Integrated Green Energy Solutions Volume 2

Scrivener Publishing
100 Cummings Center, Suite 541J
Beverly, MA 01915-6106

Publishers at Scrivener
Martin Scrivener (martin@scrivenerpublishing.com)
Phillip Carmical (pcarmical@scrivenerpublishing.com)

Integrated Green Energy Solutions Volume 2

Edited by
Milind Shrinivas Dangate
W.S. Sampath
O.V. Gnana Swathika
and
P. Sanjeevikumar

Scrivener Publishing

WILEY

This edition first published 2023 by John Wiley & Sons, Inc., 111 River Street, Hoboken, NJ 07030, USA and Scrivener Publishing LLC, 100 Cummings Center, Suite 541J, Beverly, MA 01915, USA

For more information about Scrivener publications please visit www.scrivenerpublishing.com.

Wiley Global Headquarters
111 River Street, Hoboken, NJ 07030, USA

For details of our global editorial offices, customer services, and more information about Wiley products visit us at www.wiley.com.

Library of Congress Cataloging-in-Publication Data

ISBN 9781394193660

Front cover images supplied by Pixabay.com
Cover design by Russell Richardson

Set in size of 11pt and Minion Pro by Manila Typesetting Company, Makati, Philippines

Printed in the USA

Contents

Preface

Renewable energy supplies are of ever-increasing environmental and economic importance all over the world. A wide range of renewable energy technologies has been established commercially and recognized as growth industries. World agencies, such as the United Nations, have extensive programs to encourage renewable energy technology.

This two-volume set, *Integrated Green Energy Solutions,* will bridge the gap between descriptive reviews and specialized engineering treatises on particular aspects. It centers on demonstrating how fundamental physical processes govern renewable energy resources and their application. Although the applications are being updated continually, the fundamental principles remain the same, and we are confident that this book will provide a useful platform for those advancing the subject and its industries. We have been encouraged in this approach by the ever-increasing commercial importance of renewable energy technologies.

Integrated Green Energy Solutions is a numerate and quantitative text covering subjects of proven technical and economic importance worldwide. Energy supply from renewables is an essential component of every nation's strategy, especially when there is responsibility for the environment and sustainability. These books will consider the timeless renewable energy technologies' timeless principles yet seeks to demonstrate modern applications and case studies. This volumes will stress the scientific understanding and analysis of renewable energy since we believe these are distinctive and require specialist attention.

The five most important topics covered in these two books are:

1. Education in Energy Conversion and Management
2. Integrated Energy Systems
3. Energy Management Strategies for Control and Planning
4. Energy economics and environment
5. World Energy demand

23

Energy Economics and Environment

P. Sanjeevikumar[1], Morteza Azimi Nasab[1*], Mohammad Zand[1], Farnaz Hassani[1] and Fatemeh Nikokar[2]

[1]*Department of Electrical Engineering, IT and Cybernetic, University of South-Eastern Norway, Porsgrunn, Norway*
[2]*Department of Electrical and Computer Engineering, Shahrood Branch, Islamic Azad University, Shahrood, Iran*

Abstract

Considering environmental issues and better utilization of facilities and having a superior level of service, a smart grid is one of the future requirements of the power system. The smart distribution network, which is the gateway between subscribers and the network, is one of the significant components of the smart grid that plays a vital role. Smart grids are able to confront unexpected events and they can disconnect the problematic part from the network; therefore the rest of the network can return to normal operation. Due to the fact that the smart grid is high priced, it must be implemented with great intelligence from the technical and economic and environmental perspective, to meet the requirements of subscribers, both. This chapter examines intelligent electrical systems and their function. The downside of the existing power grid system in comparison with smart power systems is presented and the main characteristics of smart grids are explained based on their capabilities. Uses of these networks are discussed regarding advanced measuring infrastructure system, meeting the demand, distributed generation and storage resources, distribution automation and comprehensive knowledge of the location of the area and electrical transportation.

***Keywords*:** Environmental, optimization, microgrid, energy economics

Abbreviations

MT	Micro Turbine
FC	Fuel Cell

**Corresponding author*: morteza1368morteza@gmail.com

Milind Shrinivas Dangate, W.S. Sampath, O.V. Gnana Swathika and P. Sanjeevikumar (eds.) Integrated Green Energy Solutions Volume 2, (1–24) © 2023 Scrivener Publishing LLC

ESS	Energy Storage Systems
CHP	Combine Heat & Power
GB	Gas Boiler
DERs	Distributed Energy Resources
DG	Distributed Generation
RLD	Responsive Load Demand
NRLD	Non-Responsive Load Demand
DSM	Demand Side Management
MCC	Microgrid Central Controller
LC	Local Controller
MAS	Multi-Agent System
SMO	Single Master Operation
MMO	Microgrid Management System

23.1 Introduction

The conventional power system faces problems such as the gradual depletion of fossil fuel resources, poor energy efficiency, and environmental pollution, around the world.

On the other hand, numerous obstacles in the development and construction of centralized production resources, limitations in the capacity of transmission lines, etc., have reduced the role of centralized production. These problems lead to a new trend of electricity generation using unconventional or renewable energy sources, including MT, FC and ESS for storage of electrical and thermal energy, EHP, GB, CHP, etc., as these devices are considered controllable units [1].

Also, Renewable energy sources such as wind and photovoltaics and solar panels are uncontrollable units that are integrated with the distribution network. This type of power generation is called distributed generation (DG) and the mentioned energy sources are called distributed energy sources (DERs).

Regulatory commissions have initiated mandatory or optional policies to use advanced measurement infrastructure to enable demand-side responsiveness. The Federal Energy Regulatory Commission, in a ruling on October 17, 2008, adopted a policy aimed at removing barriers to load-sharing participation in electricity markets.

It is explained that accountability can be activated in creating competitive pressure to reduce the wholesale price; furthermore, it increases public awareness in energy consumption and provides better efficiency for electricity markets, and improves reliability. A distribution network integrated with DGs is called an active distribution network. Nowadays, DG products have several advantages. One of the most vital advantages of DG products is its proximity to consumers and as a result, the reduction of losses and related costs in energy distribution and transmission. Other benefits include eliminating the spatial and geographical constraints of small production compared to large power plants, needless for high risk, less installation time, better environmental conditions, higher quality production capacity, greater reliability and security [2].

Due to the high permeability of DGs in the network, new challenges have arisen in terms of their safe and effective performance. In order to increase reliability and flexibility, these must have capabilities such as restructuring design and performance independent of the global network. By using these microgrids (MGs), these challenges can be eliminated [3].

23.1.1 The Concept of Microgrids

A microgrid is a small network consisting of distributed and load products that are connected to the distribution network at a low voltage level and are an important part in the development of smart grids. By forming an alliance and exchanging power with each other in two functional modes, the mode connected to the Bo network, the island mode or the network independent mode can be exploited.

This structure is in the form of a set of some small controllable and uncontrollable sources such as solar panels and wind turbines, ESS and loads capable of responding to the central controller signals called responsive load (RLD). Interruptible loads and uncontrollable loads (NRLD) are modeled as shown in Figure 23.1.

Federal Energy Regulatory Commission Advanced Metering Infrastructure [4].

The usual goal pursued to control the microgrid in independent operation mode is to supply local loads; while in the case of grid-connected operation, the main purpose is to maximize profits or minimize the cost of electricity generation. Other additional goals such as minimizing greenhouse gas emissions can also be considered using multi-objective optimization techniques. One of the main technical issues in the processes of connection, control and management of renewable energy sources is the unpredictability of the production capacity by these resources. One of the main concerns of researchers in this regard is to assess the impact of these resources on the overall security and reliability of the network. Utilization of distributed energy sources installed in microgrids can potentially increase

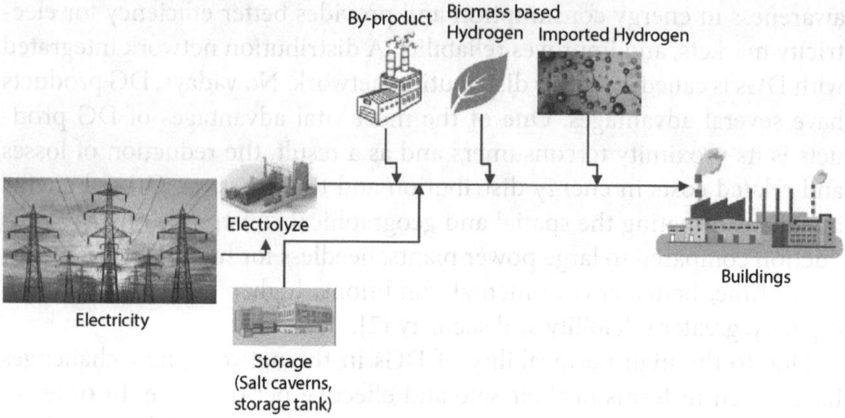

Figure 23.1 The structure of a microgrid.

energy efficiency and improve voltage profiles, reduce power density in the distribution network and stabilize the system. As an example, a microgrid is compared to conventional centralized power plants as follows [5]:

- Production capacity in a microgrid is much less than conventional power plants.
- The power generated at the distribution voltage level can be used directly to power the distribution system load.
- The microgrid sources are installed in the vicinity of the consumption site, so the microgrid load can be provided with a suitable voltage and frequency profile and with small line losses.

The microgrid's technical features make it a viable option for power supply in remote areas. Especially in places where various reasons such as topology or frequent outages due to bad weather conditions, it is difficult to supply power from the main network.

23.2 Benefits and Drawbacks of Microgrids

- Environmental issues: There is no doubt that the environmental effects of microgrids are much less in terms of gas and particulate emissions than conventional thermal power plants. In addition, the proximity of consumers to production sources can increase the level of public awareness of fair energy consumption.

- Investment and operation issues: diminishing the electrical and physical distance between production sources and consumption centers would have positive effects. These effects include optimizing the reactive power of the whole system and thus improving the voltage profile, reducing the distribution feeder density, reducing losses and reducing/delaying investment in the development of the transmission and production system.
- Power quality: Factors such as decentralization of generated power, better matching of supply and demand, reducing the effect of large-scale transmission and production exit can be effective in improving power quality and system reliability by eliminating voltage harmonics, and if properly positioned can Reduce energy supply costs [6].
- Cost saving: Resource integration can be cost effective. Because these resources are installed at the point of consumption, transmission and distribution costs are significantly reduced. In addition, the energy produced is distributed locally among consumers, which reduces power through long feeders.

Despite the potential benefits, the development of microgrids has the following challenges:

- High costs of distributed energy sources: The high cost of installing microgrids is a major challenge. This issue can be partially addressed with government funding to encourage investment in this sector.
- Technical issues: These problems are related to the lack of executive experience in controlling a large number of distributed generation sources, electric vehicles and the use of load response programs.
- Lack of Standard: Since microgrids are a new field, there are still no standards that can be used to address the protection and performance of microgrids. Therefore, standards and protocols should be used to integrate distributed generation resources, safety and protection strategies.
- Legal and administrative barriers: In many countries, despite government funding for microgrids, there are no standard rules for setting up and operating microgrids.
- Market monopoly: Microgrids may be allowed to independently supply power for essential loads when the mains

is not available due to an emergency. In this case the current electricity market will lose control of the energy rate, and microgrids will be able to retail energy at a very high rate. Therefore, designing and implementing appropriate market infrastructure for the development of microgrids is essential [7].

23.3 Causes of Increase in Power Plants

In the last decade, the number of network-connected DGs has been increasing due to the high benefits of distributed generation, and the need for a network-independent source.

Several factors have led to an increase in DGs [8]:

1. The ever-increasing demand for electricity around the world is leading to the search for new sources of energy.
2. Concerns about climate change limiting fossil fuel storage have led to increased interest in renewable energy.
3. Advances in DG technologies, such as the simultaneous generation of electricity and heat, require the production of energy close to the consumer.
4. Liberalization of the electricity market allows entry into the energy production business even with small power plants.
5. Consumer demand for high-reliability electricity has increased, necessitating storage and backup systems.

DGs include a wide range of generators such as PV photovoltaics, wind turbines, micro turbines, fuel cells, CHPs, water units, etc., whose production capacity is in the range of MW and KW.

23.4 Demand Side Management in Microgrids

Electricity companies and power grid companies have been forced to change their operation from a vertically integrated structure to a competitive market structure for various reasons. With the restructuring of the electricity industry, the philosophy of operating the power system also changed. The traditional method was to supply all power demand, no matter where it was located but the new philosophy states that system efficiency will increase if demand fluctuations are as low as possible.

Demand side management (DSM) is designed to plan, execute and monitor network activities to affect customer power consumption. As a result, the DSM can change the time pattern and amount of network load [9]. Usually, the main goal of demand-side management is to encourage users to either consume less power or shift energy consumption to off-peak times during peak hours, thus smoothing the demand curve. Reliable grid performance is primarily dependent on the perfect balance between output and load at any given time.

Assuming very little control on the demand side, the production side can be controlled according to the load. (Maintaining this balance is not easy.) This may become even more difficult as the distribution of distributed energy production increases. Since the amount of production produced by renewable sources varies according to weather conditions, it is not easy to follow the output of renewable sources from a particular load form.

Therefore, since the peak of production in renewable sources does not necessarily correspond to the peak of consumption in the demand side, energy must either be consumed artificially or stored for future consumption.

The system can rely on fossil fuels at peak times, but due to increased production diversity, the grid has to hold more reserves, which will significantly increase the total cost of electricity.

Alternative to this balance is the use of new methods and technologies that are mainly in interaction with users. Thus, unlike classical methods,

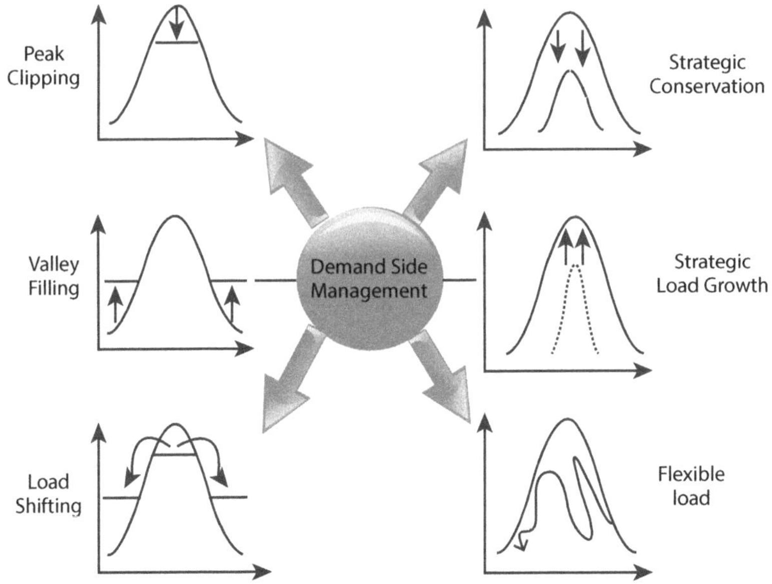

Figure 23.2 Techniques for changing the characteristic demand curve in demand side management.

which determined the amount of production capacity, the demand response (DR) can play a key role in balancing power. Due to the nature of renewable resources, it is not possible to control or demand power from these resources. The main goals of DR techniques are to reduce the peak load and create the ability to control consumption according to production.

In other words, when cheap renewable energy is available and when there is a shortage of electricity, there must be a way for end consumers to be aware of it and react. There is a significant scope for DSM to help increase system productivity and utilization. Demand side management is planned for early 2018 [10, 11].

Demand side management can be used as a tool to achieve various load shaping goals such as peak cutting, canyon filling, load shifts, strategic maintenance, strategic load growth and flexible load shaping as shown in Figure 23.2.

23.5 Centralized Control of Smart Grid

Centralized control is performed with the aim of optimizing power exchange with the upstream network, and the production levels of the smart grid change according to market prices and security constraints. Centralized control can only be implemented with controllable resources and loads.

Figure 23.3 shows the data exchange path between the central controller and the local controller of the smart grid.

According to this figure, information is exchanged in two directions. This data exchange can be through telephone lines or wirelessly. The central controller makes decisions every 15 minutes according to market

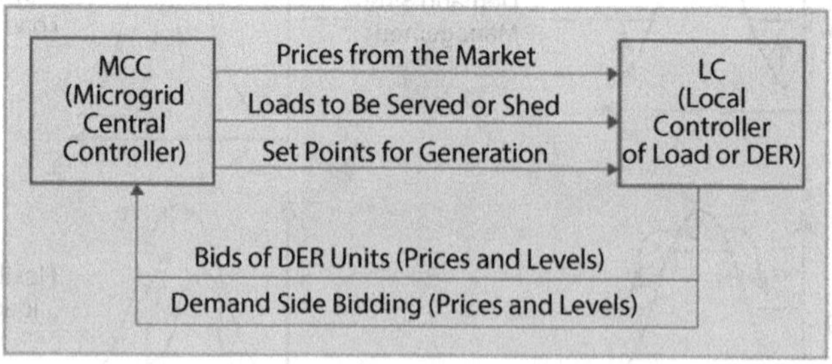

Figure 23.3 Information exchange path in centralized intelligent network control.

conditions, the capacity of the available units and the prices given by the local controller.

MCC considers the following requirements based on market policies [12]:

- DER resource prices
- Market prices
- Network security restrictions
- Load forecasts and even forecasts of renewable energy production
- Birth control points of DER resources
- Adjust loads in terms of supply or blackouts
- Market price for the next period in order to enable local controllers to bid

LCs receive control signals from the MCC to determine their generation and consumption as well as the price of electricity. To achieve this, studies are needed to predict the load, production and heat energy required.

23.6 Decentralized Smart Grid Control

Decentralized control approach enables autonomous operation of DER units and loads in the smart grid. Since each DER resource can have a separate owner, it will be difficult to control the smart grid in a centralized manner for this situation. In decentralized control, each of the local controllers intelligently controls the production operation and can communicate with each other. In this case, maximizing the revenue corresponding to each unit is not necessarily the main task of any controller, but improving the overall performance of the smart grid is of paramount importance. The use of a multi-firm system solves many problems related to the operation and performance of the system, so MAS is selected as the first candidate for decentralized control to expand the smart grid. The MAS system is schematically shown in Figure 23.4.

In fact, this system has the ability to perform decentralized control tasks using special software and methods. An intelligent network requires a strong telecommunications infrastructure to transfer information to its various parts. Strong telecommunication systems help make data transfer easier and faster [13].

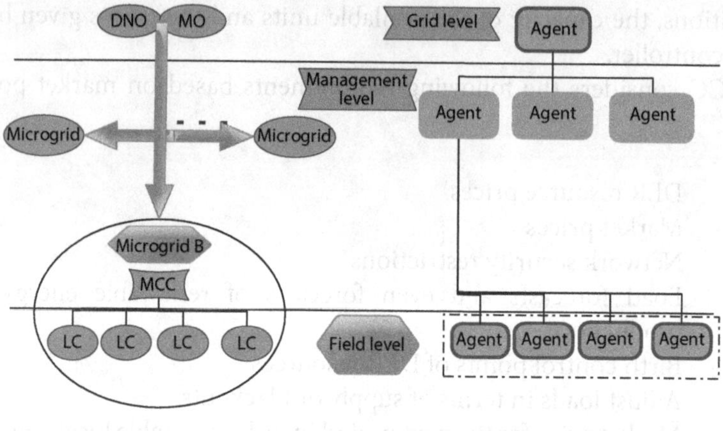

Figure 23.4 Schematic diagram of the MAS structure of decentralized intelligent network control.

23.7 DER Resource Control Strategies in the Smart Grid

DERs in smart grids are exploited with three control strategies: PQ, PV and VF. PQ control delivers constant active and reactive power to the pseudo and is especially used to extract maximum energy from DGs such as photovoltaics and wind turbines. PV control injects a constant amount of active power into the grid by keeping the bus voltage constant. This control strategy is often used for conventional synchronous DGs that are connected directly to the grid. The VF control acts like a slack bus and keeps the voltage and frequency of the smart grid constant by controlling the active and reactive power injected into the network.

It should be noted that in the case of network connection, all DG units are controlled in PQ or PV mode. In this case, DG units are involved in active power injection and voltage control. With the occurrence of the island, the intelligent reference network loses its voltage and frequency, and the task is to control the voltage and stabilize the frequency to some of its DGs. In this case, if the scattered products continue to operate in pre-island modes, we will see the instability of the smart grid in the transition to island mode. Therefore, in island mode, at least one DG unit must be in VF mode [14].

23.8 DER Participation Strategy in Smart Grid

There are two strategies, SMO and MMO, for the participation of distributed energy sources in the operation and control of smart grids. In the SMO method, a DER unit provides voltage and frequency reference for other doors, and the rest of the sources continue to operate in constant active and reactive power control mode by receiving voltage and frequency references from this unit. The SMO strategy is shown in Figure 23.5.

In the MMO strategy shown in Figure 23.6, several DERs work together to determine voltage and frequency references. Power sources for other sources are transmitted by the MGCC [15].

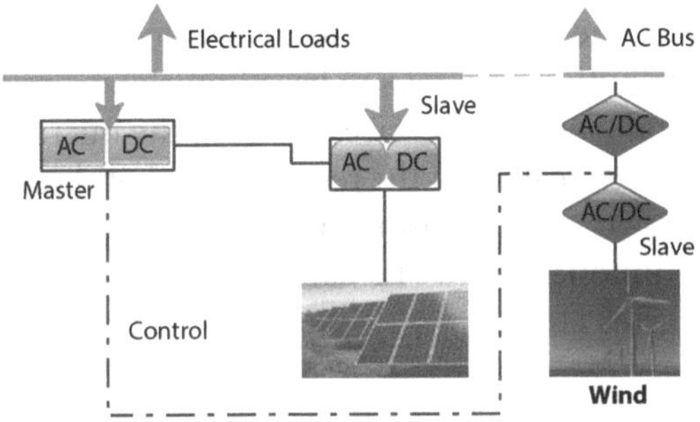

Figure 23.5 SMO control strategy.

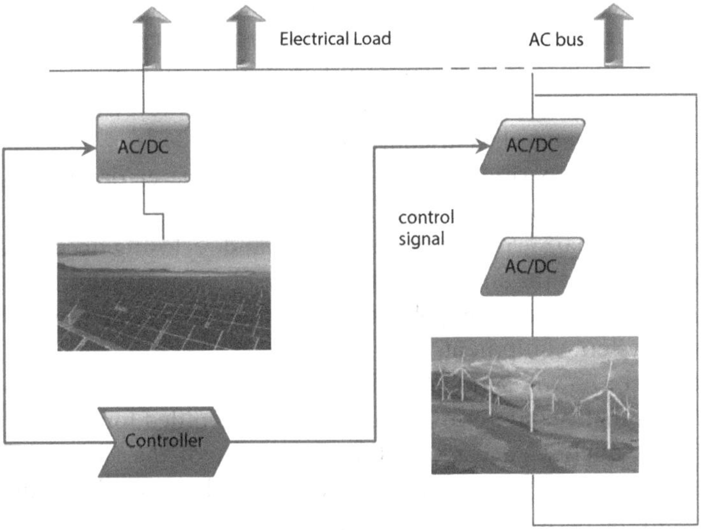

Figure 23.6 MSO control strategy.

23.9 Topics Raised in the Smart Grid

Protection, safe and economical operation, power quality, dynamics and control of smart grids are the most important issues raised in smart grid research. In this section, the most important issues discussed in the smart grid are briefly studied.

23.10 Smart Grid Protection

If an error occurs in the upstream network, the protection equipment must be able to quickly disconnect the smart grid from the upstream network in order to protect the smart grid loads. Most methods of protection of conventional distribution systems are based on short circuit current. The presence of electronic power interfaces between the micro-sources and the network prevents the creation of the required level of short circuit current. Therefore, some overcurrent sensors are not responsive to such a level of overcurrent. The unique nature of smart grids necessitates new measures to protect smart grids [16].

23.11 Detection of Smart Grid Islands

The importance of the island detection operation is due to network security reasons. It can be very dangerous to have a power feeder that is separate from the mains while network workers are in maintenance operations. Many power grids use automatic reopening. In this way, when a short circuit occurs, the network is disconnected.

After a certain period of time and by fixing the error, the switching equipment closes the circuit again.

Now, if the smart grid remains electrified and reclosing occurs between the global grid and the smart grid, frequency, phase and magnitude distortion will form between the global grid and the island, which will damage the power system equipment.

In fact, the DG will connect to the main network when it is out of sync. The result is that island performance should be avoided for safety and network power quality. As a result, the occurrence of the island should be detected as soon as possible and the power switch between distributed generation and the network should be disconnected.

Detection methods are divided into two main groups: remote detection methods and local detection methods. Local methods, in turn, are divided into three subsets: passive, active, and hybrid [17].

23.12 Smart Grid Optimization

The optimization of the smart grid is done by the energy manager. The energy manager examines the needs of his electrical and thermal loads, power quality requirements, electricity and heat generation costs, wholesale and retail service needs, upstream network specific needs, manageable load requests, density levels, etc. He performs system optimization to determine the output power of micro-resources. Some of the key tasks of an energy manager are as follows [18]:

- Determine voltage and power adjustment points for each micro source
- Supply of electrical and thermal loads
- Providing smart grid operation contracts with the transmission system
- Minimize environmental pollution and system losses
- Maximize the efficiency of exploiting micro-resources
- Provide control signal to identify the island and reconnect the smart grid

23.13 Power Quality

Due to the presence of sensitive loads in the smart grid, which must be provided with high reliability and an appropriate level of power quality (voltage drop, flicker, harmonic, etc.), the smart grid must also provide good quality power in the island mode.

23.14 Frequency and Voltage Control

In the case of connecting to the global network, the voltage and frequency are determined by the upstream network. In this case, the generating sources are involved in injecting active power and voltage control. With the occurrence of the island, the reference smart grid loses its voltage and frequency. In this case, if DERs continue to operate in pre-island modes, we will see the smart grid become unstable in the transition to island mode. In order to stabilize the smart grid in the island mode, the island must be quickly detected and then receive at least one scattered energy source to regulate the frequency and voltage of the smart grid by receiving an island command from the protection devices. If the existing DGs and DS are not able to recover the frequency and voltage of the smart grid, part of the load of the smart grid will be cut off [19].

23.15 Balance between Production and Power Consumption

The small inertia and the slow response time to the load changes of the small generators cause a large frequency oscillation in the island's smart grid for the occurrence of a small imbalance. If power is exchanged between the smart grid and the upstream network when connected to the global grid, the control strategy selected for DERs should be able to balance power generation and consumption by disconnecting the smart grid. This balance is generally maintained by rapid battery performance and the removal of unnecessary loads [20].

23.16 Ability to Easily Connect Distributed Generation Sources

It is desirable to be able to flexibly connect or disconnect distributed distribution source controllers anywhere in the electrical system without the need to redesign, which is called Plug and Play mode.

Such a system has protection and control parts that can operate in all situations.

23.17 Smart Network Security

Sometimes an error, even in a low voltage network, can lead to an increase in ground voltage.

Therefore, grounding of distributed energy sources and transformers of smart grid interface with upstream grid should be carefully considered and ground rules should be observed. LV grounding systems are defined based on the secondary grounding techniques of MV/LV transformers and within the framework of load-bearing equipment. Neutral ground connections LV are mostly classified into three samples, TI, IT and TN.

23.18 Resynchronization after Network Connection

After recovering the frequency and voltage, the control system must synchronize the smart grid and connect it to the grid so that all the loads can be supplied and power exchanges can be established between the smart grid and the global grid. The voltage range, frequency and phase angle between the mains voltage and the smart grid must be within range before reconnecting. Reconnection with improper phase difference can cause transient states and damage to network equipment by causing severe surges in transformers.

Smart grid reconnection can be done if the voltage error is less than 3%, the frequency error is less than 1% Hz and the phase error is less than 10 degrees.

23.19 Smart Grid Control Glasses

We expect to have a growing trend of smart grid presence in traditional networks in the future. As a result, the characteristics of distribution systems will be different from today's distribution systems, and this difference will become more significant as the number of smart grids increases. Therefore, it is necessary to provide appropriate control strategies for such situations. Controlling a large number of micro-sources due to communication constraints is very challenging. The transition from a connected state to a separate one is likely to lead to a loss of balance between load and output and to voltage and frequency problems. If a large number of micro-sources are disconnected or connected at the same time, the Plug and Play feature may cause a number of serious problems [21, 22].

23.20 Economic Dimensions

In the field of distributed generation resources infrastructure, a lot of investment is needed for tools and network changes for control and security. Reimbursement in open, less risky markets will be complex and profitability will be achieved by avoiding price fluctuations or punishing peak loads or unexpected consumption patterns unless scattered sources of production are the only option (costly). Otherwise, methods that are considered a kind of retreat should be used. Even so, owning one is still beyond the reach of the average person. Artificial, unlike the free market, is always there.

The question here is who will financially support this support system, because it is a heavy investment with very long-term profitability.

Normally, the network budget can be paid to the power system operator through the approved transmission tariff, which will be registered in terms of power and energy exchanged. The installation of a distributed generation unit and the application of an energy island mean that less energy will be returned from the grid and the economic contribution to capital gains will be reduced.

At the same time, the network in which the transformers and lines designed for maximum capacitance will remain the same due to the balance and emergency support; therefore in some countries the cost of disconnection is charged from the units of distributed generation units. Otherwise, the economic burden of the remaining customers will be heavier with each unit of distributed generation source. It will be levied as a hidden tax to protect customers from scattered production sources.

On the other hand, additional marginal services related to PQ, such as voltage support, can also be provided to the network operator to provide an additional source of revenue and an indirect aid to the costs of network equipment, but suitable contexts in business mechanisms and economic incentives for Profitable use of technology capabilities has not yet been provided.

In general, it is difficult to say where the limitations of the technology of distributed generation resources in distribution networks are. In fact, it is the subject of many projects that different parameters must be considered in this regard, such as voltage stability, power quality and inventory reliability.

All the issues raised depend on various features such as loads, network topology and transmission network support. The principle that the rules for connecting region to region differ makes the problem much more complicated. In practice, efforts are made to maintain inventory reliability, which is a conservative view. This is understandable in situations where reliability inventory is heavily dependent.

Generally, one should not only focus on the initial problems of developing distributed generation resources, but also the benefits of using them. In a situation with good conditions, the reliability of the whole system increases, the peak of power demand will put less pressure on the full load power system. One of the most important possibilities is to use local resources to reduce the cost of electricity by reducing transmission losses and eliminating (delayed) expensive investments in infrastructure.

It is almost certain that the central power system will undergo a revolutionary transformation in the coming years and decades, because electricity consumption has not been reduced at all and the problems of grid expansion can only be solved by installing distributed generation technology. However, crossing the boundaries and applying more or less economically and technically independent energy islands has not yet been confirmed as a good idea, and examples will be implemented in the near future that will not be without merit. Let's follow, which shows how to use networks in the not too distant future [23].

23.21 Losses

Losses in electrical energy networks in the path of transmission and consumption are divided into two components: technical losses and non-technical losses, which are described below and the factors influencing it.

Network technical losses:

Technical losses are losses that occur naturally due to the nature of the components of the power system. Many solutions have been proposed to calculate the technical losses of the network.

The most common of these are load distribution, simultaneous meter reading, statistical methods, and so on. The source of technical losses are [24]:

1. Intrinsic equipment losses (losses due to resistance of transformer unloaded lines - copper losses or pregnancy of transformers)
2. Losses due to environmental conditions (losses due to moisture and corona, losses due to dust, etc.
   Air pollution - collision and connection losses - tree branches)
3. Meter waste and measuring instruments (inaccuracy in measurement)
4. Losses due to improper power factor
5. Losses due to unbalanced load in phases and low-pressure single-phase distribution
6. Losses due to the use of substandard supplies and equipment

7. Losses due to improper designs (failure to use transformers with suitable power - failure to install transformers in the center of gravity of the load - lack of optimal fit between transmission and over-distribution voltages and medium and low pressure)
8. Losses due to obsolescence of networks
9. Losses due to network connections and not using the correct grounding system
10. Losses due to the construction of long networks to supply electricity to subscribers
11. Loss in network equipment due to current leakage

23.22 Non-Technical Network Losses

Non-technical losses are due to factors other than inherent network losses. The only solution to calculate this part of the losses is to obtain the difference between the total losses and the technical losses. Non-technical losses include unauthorized use of electricity, tampering with measuring devices and meters, incorrect reading of the officers, improper operation of the meters, and so on. Lower losses lead to lower energy costs and thus accelerate countries' economic growth by lowering the cost of production.

Lower losses make distribution companies more flexible to compete in competitive markets. Studies show that loss reduction can be considered as one of the optimization activities that will delay the development and construction operations and provide huge investments related to it. In this regard, the issue of economic evaluation and prioritization of their implementation is discussed. There are various methods to reduce losses, which according to the amount of investment can be used to reduce losses, including [25]:

1. Capacitor placement and distributed generation sources
2. Optimal placement of distribution transformers
3. Install the distribution transformer in the center of gravity of the load
4. Load adjustment of low-pressure feeders
5. Replacing the neutral wire with a larger cross section and leveling the cross section of the neutral wire with the phase wire
6. Replacement of three-phase wire with larger cross section

7. Using a three-phase system instead of a single-phase system
8. Rearrangement of medium and weak pressure network

23.23 Power System Loss Analysis

Losses can increase the capacity of power plants as well as consumption needs. To reduce energy losses, it is worthwhile to consider reducing power losses.

The huge costs of losses are a major factor in persuading companies to look for ways to reduce it. Technical losses depend on the structure of the network, cannot be removed and can only be reduced by changing the structure and equipment, but since non-technical losses are due to a factor outside the power network can be identified and removed to recover lost payout. It can be boldly said that measuring and evaluating losses is the first and most important step in related studies. Because without accurate and accurate measurement of losses, we cannot expect other studies and planning such as intelligent design of transmission and distribution system, rearrangement, voltage leveling, etc., which aims to reduce losses and improve the quality of networks. Intelligently communicate to provide an appropriate, accurate, and practical response.

Although it is very common to use power losses in peak load and loss factor to calculate energy losses in the power system, it should be noted that the loss factor in each area depends on several parameters such as peak load, transmission energy and shape of the consumption curve. And therefore, its value is mainly a function of the type of consumption that will vary from region to region. In obtaining simple loss models for different types of power systems, the electrical network topology of the power systems has been identified and there is also relevant bus data. (Branch data and consumer data obtained from the databases and databases of the country in question.) Three-phase load distribution analysis will lead to solving power system power losses and creating training set data.

23.24 The Impact of the Electricity Market on the Performance of Distribution Companies

1. Improving distribution networks and reducing losses
   In the electricity market, when regional power companies sell energy to distribution companies, they are responsible for a number of things, including wasted energy, the

improvement of the distribution network, and the non-abuse of electricity that they have. They will take these issues seriously and correct them [26].

2. Increase network reliability and reduce blackouts
   Energy purchases are regulated in the area covered by the distribution companies and they are paid according to this amount, so the distribution companies will be more sensitive and serious about increasing reliability and reducing blackouts.
3. Improving the methods of exploitation and adjustment of manpower
   Today, not only the electricity industry but also all fields should abandon traditional methods and turn to modern technologies, so electricity companies are thinking of reducing manpower and reducing costs.
4. Expanding consumption management programs and load factor correction
   In the regulations of the electricity company, they have multiplied the prices for buying electricity, so the distribution companies will be more inclined to use effective methods of consumption management and load factor correction.

23.25 Power Quality in the Restructured Electricity Market

Power quality is one of the most important issues in electrical loads, as any noise in voltage, frequency causes damage to electrical devices. Retail markets need to regulate natural electricity delivery (transmission and distribution) and other services contracted with consumers in order to meet customer needs and demands. Preparing data for general load distribution requires a large number of components such as length, line conductor size, impedance and capacity of power transformers and power consumption of all consumers. To support various storage-based power system management applications are features of system components in the database [27].

23.26 Conclusion

This chapter reviews smart grids and compares them with conventional energy distribution networks, and then examines the advantages and disadvantages of modern smart grids and offers solutions to improve operation and address unique challenges deal with smart grids. Also, the control structures of smart grids are described, and in the next step, the economic dimensions related to smart grids and the parameters affecting them are explained, and finally, solutions for better entry of smart grids into networks and their links with conventional distribution networks are presented. Finally, it can be said that by reviewing this chapter, the reader can accurately and completely deal with how smart networks enter and their detailed performance, as well as demand-side management strategies and protection in smart networks. Also, economy-oriented solutions for optimal network performance are presented, and finally a section titled Micro-network-related environment and auxiliary that

Optimal design of smart grids can improve the environment and reduce emissions.

References

1. Wang, Xubin, *et al.* "Optimal Planning and Operation of Typical Rural Integrated Energy Systems Based on Five-level Energy Hub." *2021 3rd Asia Energy and Electrical Engineering Symposium (AEEES).* IEEE, 2021.
2. Zhang, Zhenbin, *et al.* "Advances and opportunities in the model predictive control of microgrids: Part I–primary layer." *International Journal of Electrical Power & Energy Systems* 134 (2022): 107411.
3. Borghei, Moein, and Mona Ghassemi. "Optimal planning of microgrids for resilient distribution networks." *International Journal of Electrical Power & Energy Systems* 128 (2021): 106682.
4. Wu, Ying, *et al.* "Digitalization and decentralization driving transactive energy Internet: Key technologies and infrastructures." *International Journal of Electrical Power & Energy Systems* 126 (2021): 106593.
5. Lin, Wen-Ting, Guo Chen, and Chaojie Li. "Risk-averse energy trading among peer-to-peer based virtual power plants: A stochastic game approach." *International Journal of Electrical Power & Energy Systems* 132 (2021): 107145.
6. Jaramillo, Andres F. Moreno, *et al.* "Load modelling and non-intrusive load monitoring to integrate distributed energy resources in low and medium voltage networks." *Renewable Energy* (2021).
7. Nasrollahi, R., *et al.* "Sliding mode control of a dynamic voltage restorer based on PWM AC chopper in three-phase three-wire systems." *International Journal of Electrical Power & Energy Systems* 134 (2022): 107480.
8. Kumar, Avinash, Abheejeet Mohapatra, and Sri Niwas Singh. "Sequence Measurement Based Islanding Detection of DGs in Microgrid with

Enhanced Power Quality." *IEEE Transactions on Instrumentation and Measurement* (2021).

9. Alhasnawi, Bilal Naji, and Basil H. Jasim. "A new internet of things enabled trust distributed demand side management system." *Sustainable Energy Technologies and Assessments* 46 (2021): 101272.
10. Babar, Muhammad, Muhammad Usman Tariq, and Mian Ahmad Jan. "Secure and resilient demand side management engine using machine learning for IoT-enabled smart grid." *Sustainable Cities and Society* 62 (2020): 102370.
11. Gronier, Timothé, *et al.* "Iterative sizing of solar-assisted mixed district heating network and local electrical grid integrating demand-side management." *Energy* (2021): 121517.
12. Felling, Tim. "Development of a genetic algorithm and its application to a bi-level problem of system cost optimal electricity price zone configurations." *Energy Economics* 101 (2021): 105422.
13. Khan, Riaz, *et al.* "Energy Sustainability–Survey on Technology and Control of Microgrid, Smart Grid and Virtual Power Plant." *IEEE Access* (2021).
14. Rajesh, P., Francis H. Shajin, and L. Umasankar. "A Novel Control Scheme for PV/WT/FC/Battery to Power Quality Enhancement in Micro Grid System: A Hybrid Technique." *Energy Sources, Part A: Recovery, Utilization, and Environmental Effects* (2021): 1-17.
15. Aftab, Mohd Asim, *et al.* "IEC 61850 communication based dual stage load frequency controller for isolated hybrid microgrid." *International Journal of Electrical Power & Energy Systems* 130 (2021): 106909.
16. Khan, Riaz, *et al.* "Energy Sustainability–Survey on Technology and Control of Microgrid, Smart Grid and Virtual Power Plant." *IEEE Access* (2021).
17. Shukla, Apoorva, Soham Dutta, and Pradip Kumar Sadhu. "An island detection approach by μ-PMU with reduced chances of cyber attack." *International Journal of Electrical Power & Energy Systems* 126 (2021): 106599.
18. Das, Barun K., and Mahmudul Hasan. "Optimal sizing of a stand-alone hybrid system for electric and thermal loads using excess energy and waste heat." *Energy* 214 (2021): 119036.
19. Khan, Riaz, *et al.* "Energy Sustainability–Survey on Technology and Control of Microgrid, Smart Grid and Virtual Power Plant." *IEEE Access* (2021).
20. Refaat, Shady S., *et al. Smart Grid Enabling Technologies*. John Wiley & Sons, 2021.
21. Khan, Riaz, *et al.* "Energy Sustainability–Survey on Technology and Control of Microgrid, Smart Grid and Virtual Power Plant." *IEEE Access* (2021).
22. Refaat, Shady S., *et al. Smart Grid Enabling Technologies*. John Wiley & Sons, 2021.
23. Attar, Mehdi, *et al.* "Mid-term operational planning of pre-installed voltage regulators in distribution networks." *International Journal of Electrical Power & Energy Systems* 133 (2021): 107276.

24. Yu, Chih-Min, Meng Lin Ku, and Li-Chun Wang. "Joint Topology Construction and Hybrid Routing Strategy on Load Balancing for Bluetooth Low Energy Networks." *IEEE Internet of Things Journal* 8.8 (2021): 7101-7102.
25. Hadjidimitriou, Natalia Selini, *et al.*, eds. *Mathematical Optimization for Efficient and Robust Energy Networks*. Vol. 4. Springer Nature, 2021.
26. Rathore, Arun, and N. P. Patidar. "Optimal sizing and allocation of renewable based distribution generation with gravity energy storage considering stochastic nature using particle swarm optimization in radial distribution network." *Journal of Energy Storage* 35 (2021): 102282.
27. Wang, Shuren, *et al.* "Comprehensive assessment of fault-resilient schemes based on energy storage integrated modular converters for AC-DC conversion systems." *IEEE Transactions on Power Delivery* (2021).

24

Stringent Energy Management Strategy during Covid-19 Pandemic

Nagajayanthi B.

Vellore Institute of Technology, Chennai Campus, Chennai, Tamil Nadu, India

Abstract

The Indian economy depends on energy. The Covid-19 pandemic has presented a lot of challenges to work with protective equipment, less commuting, and social distancing. With Covid-19, there are no more peak hours' energy usage patterns during weekdays. There was delayed early energy pattern and peak midday pattern. People spend a lot on the electricity and energy management system at home. Residential load has increased but commercial load has decreased. This energy dynamic needs to be balanced. Electrical utilities and facilities were managed efficiently in India during this pandemic. Moreover, energy management needs to be done on campuses and industries with clean air and reduced energy setting so that the services could be resumed based on requirement. The path towards carbon reduction needs to be paved for a climate-neutral economy. Information Communication Technology improves energy efficacy and energy management. Renewable energy reduces global warming. Monitoring energy using data driven technologies prevents future energy shortages. Energy should be available uninterrupted with reliability at an affordable price to the common citizens. India should not depend on imported energy resources. This chapter focuses on energy management, energy efficiency, and energy utility techniques required to face unexpected catastrophic crises such as Covid-19. In addition, due to the burgeoning electrical devices, there is an increasing demand for electrical energy. During the Covid-19 pandemic, we spend a lot of time with the devices. Devices would get damaged if there are power variations. A smart grid was implemented and tested to balance energy usage in urban and rural areas. This design focuses towards monitoring power usage and prevents overloading.

***Keywords*:** IoT, energy, overload, power management, MQTT

Email: nagajayanthi.b@vit.ac.in

Milind Shrinivas Dangate, W.S. Sampath, O.V. Gnana Swathika and P. Sanjeevikumar (eds.) Integrated Green Energy Solutions Volume 2, (25–48) © 2023 Scrivener Publishing LLC

24.1 Introduction

Covid-19 has changed the lifestyle of people [11]. People spend more quality time at home [17] by reducing their business trips. People use more devices, ranging from mobiles in online classes to laptops for students. Information and Communication Technology provides energy management solutions [8]. Focus is shifted towards renewable [1] energy sources and energy conserving technologies. Due to social distancing and manpower issues, sustainability is required in terms of financial, environmental and social means. Covid-19 was identified in Wuhan, China, in December 2019. There was a major breakdown in the economy [15]. For post-pandemic development, energy sector development is required due to consumer spending patterns [7]. Pandemic spread is uncertain across various parts of the world. Vaccination is indeed a short-term immunity in the near future [10]. Stringent energy measures are to be followed to balance and maintain services from healthcare to home [13]. A smart grid system has hardware like smart meters, sensors to gather data and monitor the usage [20]; software to extract data relating to the usage and services to manage energy usage. By saving energy, we save money, we reduce pollution caused by non-renewable sources of energy. We have proposed a stringent Smart grid System that supplies energy to the customers as per need, thereby eliminating overhead cost and wastage. Power distribution is digitalized. An industry might require a lot of power whereas a household might require less electrical energy. Our system analyses the utilized power and actuates the supply mains in case of overloading or power line fault [5, 12].

24.2 Energy Management

There are different methodologies to harness energy. Energy sources are of different types: (i) primary energy sources [14] are found in nature such as wind and coal; (ii) secondary energy sources, such as electricity, are derived from primary energy sources; (iii) commercial energy sources, such as electricity, are available for usage with a price [19]; (iv) non-commercial energy source, such as cow dung, is not chargeable; (v) renewable energy source, like solar energy, is sustainable and does not get depleted; (vi) non-renewable energy like oil gets depleted over a course of time.

Depending on the application, the energy resources and the appliances are chosen accordingly. Energy-efficient smart appliances consume low energy. Deep energy retrofit is achieved using building design with heating, ventilation and air-conditioning (HVAC) facility. Thermal throttling is reduced if energy is utilized efficiently. By monitoring energy usage, cost is reduced, health hazards due to exposed radiation is reduced. As per the UN report [16], sustainable development goals have been identified, developed and followed. Currently energy management systems are expensive and hence the need for a smart grid to deliver power economically, securely and efficiently [5].

24.3 Smart Grid Design

Smart grid comprises both software and hardware. This has advanced metering infrastructure, demand response, and distributed grid management. The smart grid is an electrical grid that is optimized to deliver and control distribution of power remotely. This aims at optimized power regulation based on the needs of the end users. Communication is set between the generating source and the consuming nodal distribution center.

The proposed smart grid infrastructure is based on the ground station, cloud and the app [3]. Message Queuing Telemetry Transport protocol is a lightweight IoT messaging protocol used between the microcontroller and the things.

24.3.1 Ground Station

The ground station involves the preparation of the smart grid as shown in Figure 24.1. This measures the power from the current sensing transformer and the voltage from the voltage sensing transformer in analog form. This is digitized using the microcontroller and is displayed in the digital display. If there is an overload, the LED glows and alerts accordingly.

The functional flow involved in building the ground station is as shown in Figure 24.2. This explains the sequence of steps involved in designing the breadboard for smart grid.

24.3.2 Gateway

The digital data from the microcontroller is sent through a gateway such as the Wi-Fi module to the cloud as shown in Figure 24.3.

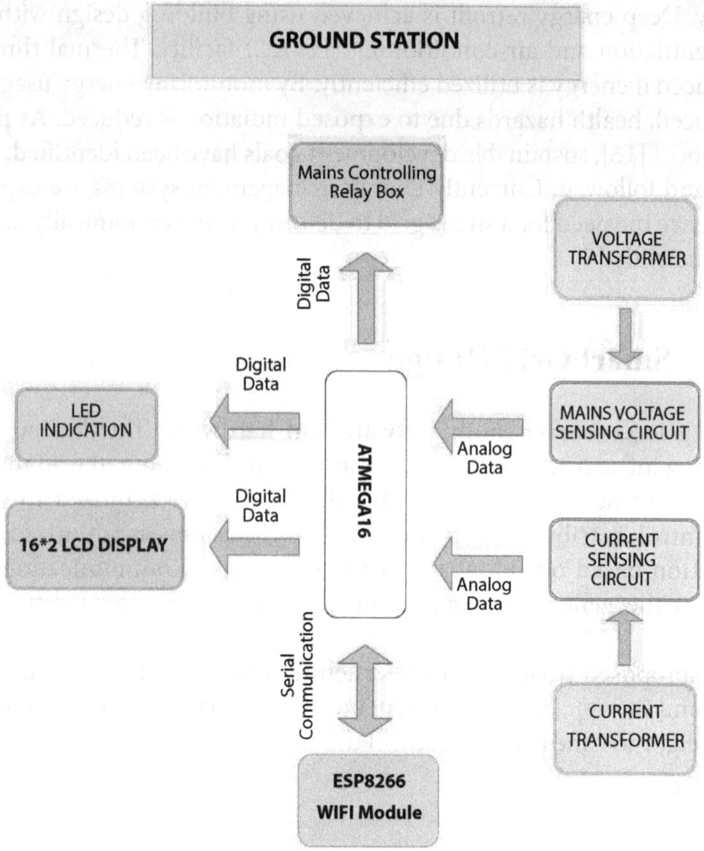

Figure 24.1 Smart grid in the ground station.

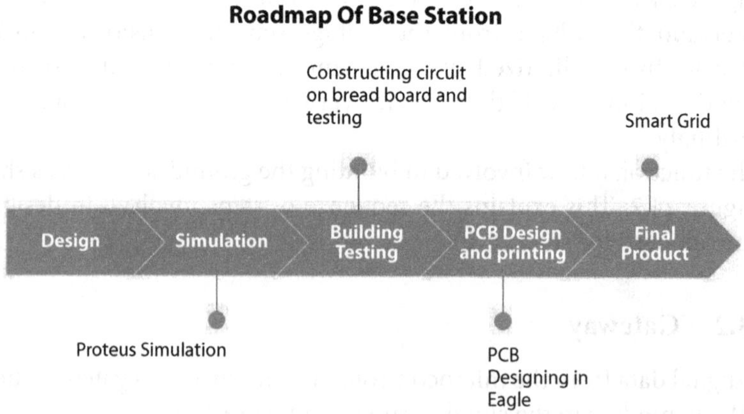

Figure 24.2 Steps involved in smart grid design.

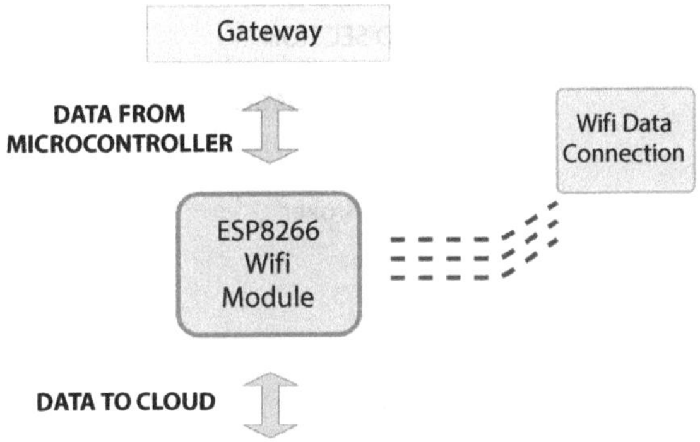

Figure 24.3 Digital data is sent to the cloud through a gateway.

24.3.3 Cloud

The data flow, data analysis and data visualization involving the data from smart grid to the cloud is as shown in Figure 24.4. The ground station provides the data through the gateway. This is encoded and sent to the cloud. Data is visualized in the dashboard as shown in Figure 24.5.

The steps involved in displaying the data in the Dashboard or in the App as per customized requirement is as shown in Figure 24.6.

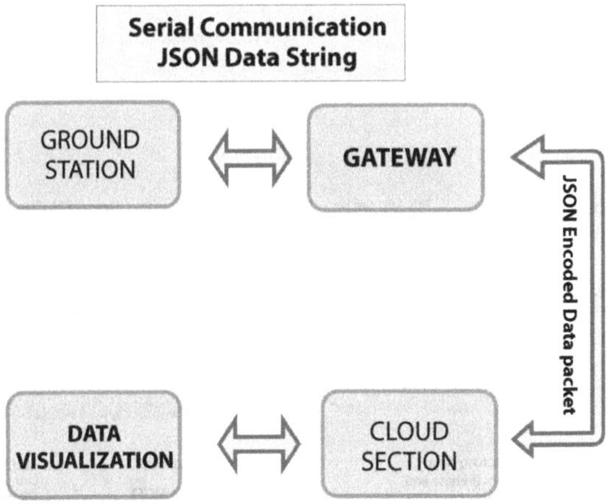

Figure 24.4 Data flow from the smart grid to the cloud.

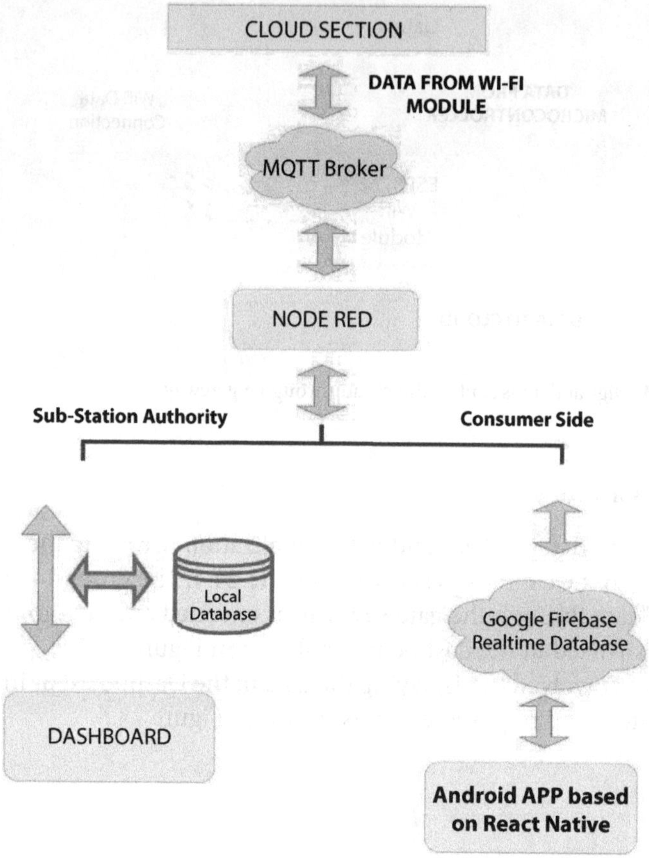

Figure 24.5 Data visualization.

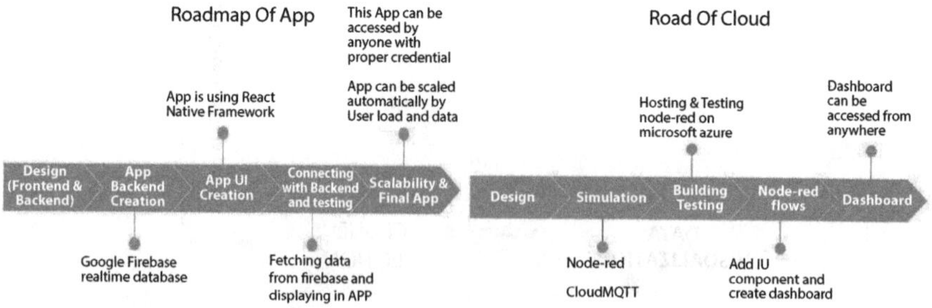

Figure 24.6 Steps involved in data display.

24.4 Smart Grid Design and Testing

Smart Grid Pilot Project Was Designed Using Proteus Simulation as shown in Figure 24.7. This circuit was simulated with input voltages ranging from 90V - 350V to replicate the scenario of the actual sub-station. To analyze overload, a variable 10 K potentiometer was added. For Serial communication, there is a virtual box for data communication.

This circuit was tested and the data was displayed for power consumed every 2 seconds as shown in Figure 24.8. Voltage and current measurement using Proteus is shown in Figure 24.9.

Offset circuit shown in Figure 24.9 is used for current measurement. Output from Current Transformer is in millivolt. This is further converted to digital for display.

Circuit shown in Figure 24.10 is used for voltage measurement. Mains voltage is stepped down to around 12V. Then the rectifier converts it to DC. The ripples contained in the DC is removed using capacitor. Voltage is regulated by the Zener diode.

Node-RED is a browser-based editor used to connect things, APIs and provide online services. Cloud MQTT is a lightweight messaging protocol

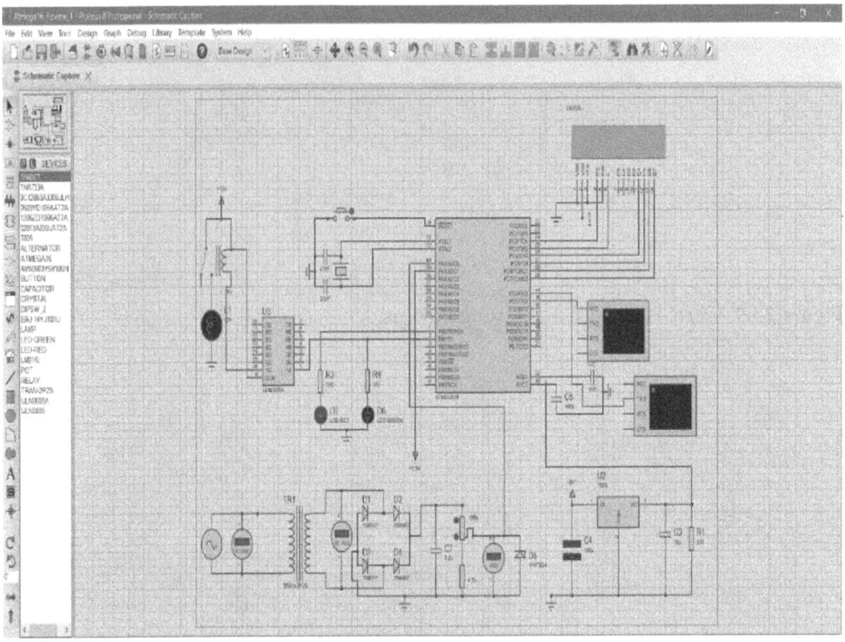

Figure 24.7 Smart grid simulation using Proteus.

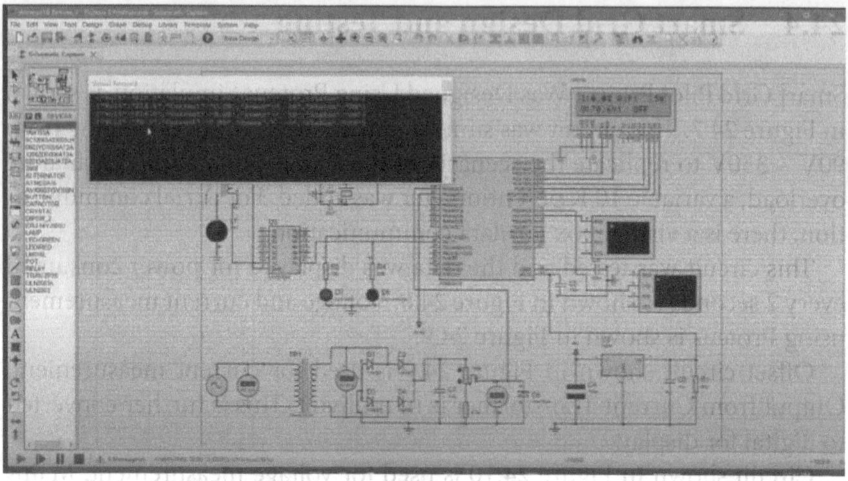

Figure 24.8 Power consumption display in Proteus.

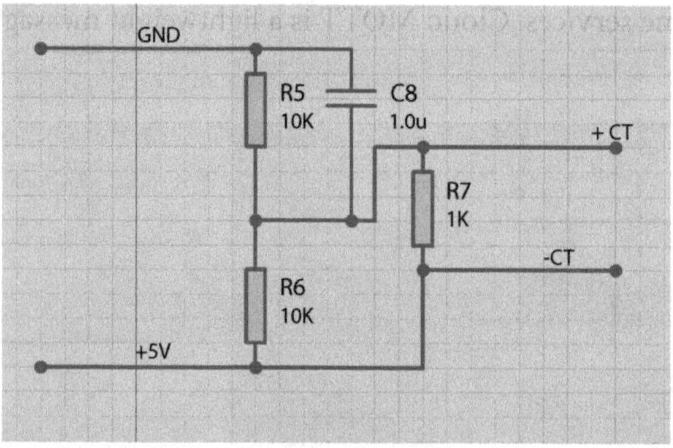

Figure 24.9 Current measurement using Proteus.

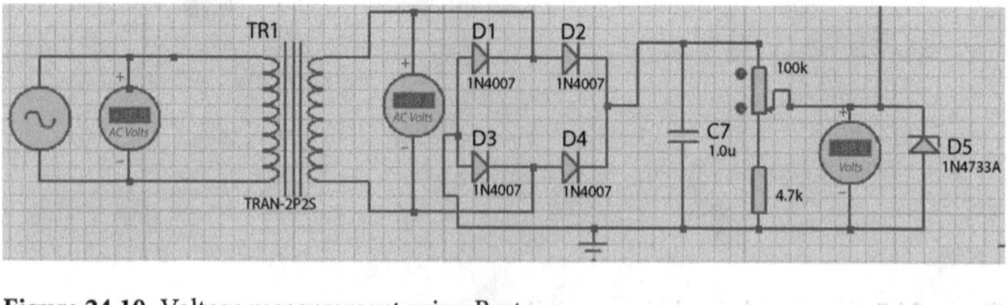

Figure 24.10 Voltage measurement using Proteus.

used between devices for messaging. Microsoft Azure provides platform as a service to host our Smart Grid Application. It is used as the virtual machine. Smart Grid circuit is simulated using Node-RED [9] as shown in Figure 24.11 and Cloud MQTT; tested using Microsoft Azure as shown in Figure 24.12; with Voltage and power consumed displayed in the Node-RED dashboard as shown in Figure 24.13.

For the clients, an App is created using Google Firebase [18] as the back-end. The client app is connected with Google Firebase online database. It streams live data from the database and keeps on updating the client regarding voltage, power, current, and the status. Also, we have customized

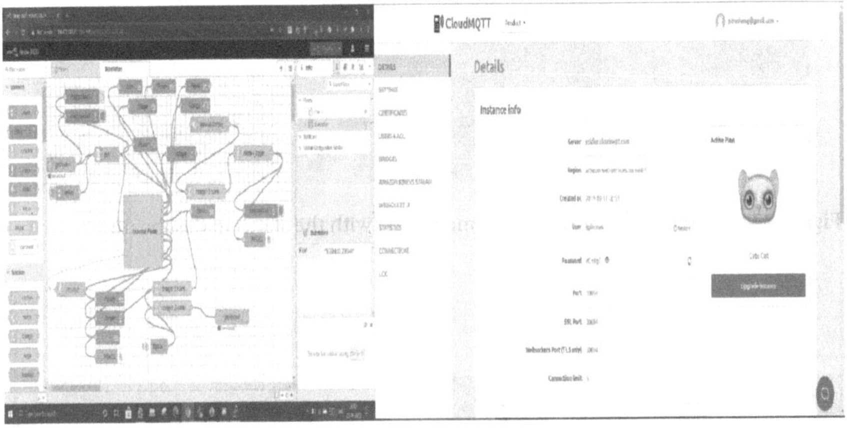

Figure 24.11 Circuit is tested using Node-RED and MQTT Implementation of smart Node-RED.

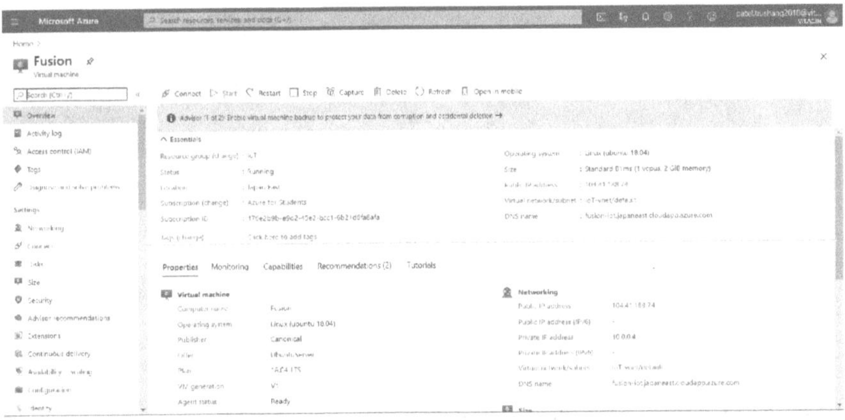

Figure 24.12 Microsoft azure as virtual machine.

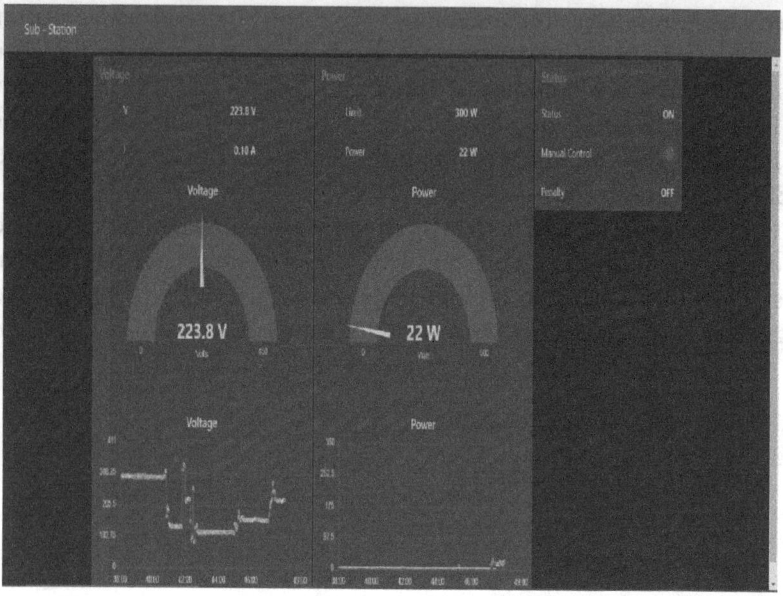

Figure 24.13 Voltage and power consumed along with the status in dashboard.

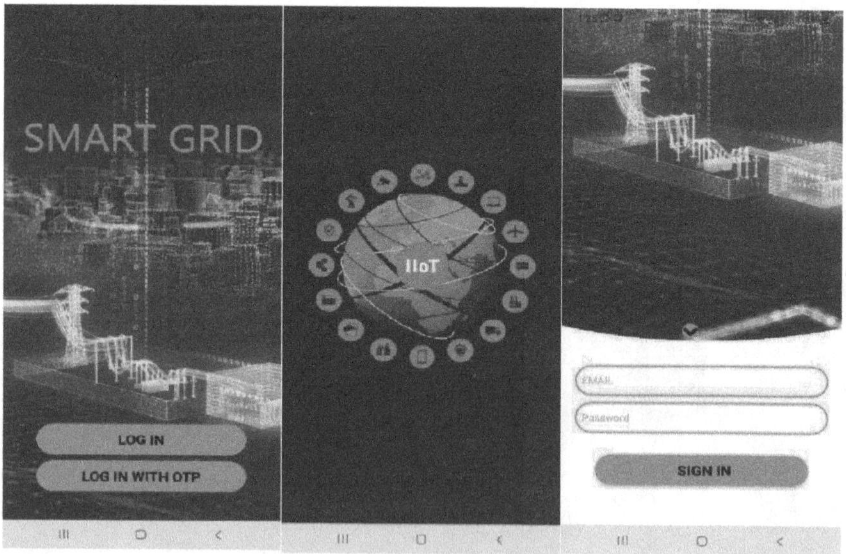

Figure 24.14 Smart grid app in the client side.

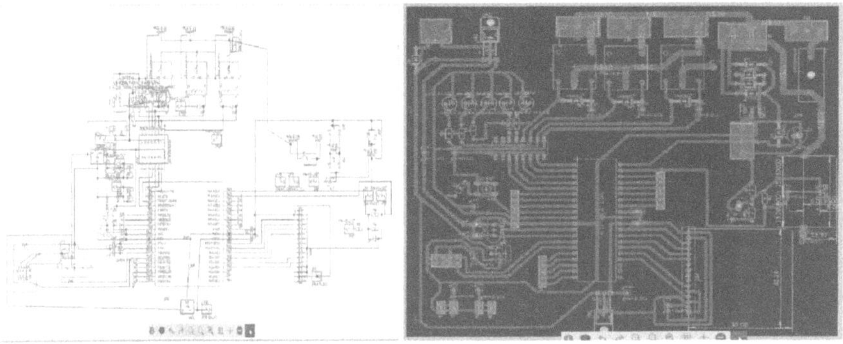

Figure 24.15 Fusion 360 software to build the PCB for smart grid.

the client with a provision to control the main supply connection remotely from his mobile app. Otherwise, based on the limit set by the client, the app decides and actuates the main supply accordingly. The customized app using React Native Framework is as shown in Figure 24.14.

24.5 Implementation of Smart Grid

The PCB Design for Smart Grid was implemented using Fusion 360 software as shown in Figure 24.15.

Smart Grid is implemented using ATMEGA 16 L power constrained high performance Microcontroller as shown in Figure 24.16. This processes 8 bits of data at a time. The main voltage sensing transformer will step down the main voltage from 230 V to workable 12V. A current transformer steps down thousands of current to 2 Amp or as required. A relay Driver 2LN 2004 IC has seven bipolar dual Darlington pairs. For each relay driver, the output current is 500 mA and the output voltage is 50V. An ESP 01 Wi-Fi module networks the ground station to the cloud station.

The voltage form the voltage sensing transformer and the current from the current sensing transformer gives the power. This is digitalized by the ATMEGA microcontroller. If it exceeds 220 V, an alert signal is displayed in the LED and this is also simultaneously displayed in the LED display. This module is connected to the cloud using a Wi-Fi module.

Cloud MQTT [2] is used as the broker. Node-RED is a flow-based GUI Editor. Node-RED is an open-source software-based visual programming tool. It wires the devices, processes the data, controls the devices and sends alerts to the devices based on the data received. This is used to manage multiple users in the cloud. This is connected to the cloud using MQTT protocol.

Google Firebase is the real-time database storing the data from the smart grid. Client side Mobile App is implemented using React Native Java App.

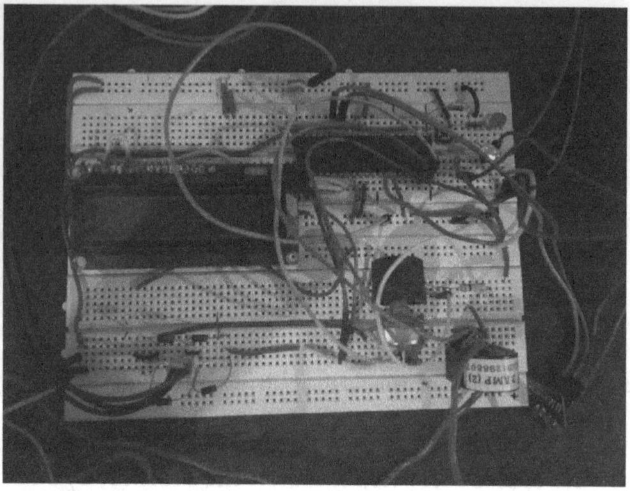

Figure 24.16 Smart grid.

To host our application, Microsoft Azure is used to provide platform as a service. A virtual machine is created and a static IP address is assigned. Putty software as shown in Figure 24.17 is used to provide a secure SSH channel between the client and the machine using command line interface (CLI). Client is authenticated to access using his credentials as shown in Figure 24.18.

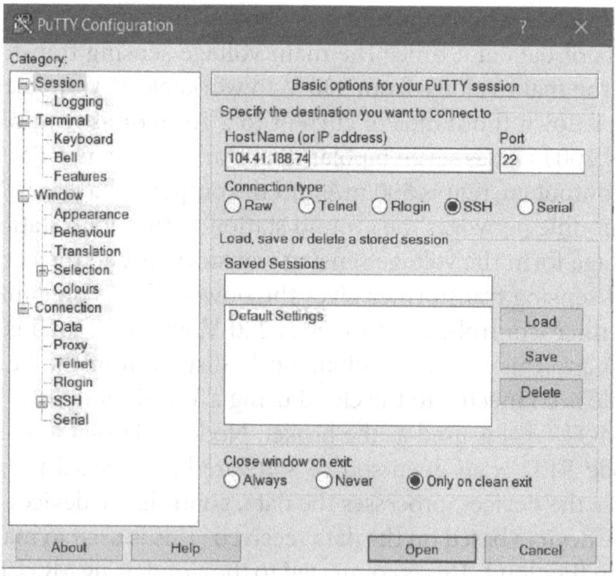

Figure 24.17 Putty configuration.

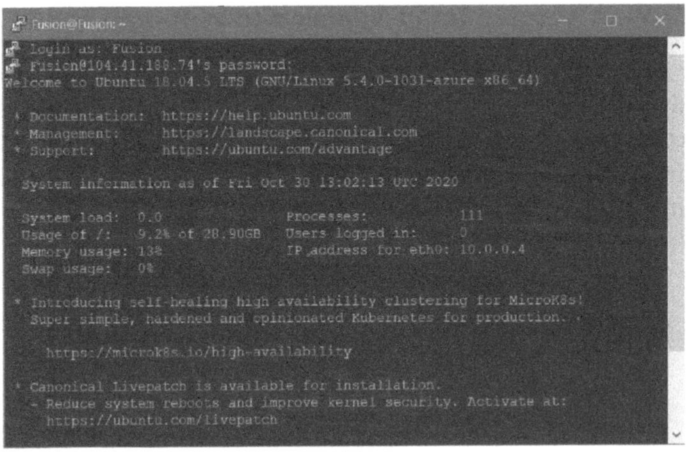

Figure 24.18 CLI for secured access.

24.6 Energy Management to Check Overload Conditions

Overload Condition is checked with varying load of 18W, 60W, 200W and 260 W. This is done by adding bulbs in the socket as shown in Figure 24.19. Bulbs of varying power were used for load to check varying load conditions. Electrical installation cannot be adjusted manually for varying loads [6]. With the Covid-19 pandemic, residential load has increased [4]. The proposed method tracks the varying overload conditions in real time.

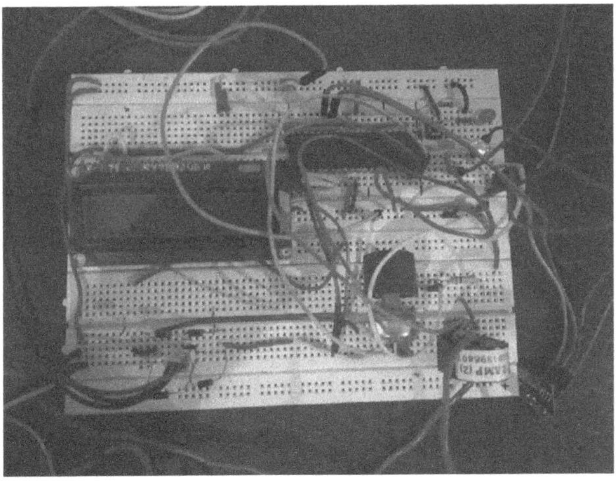

Figure 24.19 Smart grid with varying load.

24.6.1 With Varying Input Voltage and Without Load

Initially the input voltage was varied using a potentiometer and the output was analyzed using the dashboard, Client App, and data in Google firebase. This was tested without load. Cloud MQTT displayed the voltage values as shown in Figure 24.20. Power values were visualized in the Node-RED as shown in Figure 24.21.

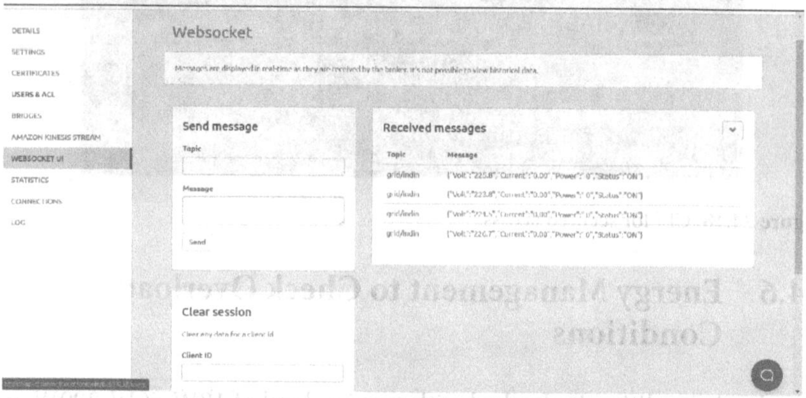

Figure 24.20 Power displayed in Cloud.

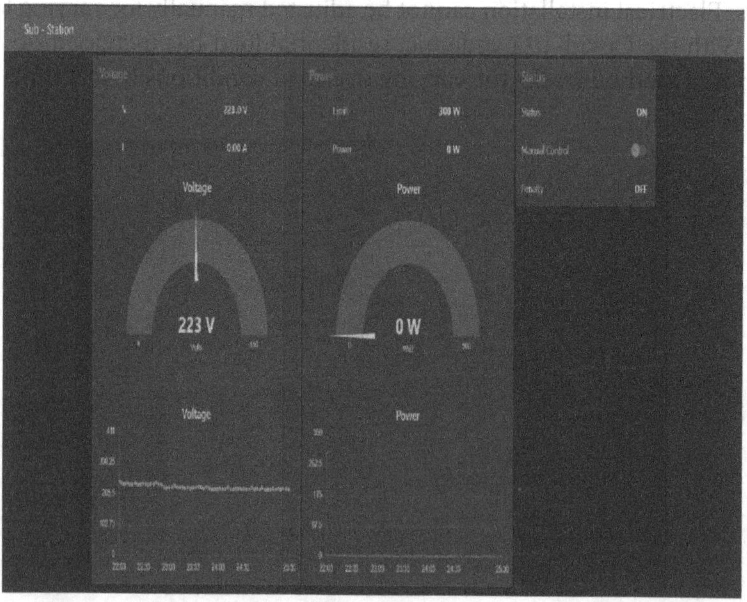

Figure 24.21 Optimum power values displayed in Node-RED dashboard.

By using the potentiometer to vary the input voltage, the dashboard displayed the corresponding output in real time as shown in Figure 24.22.

The real-time data was fed from the ground station to the Google Firebase as shown in Figure 24.23.

The Client Side App as indicated in Figure 24.24 indicates voltage, current and power consumed by the client. If the value is below or above the

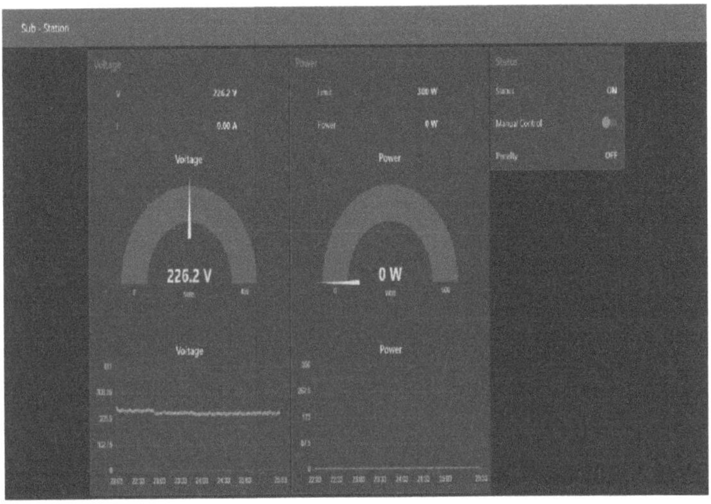

Figure 24.22 Varying voltage in Node-RED dashboard.

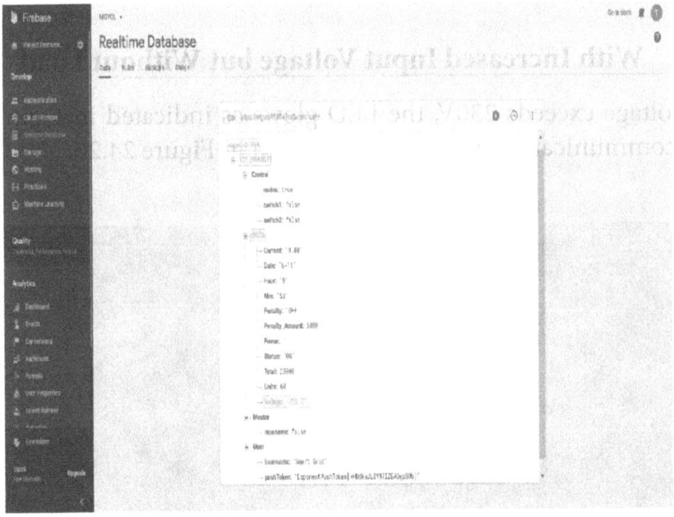

Figure 24.23 Google Firebase as the database.

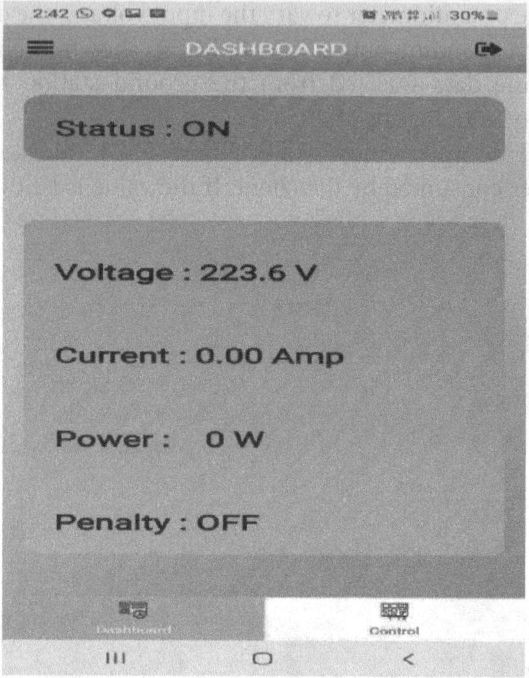

Figure 24.24 Client app.

optimum value '230V', it is indicated in the app. If the power consumed is more, the mains is turned off automatically.

24.6.2 With Increased Input Voltage but Without Load

If the voltage exceeds 230V, the LED glows as indicated in Figure 24.25. This is communicated to the cloud as shown in Figure 24.26.

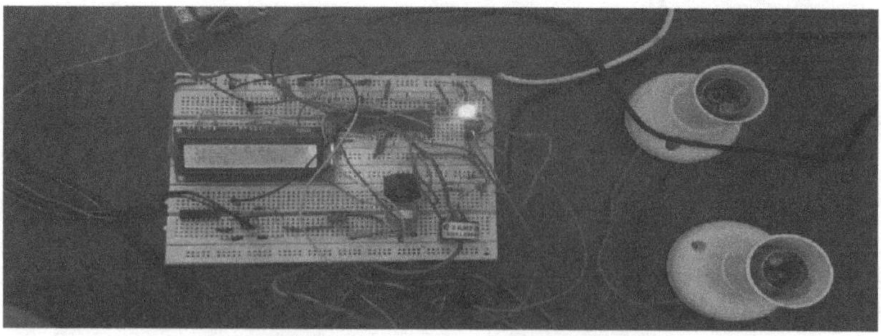

Figure 24.25 Overload condition in the power without load.

Figure 24.26 Cloud MQTT message for overload without load.

This is displayed in the dashboard and is notified as 'RED' for over-voltage and under-voltage load conditions as shown in Figure 24.27 and Figure 24.28. If status is 'OFF' then the mains supply is turned 'OFF'.

24.6.3 With Optimum Input Voltage and Load

With optimum load of two bulbs each of 9 W, the output was displayed in Node-RED and Google Firebase as optimum. This is indicated by green color as shown in Figure 24.29.

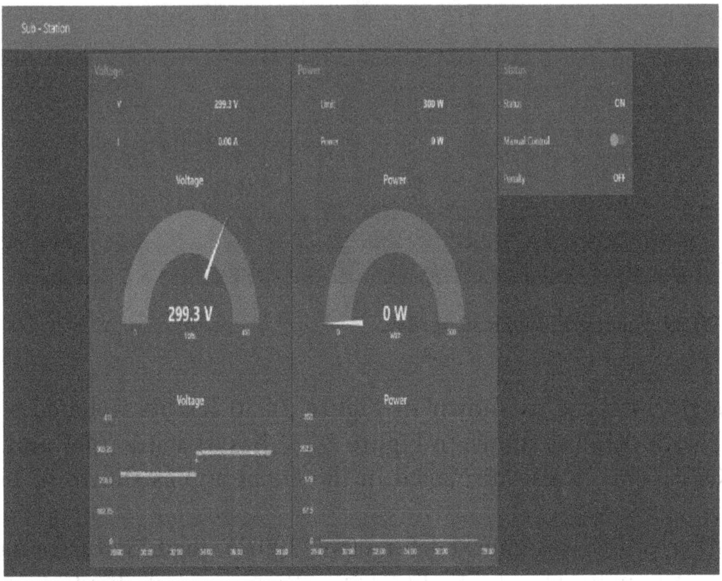

Figure 24.27 Node-RED indicating over-voltage.

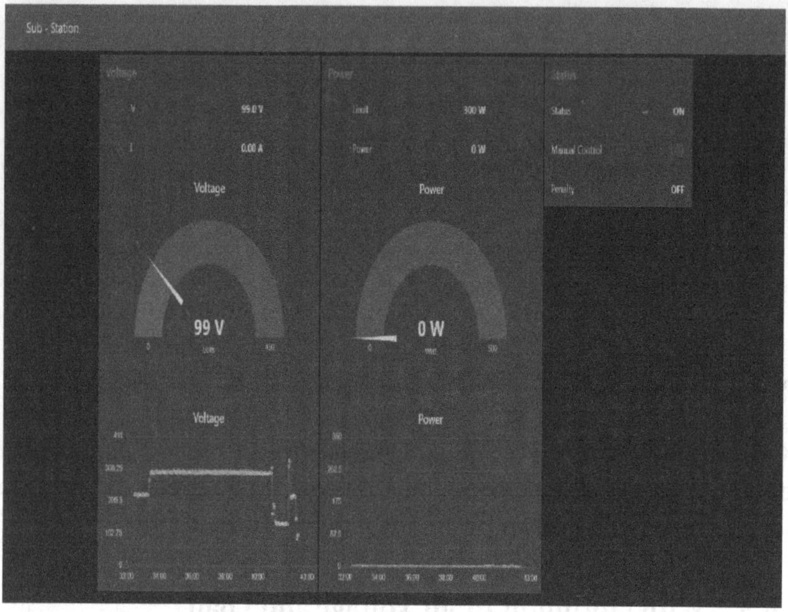

Figure 24.28 Under-voltage indication.

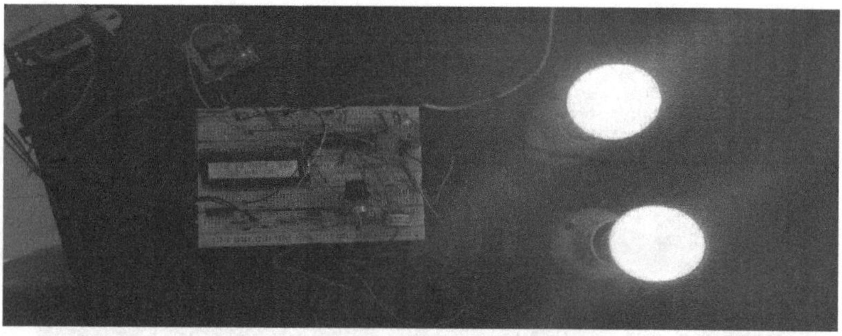

Figure 24.29 Smart grid with load of 18W.

Google Firebase as shown in Figure 24.30 has its updated power value. Node-RED as shown in Figure 24.31 has its status 'ON' and penalty 'OFF'. This is also displayed in the client app as shown in Figure 24.32.

Similarly this was checked for 60W, 200 W load. Status remained 'ON' and penalty was 'OFF'. But when there was overload of greater

Received messages

Topic	Message
grid/indin	{"Volt":"221.5","Current":"0.06","Power":" 14","Status":"ON"}
grid/indin	{"Volt":"222.0","Current":"0.08","Power":" 18","Status":"ON"}
grid/indin	{"Volt":"222.0","Current":"0.08","Power":" 16","Status":"ON"}
grid/indin	{"Volt":"220.1","Current":"0.10","Power":" 21","Status":"ON"}
grid/indin	{"Volt":"218.9","Current":"0.10","Power":" 21","Status":"ON"}
grid/indin	{"Volt":"218.7","Current":"0.08","Power":" 17","Status":"ON"}
grid/indin	{"Volt":"215.3","Current":"0.11","Power":" 22","Status":"ON"}
grid/indin	{"Volt":"215.4","Current":"0.10","Power":" 21","Status":"ON"}

Figure 24.30 Status 'ON'.

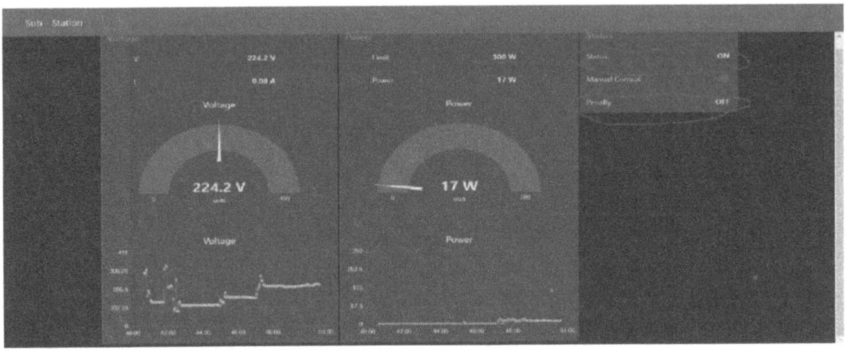

Figure 24.31 Node-RED-Status-'ON' and Penalty –'OFF'.

than 260 W as shown in Figure 24.33, Cloud MQTT displayed Status 'OFF' as shown in Figure 24.34; Node-RED displayed Status 'OFF' and Penalty 'ON' as shown in Figure 24.35; Google Firebase was updated in real time as displayed in Figure 24.36; and the Client App was updated in real time with power consumed and with Penalty turned ON as shown in Figure 24.37.

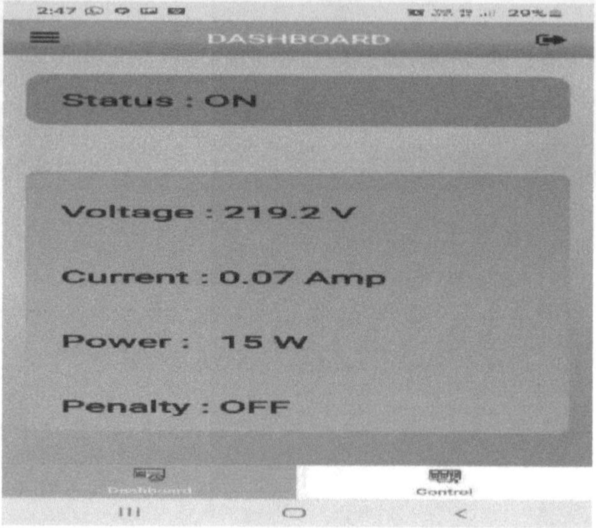

Figure 24.32 Client app with power displayed and Penalty –'OFF'.

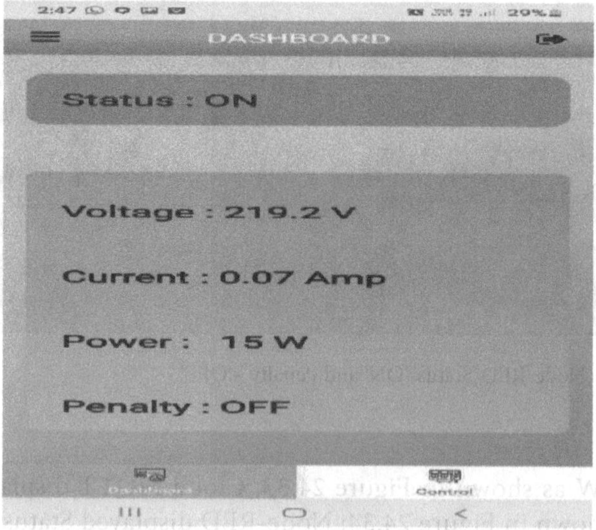

Figure 24.33 Smart grid with overload.

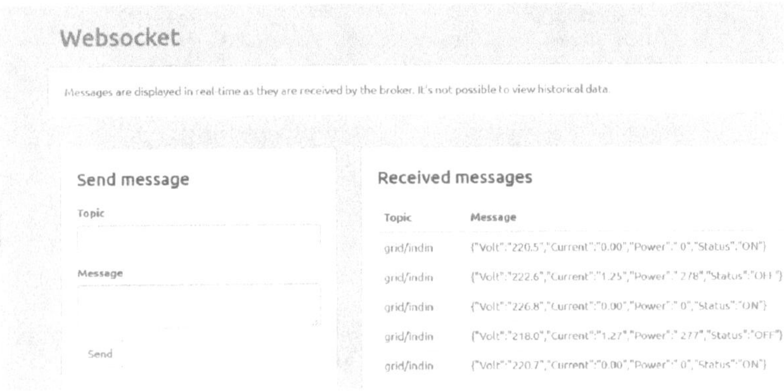

Figure 24.34 Cloud MQTT with overload.

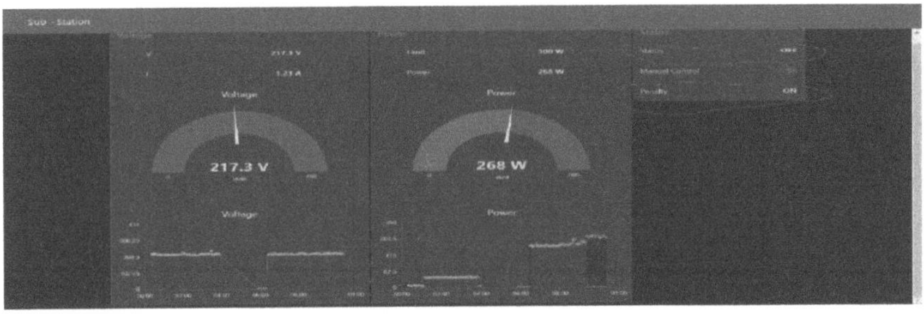

Figure 24.35 Node-RED with Status -OFF and Penalty -ON for overload.

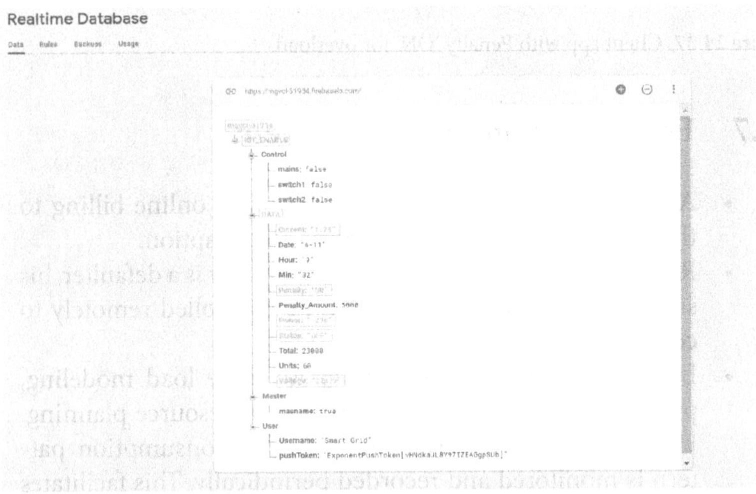

Figure 24.36 Google firebase update for overload criteria.

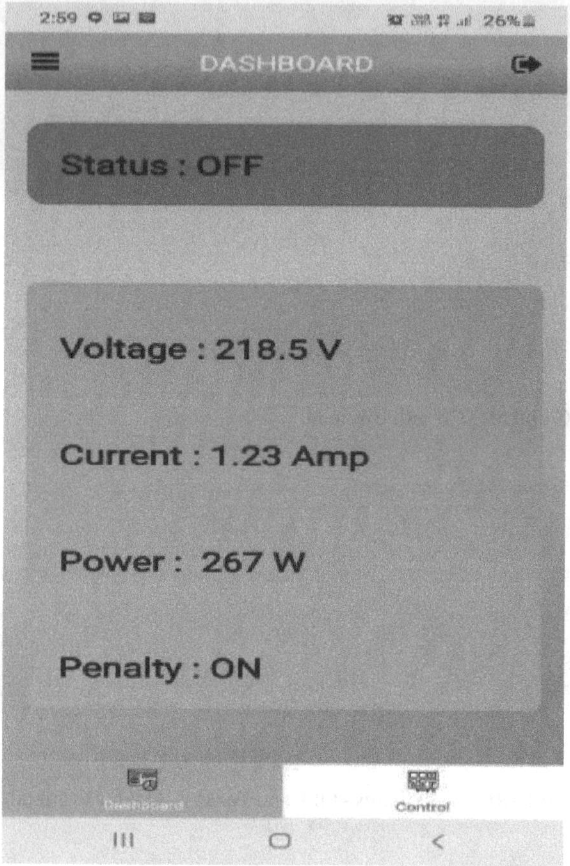

Figure 24.37 Client app with Penalty 'ON' for overload.

24.7 Features of Smart Grid System

- Automated billing: Smart meters provide online billing to the customer based on their power consumption.
- Remote Connect/Disconnect: If a customer is a defaulter, his smart meter consumption could be controlled remotely to connect and disconnect.
- Load modeling and forecasting: Accurate load modeling, predicting and forecasting is crucial for resource planning. Using data from smart meters, energy consumption pattern is monitored and recorded periodically. This facilitates resource planning.

24.8 Conclusion and Future Work

Using the displayed values in the Client App, the user is alerted and the mains are turned 'OFF' automatically. If there is a fluctuation in the input voltage, this causes the mains to be turned 'OFF' automatically. This helps to identify the fault easily. From a history of the usage pattern, power usage could be calculated and managed efficiently. With economic crisis in this pandemic, power consumption needs to be managed stringently. The percentage demand of electricity has changed in most of the sectors and therefore, the new load mix in different countries was observed and hence this solution would meet the need of the hour. Covid-19 pandemic healthcare depended on the individual person's access to energy and its reliability [13]. The proposed Smart Grid, analyses, proposes and manages electricity demand. The energy sector has had a vital role in ensuring the availability of healthcare, home-schooling, remote work and digital services during the Covid-19 pandemic. Using Smart grid load balancing, meters could charge the consumers based on usage. Unused power could be shared with the needy. The currently proposed system supports single-phase supply. This could be extended to three-phase supply protecting high load. Smaller grids could be connected to form a large grid. Data could be encrypted with block chain to ensure security.

References

1. Aviad Navon, Ram Machlev, David Carmon, Abiodun Emmanuel Onile, Juri Belikov and Yoash Levron [2020], "Effects of the COVID-19 Pandemic on Energy Systems and Electric Power Grids—A Review of the Challenges Ahead", *Energies*, Vol. 14.
2. Ali M.Hasan, Abdulkareem A. Kadhim [2020], "Design and Implementation of Smart Meter for Smart City", *Iraqi Journal of Information and Communications Technology (IJICT)*, Vol. 3(3).
3. Alireza Ghasempour [2019], "Internet of Things in Smart Grid: Architecture, Applications, Services, Key Technologies, and Challenges", *Inventions*, Vol. 4(22), doi: 10.3390/inventions4010022.
4. Bilal Naji Alhasnawi, Basil H. Jasim, Zain-Aldeen S. A. Rahman and Pierluigi Siano [2021], "A Novel Robust Smart Energy Management and Demand Reduction for Smart Homes Based on Internet of Energy", *Sensors*,Vol. 21 (4756), pp.1-28.
5. Claude Ziad El-Bayeh, Khaled Alzaareer [2019], "Energy Management in Smart Grid", *IEEE Smart Grid*.

6. I W Sutaya, K U Ariawan, I G Nurhayata [2021], "The design of automatic three phases load balancing for dynamic electrical installation", *Journal of Physics Conference Series IConVET* 20, IOP Publishing, 1810-012002.
7. Jiri Jaromir Klemes, Yee Van Fan , Pen Jiang [2020], "COVID-19 pandemic facilitating energy transition opportunities, *International Journal of Energy Research*, Vol. 45(3), pp. 3457-3463.
8. Kaile Zhou, Shanlin Yang [2018], "Comprehensive Energy Systems", *Science Direct*.
9. Mirza Jabbar Aziz Baig, M.Tariq Iqbal, Mohsin Jamil, Jahangir Khan [2021], "Design and implementation of an open-Source IoT and blockchain-based peer-to-peer energy trading platform using ESP32-S2, Node-Red and, MQTT protocol", *Energy Reports*, Elsevier, Vol. 7, pp. 5733-5746.
10. Megan Scudellari [2020], "The Pandemics Future", *Nature*, Vol. 584, pp. 22-25.
11. Rajvikram Madurai Elavarasan, GM Shafiullah, Kannadasan Raju, Vijay Mudgal, M. T. Arif, Taskin Jamal, Senthilkumar Subramanian, V.S. Sriraja Balaguru, K.S. Reddy, Umashankar Subramaniam [2020], "COVID-19: Impact analysis and recommendations for power sector operation", *Applied Energy*, Vol. 279.
12. Shama Siddiqui, Muhammad Zeeshan Shakir, Anwar Ahmed Khan, Indrakshi Dey [2021], "Internet of Things (IoT) Enabled Architecture for Social Distancing during Pandemic", *Frontiers in Communications and Networks*.
13. Vanesa Castan Broto, Joshua Kirshner [2020], "Energy health is needed to maintain health during pandemics", *Nature Energy*, Vol. 5, pp. 419-421.
14. Wadim Strielkowski, Irina Firsova, Inna Lukashenko, Jurgita Raudeliunie, and Manuela Tvaronavičiene [2021], "Effective Management of Energy Consumption during the COVID-19 Pandemic: The Role of ICT Solutions", *Energies*, MDPI, Vol. 19 (893).
15. International Finance Corporation [2020], "Impact of Covid-19 on the power sector".
16. UN Secretary-General. [2019] Special Edition: "Progress Towards the Sustainable Development Goals". UN Economic and Social Council, 2019.
17. Impact of Covid 19 on Electricity [2020], https://iea.org.
18. Ravishankar Velladurai [2019], "Interacting with Google Firebase data using Node-Red and Raspberry Pi', https://www.c-sharpcorner.com/article/how-to-interacting-with-the-google-firebase-using-a-raspberry-pi-and-node-red/.
19. https://www2.deloitte.com/us/en/insights/industry/power-and-utilities/energy-study-of-businesses-and-residential-consumers.html
20. https://www.centricabusinesssolutions.com/us/sites/g/files/qehiga201/files/documents/WP_Energy%20Management%20in%20Healthcare%20Facilities_NA_0.pdf

25

Energy Management Strategy for Control and Planning

Anmol D. Ganer

G H Raisoni Institute of Engineering and Technology, Nagpur (MH), India

Abstract

Energy is one of the primary sources of economic and technological development in any area. Energy is a leading requirement like food, cloth, and shelter just like illumination, electrical appliances like fan, television, air-conditioning and other things. So, as the demand of this all box of tricks the planning must be there on various strategies of energy management systems. To conserve energy, the investigation of energy supplied where it uses, conserves, and wastes. The audits improve the performance of activities from generation to the receiving end.

This chapter is not only to pioneer the management but also to preserve. The preservation squanders the energy where and when it will be. Sundry action is planned for organizing state of the art policies for plants and also energy security. The planning is smart energy management and the smart city, including smart town planning, smart security, smart transportation/mobility, smart governance, smart building/home, smart energy, and smart grid. The propound creep up on the impact of deputy applications and high-power rating loads in magnitude hours to energy expending expenditure of consumers. Replica results show that the propound creep up on the efficiency in reducing vertex load demand and electricity spending charges with a rising in the comfort level of patrons. In the case of the smart grid, the main application is a smart city; the features are taking under the topic, i.e. need to enhance for consumers for their benefits are developing the nature of IoT in electrical appliances, reducing the maintenance cost, improve public safety via all-time surveillance, aid economic development, enhancing consumers service and obvious conservation.

IoT can play a vital role in energy management strategy in smart city and grid networks; it helps to contribute to various applications in which the consumers can benefit from economics and audit. The benefits are smart metering networks,

Email: anmol.ganer@raisoni.net

Milind Shrinivas Dangate, W.S. Sampath, O.V. Gnana Swathika and P. Sanjeevikumar (eds.) Integrated Green Energy Solutions Volume 2, (49–70) © 2023 Scrivener Publishing LLC

smart city lighting, smart grid automation, ecological monitoring solutions, efficient public service tracking. The main sub-assistant benefit to smart energy is a microgrid. A microgrid is a standalone structure. The Microgrid is a widely founded research in the current curriculum that has propounded the centralized and distributed scheme application. In this research, we are concentrating on energy security. The main concept about energy security is to, before expansion and delivery; we must go through the queries that why we need security, on which ultimatum, from whom, and security on which merit. Energy security is a global confusion issue for the general public, policymakers, or legislators.

Strategies must be planned to increase and improve the individual renewable type of energy to use in the industrial premises to reduce consumption. Planning is to strengthen the renewable energy source, which is the only way to reduce greenhouse emissions—a blueprint for the Microgrid at an individual level in domestic and commercial consumers. Blueprint to replace the running energy infrastructure so advance new better consumption free structures develops for consumers. Strategies to reduce fuel squander. So, it will directly reduce consumption and indirectly greenhouse emissions and air pollution. This chapter is a case study of different energy management proposals to conserve energy investments via organizing online, offline training programs and evaluating energy performances, action planning, and economics.

Keywords: Energy, micro-grid, smart city, smart security, smart mobility, smart governance

25.1 Energy Management and Audit

Energy is one of the primary sources of economic and technological development in any area. Energy is a leading requirement, like food, clothes, and shelter, like lighting, and electrical appliances like fan, television, air-conditioning, etc. [1]. To meet the demand of this box of tricks the planning must be there on various strategies of energy management systems. [1] To conserve energy, the investigation of energy supplied information about where energy is used, conserved, and wasted. The audits improve the performance of activities from generation to the receiving end. Energy management is a system in which we have to manage the energy to conserve in terms of many applications in the residential and commercial levels. [1] The management of energy is used to improve the cost as well as give benefits to consumers. So, as per the energy management requirement, it needs to conserve more to improve the audits reports. [1] The administration intends to fulfil and continue prime energy acquisition, and use it all over the system to shrink energy cost/misuse without influencing the generation and superiority.

25.1.1 Steps for Energy Audit Management

Step no.	Step for energy audit management
01	Plan and organize the resources. Macro data collection for audit reports. Observation on energy consumption.
02	To convene a cursory conference/cognizance campaign with all regional leaders and other covered parties. Awareness creation orientation programs are to be held.
03	Collection of historic and baseline data collection for previous waste energy consumption to analysis the previous issues.
04	Analysis of energy consumption with also energy and material balance in system analysis with energy loss/waste analysis
05	Analysis of the Capital benefits.
06	Reporting and presentation to top of management.
07	Implementation and follow up on or regular energy utilization and conservation.

25.1.2 How An Energy Audit can be An Effective Energy Management

Continuous power consumption audits aid in the recognition of intentional and unintentional wastage of electrical power. [25] Audits also aid in recognizing moments for power control and custom for justification of ruin. [25] Fitful reduction in power misuse helps to lessen greenhouse gas, and constructs shielded functioning territory and other protective elements. [25] The practical of decay also clinch the health and welfare of independents. Chronicle of misuse of power can be accomplished by conferences recommendation that ameliorates imparting in the middle of assets administration and society functioning in these sites. [25]

25.1.3 Power Conservation through Energy Audit

Gradually Power ultimatum retains revolting so that it is necessary to diminish power misuse; for that, energy conservation is required. [26] The most acceptable choice to save power is the summons to dictate when, where, why, and how power is utilized in plants and premises. [26] These statistics aid in recognizing the circumstances where there is the requirement to improve power efficiency and reduce manufacturing capital. [26] Usually, a power inspection is carried out by certified power auditors. [26] By managing power inspection tasks in the company, artisans initiate given authority as a feasible levy and try to save it in day-to-day activity. [26]

25.1.4 Study of Energy Management and Audit

The expansion of wealth in a country is often closely linked to its consumption of power. [27] The government has taken new strides to develop renewable energy roots and minor steps in public conservation of electrical energy. [27] As per the current scenario, the dictate of power has increased and become a scheduled work in our lifestyle. [27] Why is electrical auditing and power management necessary? An energy audit is a study of power wastage in premises like domestic areas, commercial flats and industries, etc. [27] Energy auditing is the only path that will lead to a solution regarding the conservation of electrical energy. [27] It is observed that in India the demand of electrical power rises at the rate of 10 to 11% every year while the generation of electrical power rises at the rate of only 5 to 6% every year. Eventually, the gap between demand and production of power extends at the rate of 3 to 4% every year. [27] An electrical energy auditing and management project can act as a motivation to the public to go through conservation actions in all forms.

25.2 The Different Steps of an Energy Management Approach

The purpose of power management is to control and optimize the consumption of your organization and increase the amount of conservation. [2, 3] Energy management is a process to plan and organize how to improve conservation and control unnecessary consumption. The procedure is to plan and produce power in renewable and non-renewable sources and develop the microgrids system for local heads [www.energylens.com]. The main objective of energy management is to minimize the environmental effect. [2, 3]

It is necessary to sustain prime energy procuring and usage all over the organization and reduce the cost of energy waste while maintaining the quality of energy and its production quantity [www.beeindia.gov.in]. Before learning about energy management approaches, we must know about every states' generation capacity as per the plants. [2]

The commitment should be made in Energy Management Strategies, i.e.:

1. Assess Performances and Set Goals.
2. Create an Action Plan.
3. Implement an Action Plan.
4. Evaluate Progress.
5. Recognize Achievements.
6. Reassess Performance and set goals, and then the process and cycle will continue.

The following flow chart shows the process.

25.2.1 State-Wise Generation Capacity till 2019

Sr. no.	States	Hydro	Thermal	Nuclear	Res (MNRE)	Total
01	Jammu & Kashmir	2369.48	810.47	67.98	180.39	3428.32
02	Himashal Pradesh	2910.48	245.41	28.95	853.84	4038.68
03	Punjab	3701.05	9004.09	190.01	1202.42	14204.97
04	Chandigarh	101.71	53.17	0.01	25.20	100.09
05	Haryana	1948.21	8781.12	109.94	411.75	11424.01
06	Uttrakhand	1815.69	961.90	31.24	547.40	3356.23
07	Rajasthan	1930.97	11763.15	556.74	6773.63	21024.50
08	Delhi	723.09	6937.35	102.83	121.57	7854.84
09	Uttar Pradesh	3421.03	18623.21	289.48	2677.01	25010.73
10	Bihar	110.00	3905.33	0.00	326.15	4341.47
11	Gujrat	772.00	22168.00	559.00	7295.32	30794.32
12	Madhya Pradesh	3223.66	128.05.41	273.00	4019.80	20321.87
13	Chhattishgarh	120.00	12723.44	48.00	535.35	13426.79
14	Jarkhand	191.00	1544.74	0.00	29.72	1764.46
15	DVC	186.20	6985.04	0.00	0.00	7171.24
16	Meghalaya	387.19	140.09	0.00	31.05	558.33
17	Manipur	88.93	138.97	0.00	5.51	233.41

(Continued)

(Continued)

Sr. no.	States	Hydro	Thermal	Nuclear	Res (MNRE)	Total
18	Assam	431.22	1027.53	0.00	49.50	1505.32
19	Nagaland	53.37	70.33	0.00	31.16	155.37
20	West Bengal	1396.00	8805.77	0.00	435.82	10637.59
21	tripura	62.38	643.85	0.00	21.10	727.33
22	Mizoram	36.67	61.16	0.00	94.19	192.02
23	Daman & Diu	0.00	169.97	7.00	10.61	187.58
24	Odisha	2150.92	4992.90	0.00	194.60	7338.42
25	Maharashtra	3331.84	30473.48	690.00	8578.88	43074.19
26	Telangana	2449.93	7876.07	148.73	3659.52	150087.13
27	Dadra & Nagar Haveli	0.00	240.78	0.00	5.46	255.24
28	Goa	0.00	522.45	26.00	0.96	549.41
29	Karnataka	3599.80	9960.82	698.00	12438.85	26697.47
30	Andhra Pradesh	1673.60	14524.76	127.27	6725.88	23051.51
31	Lakshdweep	0.00	0.00	0.00	0.75	0.75
32	Kerala	1881.50	2451.76	362.00	379.46	5074.72
33	Tamil Nadu	2203.20	15086.07	1448.00	11165.41	29902.68
34	NLC	0.00	100.00	0.00	0.00	100.00
35	Puducherry	0.00	280.90	86.00	0.16	367.06

[www.mospi.nic.in]

25.2.2 The Effective Plan should Incorporate Four Basic Steps

1. Measuring energy: Collect energy consumption statistics from buyers within a service & examine individual's jolt on full utilization (Figure 25.1.A).
2. Fix the basic low consumption equipment: Finding energy-efficient devices, like LED lighting and fixing power quality issues are important parts of energy management.

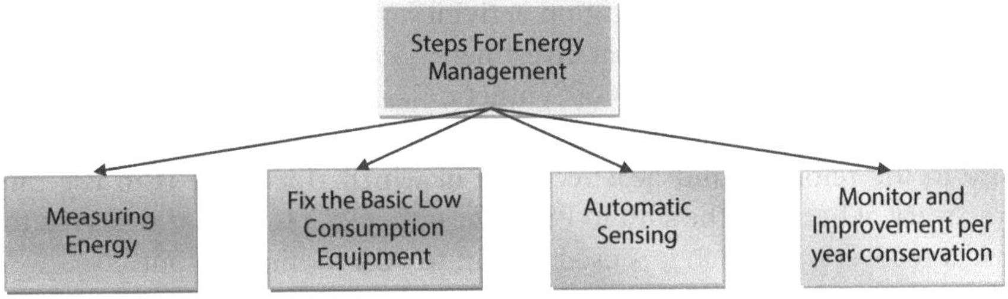

Figure 25.1.A Four basic steps for energy management.

3. Automatic/Sensing: Controlling the lighting and other electrical appliances with residence sensors automatically turns lights when they are needed.
4. Monitor and improvement for per year conservation: It can be said that loss of conservation due to an unplanned, unmanaged shutdown of energy equipment and processes occurs each year. This can be attributed to inadequate automation and regulation, substandard maintenance and a lack of behaviour continuity. [www.govtech.com]

25.3 Preliminary Technical and Economic

25.3.1 Assessment of Synthetic Gas to Fuel and Chemical with Emphasis on the Potential for Biomass Derived Syngas

Syngas can theoretically be made from any hydrocarbon feedstock, including natural gas. [19] However, the most cost-effective syngas production paths are based on natural gas, with remote or stranded deposits being the most cost-effective choice. [19] The new synthesis of liquid fuels from syngas necessitates using natural gas as the hydrocarbon source due to economic reasons. [19] Nonetheless, the syngas production process accounts for more than half of the capital cost in a gas-to-liquids plant. [19] The magnitude of the synthesis process influences the technique used to produce syngas. [19] With the addition of feedstock handling and more complex syngas purification activities, solid fuel syngas production can demand even more capital investment. [19]

The most significant influence on improving the economics of gas-to-liquids plants is achieved by (1) lowering capital costs connected with syngas production and (2) increasing thermal efficiency through greater heat integration and use. [19] With the addition of feedstock handling and more complex syngas purification activities, solid fuel syngas production can demand even more capital investment. [19]

The Economic Assessment of Synthetic Gas: Natural Gas is released into the Zinc Oxide and the Fuel gas to the steam reformer. The air and water are to the reformer and heat recovery in which zinc oxide is to use the recovery of heat and the shift to PSA. The out of PSA is again transferred to Heat recovery, in which it is used to shift to PSA for final output.

25.3.2 Natural Gas Storage/Co-Fired Retrofit System

It has been argued that utilities may boost their usage of gas by acquiring it while it is cheap and storing it for later use, rather than buying it from the source, which is expensive. [20] The motive of this project was to look into the technological practicability of this proposition. [20] Combustion of gas in coal-fired power stations has the potential to help these plants meet Title IV of the Clean Air Act's criteria. [20]

Biomass combustion renewable organic material is primarily observed as an extremely economical and freely installed way to alleviate the coal power section's carbon dioxide excretion. [30] Aside from scheme and market profits and congestion, the accomplishment of the co-firing in a steam power plant is pretentious. [30] Several technical solutions have been developed and illustrated for co-firing schemes, from naturally immediate steam schemes to the more experienced aligned and indirect co-firing systems. [30]

25.4 Evaluation of Energy-Saving Investments

In the evaluation of power conserving investments, we need to suggest a uniform disclosure method to consumers. [3] Revelation for energy saving is associated with the use of revelation policies in Truth-In-Lending and other sectors. [3] Five methods of evaluating energy-saving investments are analyzed: [3]

The methods combined with policy revelation, education, and product standards would be the perfect alternative to reach energy-saving goals [3].

The government has opened (Figure 25.2.A) the awareness of conservation of energy in recent years. [4] Consumers are not aware of energy savings and their investments; the government has granted special subsidies and loans to decrease energy waste usage in the construction sector, which has been researched from an energy-saving point of view. [4] A

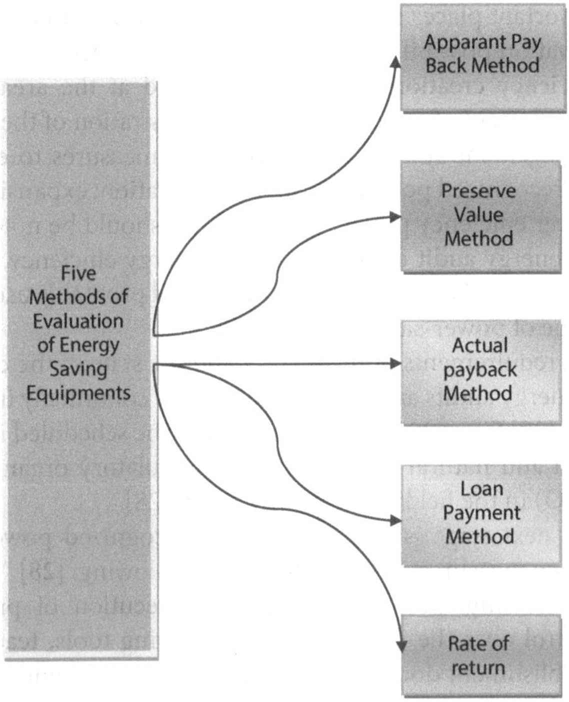

Figure 25.2.A Evaluation of energy-saving investment.

blueprint is fabricated, which shows how various elements are organized in the profitability of Energy-saving investment. [4] Tactful research came out to authorize how taxes, subsidies, energy tariff trends, and modern technology affect the solution achieved. [4]. Energy-saving investments should be carefully selected for the national economy to generate supreme results with desired interest [5]. Further in future cost, it may decide what formation of investment is going to be planned out of energy-saving which is being replaced. Furthermore, in attending and future interest, the tariff will recreate a major part. [5]

25.4.1 Power Survey – Energy Inspection

Energy Inspection is a layer of technical, administratiove, and ecoproductive activities targeted at recognized possible capital effective optimizing the consumption of power assets. [28] To accomplish the aim, the following points should be observed: [28]

1. The expansion of power certificates; evaluation of the share of power in manufacturing costs; recognition of most

important places of power-saving; evaluation of power conservation possibility in selected places; inspection of power efficiency creation ongoing or planned at the area audit; enlargement of schemes for the administration of the power management system. Expanding tacit measures to execute the recognized potential power conservation; expansion of a power efficiency project object. [28] It should be noted that the energy audit is the first step of energy efficiency, whose main target is recognition, technical and profitable use of the scope of power-saving measures. [28]
The requirements of the market and the state of the content of energy audits and energy auditors are continually increasing. [28] Since 2010, power audits may be scheduled in both legal and natural persons in a self-regulatory organization (SRO) in the field of the energy audit. [28]

2. The next stage is carrying out the recognized power efficiency measures, which includes the following: [28]
Engineering, project arrangement, execution of projects, control over the delivery of power-saving tools, teachings, establishment documentation, power management, observing, helping to sustain power-saving tools. [28]
3. The last stage is the definition of routine personnel portion power-saving tools and technical schemes. The power-saving audit which was managed jointly with the Norwegian energy group under social Kirvovsk, produced power saving in the range of 30% to 45%. [28] To design methodological equipment for administration direction on the power efficiency in the undertaking is compulsory to execute the detailed modelling schemes to scrutinize power consumption, the structure of which should depend on the theme of unity methodology, software, details, technical and administrative support. [28]

25.5 Off-Line and On-Line Procedures

25.5.1 Concept

This topic describes two broad strategies for dealing with multi-stage optimization issues under uncertainty, including off-line and online decisions. When 1) the uncertainty is exogenous and 2) there is a heuristic for the online period that can be treated as a parametric convex optimization problem, the approaches can be used. [21] The first technique uses a systematic procedure to replace online heuristics with an anticipatory solver. [21] We use two case studies to illustrate our methods: an energy management system with unpredictable renewable supply and load demand and a vehicle routing problem with unknown journey times. [21]

Dealing with uncertainty in optimization problems is difficult, but it is becoming increasingly acknowledged as a requirement for obtaining practical outcomes. [21] However, as the ambiguity is gradually revealed, online algorithms have the potential to utilize further knowledge. [21] A rising number of works have been addressing online problems using techniques originally established for stochastic programming, such as sampling and the Average Sample Approximation, to tap into this potential. [21] Sampling refers to acquiring realizations (scenarios) of the random variables used to model uncertainty; it is feasible to enrich an online method with some degree of anticipation by solving deterministic optimization problems over multiple scenarios and computing averages. [21]

25.6 Personnel Training

Given the trend in worker training toward specified and the application of technologies for self-learning [22], we investigate the issue of complexity as a factor in the functioning and preservation of electricity-generating equipment (Figure 25.3.A) [22].

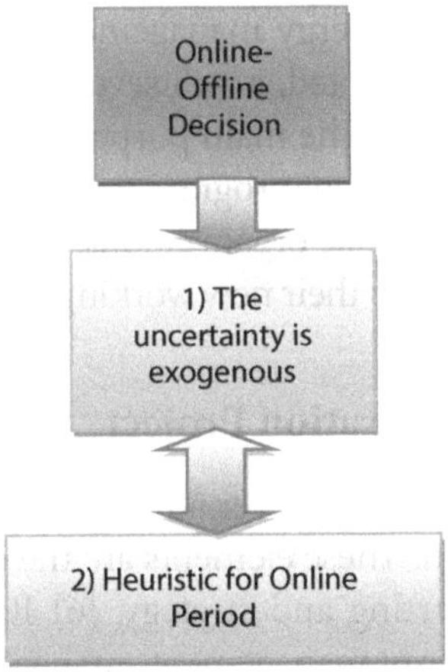

Figure 25.3.A Flow chart.

25.6.1 Training Method for Electricity Work Safety

For technical and non-technical abilities, a simulation is a beneficial approach for improving learning and increasing the safety of work processes. However, seeing, assessing, and providing feedback on these skills is extremely difficult because the procedure requires experienced observers' employment, and comments may be judgmental and ineffective. As a result, a disciplined approach for developing compelling simulation scenarios and instruments for performance observation and feedback is critical. To do this, we built a high-fidelity simulation-based training model for electricity distribution personnel in the current study.

25.7 A Successful Energy Management Program

25.7.1 Introduction

As perceived in the data, companies are becoming more efficient in their use of energy. They have made it possible to conserve energy with more positive approaches towards the system. [6] The superior Management of energy is feasible, so for various companies, there is an inclination to set up the energy management program. [6] Companies realize that to get a good energy management plan, they need to initially leave a lot of capital on the table in the beginning. [6] Companies can reduce energy consumption by 50% due to the availability of new technologies and alternative energy plans. [6] By the way, this as energy management is not just a technological challenge but also a technical change within economic limits. [6] Energy management is nothing but energy security, and energy security is also a part of conservation, which is only done with a successful energy management program. [6] By default, if a management plan is not adopted, it will severely affect energy security and show future discrepancies. [6] The main purpose of this chapter is to give the outlines of an energy management program that can be and has been adopted by small, medium and large-scale organizations. [6] While adopting new programs, companies can develop their new working organizational structure. [6]

25.7.2 Power Administration Project

All the elements of the inclusive power administration project are illustrated in the given block diagram. These elements are the corporate structure, education, audit, policy, reporting and strategy. [6] It is an aspiration that by knowing the bases of energy management, managers can adapt good skills of an energy management working program to the existing corporational structure. [6] The detailed corporate structure is considered in the diagram below.

25.7.3 Corporate Structure

The corporate blueprint for energy management is shown in block diagram 25.7a. The chart is designed to take over the existing structure for all small, medium and large-scale companies. [6] In the diagram, the President block may be general manager, and the VP block may be divisional managers of specific areas and other designation, but the foundational principles are the same. [6] The main characteristic of the blueprint is the site of energy managers. [6] The availability funding proposals of energy management priorities should be known and acknowledged. [6] The corporation of energy management managers is also dedicated to the support administration to give to the position (Figures 25.4.A and 25.4.B) [6].

25.7.4 Energy Management Managers

The critical element for an Effective and Successful Energy Management Program is to have top-level management support; all the main decision-making authority depends on that top-level management. Most important is the selection of Energy Management managers, who can

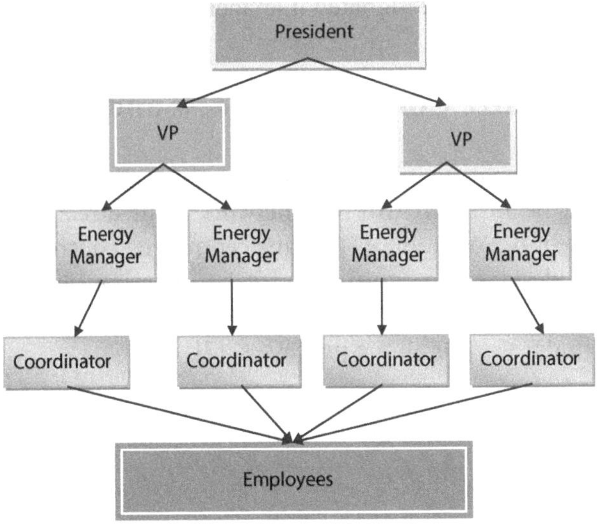

Figure 25.4.A [6] Energy management program.

Figure 25.4.B [6] Flow chart: energy management program.

secure this reinforcement among all others. [6] Whoever is selected for this managing designation must understand the vision and mission of what managing energy can do for the organisation. [6] There is an excellent motivation for energy managers to be energy engineers and endeavour to perform the whole effort alone. [6]

As the block diagram shows, the various administrative and educational authorities, duties, audits, strategies, reporting and organisational work are to be handed over to the designated managers. [6]

25.8 Centralize Control of Process and Facility Plants

In the Power management control system, the centralized management plants should be installed so other networks will directly or indirectly connect to the junction point. [7] In this centralized control model, single equipment is designed to control, and it has a responsibility to manage other equipment to control it in a centralized way. This model can be executed in a similar way for all the processes. [7]

25.8.1 Centralized and Decentralized Waste Water Management

In highly populated areas of industrialised countries, the classic wastewater management approach has been successfully utilised for decades. [23]

The issue is the expense of installing a centralised system [23]. However, there is disagreement over the level of technological sophistication that should be used. [23] We argue in this study for the creation and use of contemporary local administrate foundries that are developed, including those built using the latest commercial processes. [23] Manufacturing expenses for such package plants are likely to be kept low when mass-produced. [23] The plant should create a sanitary effluent that can be used for toilet flushing, laundry, floor cleaning, and yard watering. [23] Remote sensing and specialised service firms should be used to keep the plants running smoothly. [23]

25.8.2 Central Jurisdiction System

A Unite jurisdiction represents a system in which all equipment like actuators, sensors, and other amenities are joined to a sole handler or array of handlers found in a frequent control room. [23] Detecting all controls, operator interfaces and indicators in a frequent room ameliorate operator knowledge of system conditions and speed retaliation to eventually. [23] Streamlined jurisdiction system should only be contemplated for small commercial amenities and if, applied must have fully unnecessary processors. [23] When unessential is provided in a Streamline jurisdiction system, shielded wiring traces must be provided to self-confident that jurisdiction signal and form of instruments or protected design are not subject to usual defeat from electrical fault, physical and environmental request. [23]

25.8.3 Centralized Process Control System

Centralized Process Control System continues to manufacturing lessen, capital and employees results become progressively difficult [29]. A sturdy administrator jurisdiction & Data Acquisition (Figure 25.5.A) system also makes sure flexible & firm instantaneous online observing, manufacturing data reporting, and remote troubleshooting. [29]

25.9 Energy Security

25.9.1 Energy Security Concept

Energy Security is the continuous accessibility of energy sources at rock-bottom cost. [8] Electrical Energy Security has many aspects: (a) the first aspect is long-term security mainly dispense with felicitous stakes to supply

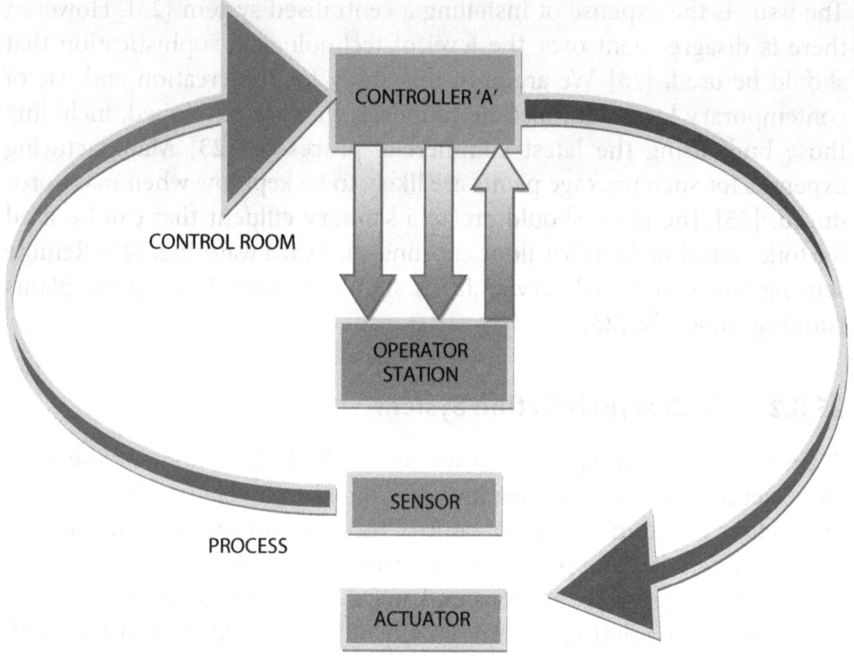

Figure 25.5.A Centralized control system.

energy in line with remunerative growth; (b) the second term is short-term security that concentrates on the potential of the electrical system to respond swiftly to instant changes in the supply-demand balance. [9] Energy Security is one of the main goals in Energy Management Strategies and Control and Planning. Energy Security can be distinguished as per the source of risk, the extent of the impact, the sternness filters in the form of speed, suspension, spread, singularity, size and sureness of implications. [8] Electrical Energy security is constrained by the innumerable element in the middle of results, and energy conservation patterns play a vital role. [9] Electrical Energy storage is the summons by which energy is stored or conserved from the power grid from which it will be used when it will transform back to electrical energy [9]. There are many other techniques to store electrical energy in future scope. [9] Nowadays, electrical energy is stored in hydroelectric energy form that incorporates about 99% of electrical energy storage worldwide. In another way, the battery electrical energy storage uses chemical energy with both methods its relent its merits and demerits. [9] As per a survey, electrical energy is mainly conserved when low demand or peak energy generation at small costs. [9] The electrical energy is used in that season when there is any scope of power energy, and the electrical generation capital is high. [9] The stored energy has many uses, such as in electrical equipment, motor conveyance and non-rotary energy resources. It is obtaining distinct heed with the worldwide usage of renewable energy sources. [9]

25.9.2 Smart Grid Security

In the case of Power Grid, this point is an aspect that seems to be a Smart Grid is a very loaded structure with numerous meanings. [10] Before making it brighter, first, we must learn the electrical Grid. It became a smarter one. [10] The orthodox Electrical Grid gives finite mechanisms to control and surveil. [10] In these Grid systems, the Power suppliers are also aware that some consumers are afraid to pay the energy bill. Some of those are taking the risk of theft of the electricity when they needed it. [10]

25.10 Evaluate Energy Performances

25.10.1 Concept

As India is the largest energy industrial and residential consumer, the population is growing. [11] The new techniques possess structure and generation executive to enhance the design to magnify the electrical energy performances. [11]

25.10.2 Building Energy Performance

Standardized accelerative procedures for making energy performance usually begin by identifying many different (Figure 25.6.A) characteristic features and then assigning weights to those features to arrive at a predominant metric measuring overall building energy performance. [12]

25.10.3 Illumination and Energy Performance

Artificial illumination is the prime electricity element in most offshore structures. [13] Using a photoelectric dimming set-up, this open plan office ensures complete air conditioning in the offices while using significantly less energy than the traditional set-up. [13] Indoor Illumination and electric lighting load levels and sunlight accessibility were comprehensively measured and scrutinized. [13]

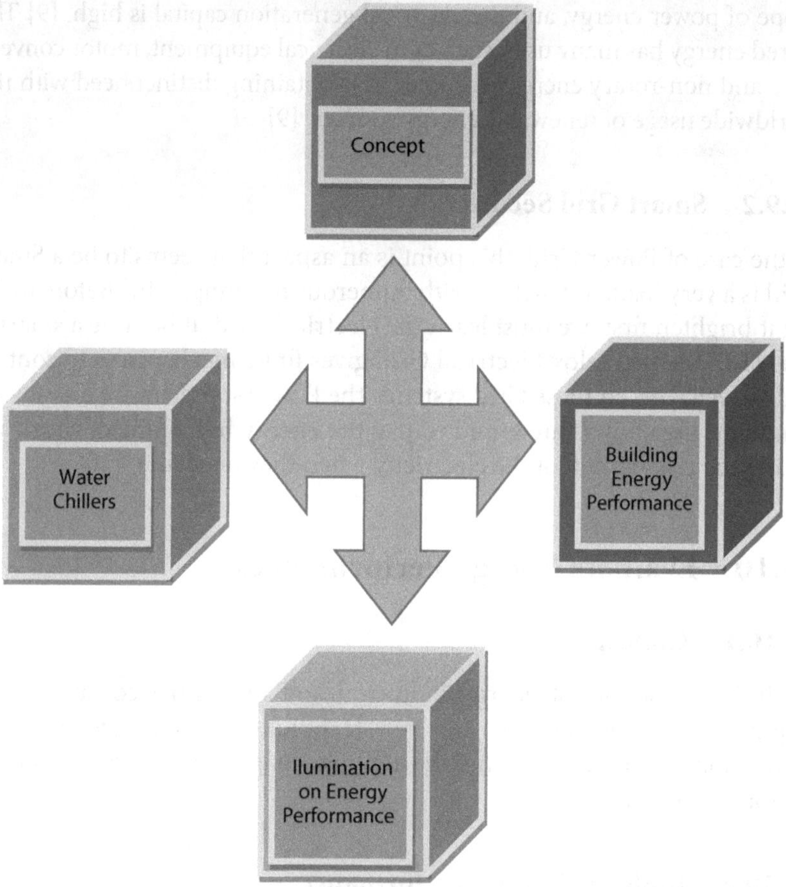

Figure 25.6.A Schematic diagram evaluate of energy performance.

25.10.4 Energy Performance of Water Chillers

The cunning or authorization of the power grasping of building furnished with freezing or heater running on vapour squeezing cycles require a précised assessed under different operating conditions, they demonstrated the staging of full and part load operation. [14]

25.11 Energy Action Planning

The desperate worldwide need to fortify the environment, conserve energy and diminish greenhouse gas emissions global has compelled enterprises to execute both single energy-saving measures and a more structured approach to upgrade the complete enterprise's energy performances. [15] Energy Action Planning and Management strategy is being given precedence as enterprises strive to diminish energy costs, fulfil official requirements, and facilitate their collective image. [15]

However, as previously researched, city and territory plans with over 50,000 use similar long-term model-based energy planning processes [16].

Structures are the substantial citified power users, but their effect can be the most significant retrenchment by refinement ability. [31] Decision frolic an essential role in synchronizing civic and city inducement strategies. [31]

25.12 Energy Economics

In the future decades, the worldwide energy system will face three significant strategic challenges: the increased risk of interruptions to the energy supply, the possible harm to the environment, and ongoing energy poverty [17]. As part of the more extensive process of human growth, concrete governmental action is required promptly to change this trend. [17] Basic human necessities, like food and shelter, must be prioritised. [17] These requirements can be addressed with the help of modern energy services. [17] In practice, tangible increases in human welfare can be achieved fast and at a low cost in the short term. [17]

Energy Economics: Environmental Effect of the renewable energy system: Using data from the literature, sustainable energy production systems were evaluated in opposition to various defendability indices. [18] Evaluating different renewable energy sources: This refers to greenhouse gas emissions, the payback time for each renewable energy project, and the cost of the electricity generated by these projects. [18] Hurricane and compact water energy generation have been discovered to be the most sustainable sources of electricity generation. [18]

25.13 Case Study

The current study proposes an excellent sustainable spirit administration approach for crude oil centralized power stratagem. [24] Thus, dictate consignment and large cloud cover data sets were analyzed using an optic grid and examined variance. [24] Using the K-means method, the data cast-off classifies countries into different groups created on its ace green energy potential sustainable power. A point examined (FA) in all Iranian provinces was conducted utilizing a large cast of 13 communal, profitable, ambient, and technological adaptable. [24] Then, using factor score coefficients, a fully-fledged numerous yardstick verdict scrutinized norm was used to ascertain the mandate preferences to optimal tincture. [24] Such requirements have been incorporated into the renewable energy management strategy. The best blend of sustainable power was created. [24]

References

1. R. Kannan, W Boie. "Energy management and practises in SME: Case study of a bakery in Germany", Vol. 44, Issue 6, April 2003.
2. Ms. Shradha Chandeakant Deshmukh, Ms. Varsha Arjun Patil, "Energy conservation and Audi," Electrical Engineering Shivaji University, Islampur, Tal-Walva, Dist-sangli, Maharashtra, India.
3. Sherman Hanna, Evaluation of Energy Saving Investments, *Journal of Consumer Affairs* Vol. 12, no. 1, (Summer 1978): https://doi.org/10.1111/j.1745-6606.1978.tb00633.x, http://www.jstor.org/stable/23858453
4. Birger Rapp, Jan Selmer. A model for the evaluation of energy conservation investments: *Energy and Building*, Volume 2, Issue 3, August 1979. http://doi.org/10.1016/0378-7788(79)90005-7
5. Walter L. Lom, The Economic Evaluation of Energy Saving Investments: Innovation for Energy Efficiency. *Proceeding of the European Conference, Newcastle upon Tyne, UK, 15-17 September 1987.*
6. William H. Mashburn, P.R, *CEM Energy Management Handbook*, 5th ed. Edited by Steve Doty, Wayne C. Turner.
7. http://ifs.host.cs.st-andrews.ac.uk
8. Christian Winzer, Conceptualizing Energy Security, *Energy Policy*, Volume 46, July 2012.
9. Strielkowski, Wadim; lisin, Evgeny; Tvaronaviciene, Manuela. *Journal of Security & Sustainability Issues*, 2016, Vol. 6 Issue 2, pp. 235-244. 10p.
10. Gilbert N. Sorebo, Michael C. Echols, Smart Grid Security: An End-to-End View of Security in the New Electrical Grid.
11. Aurelie Foucquier, Sylvain Rbert, Frederic Suard, Louis Stephan, Arnaud Jay, State of the art in Building modelling and energy performance prediction: A review.
12. Wen-Shing Lee, Chung-kuan Kung, Using Climate Classification to evaluate building energy performance, *Energy*, Volume 36, Issue 3, March 2011, pp. 1797-1801.

13. Danny H. W. Li, Tony N. T. Lam, S. L. Wong, Lighting and energy performance for an office using high frequency dimming controls: *Energy Conservation and Management* Volume 47, Issue 9-10, June 2006.
14. Luca Cecchnato, Manuel Chiarello, Marco Corradi, A simplified method to evaluate the seasonal energy performance of water chiller, *International Journal of Thermal Sciences*, Volume 49, Issue 9, September 2010.
15. Paunescu, Carmen. Effective Energy Planning for Improving the Enterprises Energy Performance, *Management & Marketing*, Vol. 11, Issue 3, (2016): 512-514, 516-531. DOI:10.1515/mmcks-2016-0013
16. Atom Mirakyan, Roland de Guio, Integrated energy Planning in Cities and territories : Areview of methods and tools, *Renewable and Sustainable Energy Reviews*, Volume 22, June 2013.
17. Fatih Birol, Energy Economics: A Place for Energy Poverty in the Agenda? *Energy Journal*, Vol. 28, No. 3 (2007), pp. 1-6. DOI: 10.5547/ISSN0195-6574-EJ-Vol28-No3-4
18. Varun, Ravi Prakash, Inder Krishnan Bhat, Energy, economics and environmental impacts of renewable energy systems; *Renewable and Sustainable Energy Reviews,* Volume 13, Issue 9, December 2009.
19. Spath P L; Dayton, D C National Renewable Energy Lab, Golden, CO (US): 2003-1-01: U S Dept of Energy Lab., Golden, CO (US).
20. Cobb, Jr, J T; Johnson , H R (1.K& M Engineering and Consulting Corp., Washington DC (United States) Preliminary Technical and Economic Feasibility of a natural gas Storage / Co-fired retrofit system. 1995-05-01: USDOE Morgantown Energy Technology Center (METC) WV (United States).
21. Allegra De Filoppo, Michele Lombardi and Michela Milano, Method for Off-Line/On-Line Optimization under Uncertainty.
22. Guillermo Rodriguez; Yasmin Hem'ndez, Israef Paredes-Rivera, Emerging Technology for Electric Power Generation Personnel Training, *2011 IEEE Electronics; Robotics and Automotive Mechanics Conference, 15-18 Nov. 2011.*
23. P. A. Wilderer; D. Schreff, Decentralized and Centralized Wastewater Management: A Challenge for technology Developers; *Water Sci Technol* (2000) 41 (1):1-8.
24. Pouya Ifaei, Alireza Farid, Chang Kyoo Yoo, An Optimal Renewable Energy Mangement: Case Study, *Energy.* Volume 158, 1 September 2018.
25. Aedah M J Mahdi, Energy Audit a step to effective Energy Management.
26. Vivek Jadhav; Rushikesh Jadhav; Pramod Magar; Sandip Kharat; U.S. Bagwan. Energy conservation through energy audit – *IEEE Explore.*
27. G. Joga Rao, K Pavan Srihari, Associate professor, Department of EEE, Raghu Institute of Technology, Vishakapatnam, Andhra Pradesh. P. G. Diploma in Energy and Environment management, A P Productivity Council, Andhra Pradesh, India. Study of Energy Management and Audit.
28. N. M. Kuznetsov, Power Survey – Basic of Energy Saving.

29. Deny Kalfarosi Amanu, Cindy Chairunissa, Centralised Process Control system: "A Choice for Optimising Mature Field Facilities". October 2013. DOI: 10.2118/165748-MS
30. Emmanouil karampinins, Panagiotis Grammelis, Michalis Agraniotis, LoannisViolidakis, Ammanuel Kalaras. Co-firing of Biomass with coal in thermal power plants, technology schemes, impacts, and future perpectives. *Wires Energy and Environment.* First published: 19 November 2013. http://dpi.org/10.1002/wene.100
31. Giancarlo Caponio, Vito Massaro, Giorgio Mossa, Strategic Energy Planning of Residential Buildings in a Smart City: A System Dynamics Approach. https://doi.org/10.5772/61768

26

Day-Ahead Solar Power Forecasting Using Statistical and Machine Learning Methods

Aadyasha Patel[1] and O.V. Gnana Swathika[2*]

[1]*School of Electrical Engineering, Vellore Institute of Technology, Chennai, India*
[2]*Centre for Smart Grid Technologies, School of Electrical Engineering, Vellore Institute of Technology, Chennai, India*

Abstract

A viable source of energy for the future is renewable sources of energy. Expansion and evolution of this field is occurring very swiftly. With progressive change in climate taking place, there is an urgent need for clean energy to replace the conventional energy sources. The repercussions of global climate change are influencing administrators to adopt clean, green, non-polluting and unlimited sources of energy. Harnessing these renewable resources will resolve the environmental concerns of many countries. Among the multiple alternative sources available, the most promising alternative to harness is solar energy. Solar energy has the power to alleviate energy security and climate change issues worldwide. The precise prediction of solar output power is a critical criterion to confirm the reliability, stability and efficiency of the photovoltaic system. The impulsive nature of the photovoltaic power generation can be overcome by forecasting the output power.

For the economic operation of a photovoltaic power plant, the forecasts are treated as references. The benefit of forecasting is that it permits time to the power plants to do the essential adjustments beforehand such that during peak hour the power plant is under minimal stress. An integral part of the solar power system is its power forecasting tool, which is responsible for delivering quality power output. The primary prediction techniques of solar power generation are physical, statistical and machine learning methods. The physical method calculates output power by employing mathematical models on the meteorological data. The process is comprehensive and complex. The statistical method gives better-quality results for the time ranging from one hour to six hours and depends upon the historical data (meteorological and output power) for prediction. The implementation of

**Corresponding author*: gnanaswathika.ov@vit.ac.in

Milind Shrinivas Dangate, W.S. Sampath, O.V. Gnana Swathika and P. Sanjeevikumar (eds.) Integrated Green Energy Solutions Volume 2, (71–102) © 2023 Scrivener Publishing LLC

this method is comparatively simple and is appropriate for linear series data with steady variations. The machine learning methods use the trained sample data to make the predictions. Machine learning algorithms are capable of resolving complications that certain methods cannot. They have the capacity to create a bond between the multiple variables of the sampled data for making precise and accurate predictions of hours or days ahead of time. Currently, statistical and machine learning methods are preferred widely due to their ability to predict quality output from the historical data.

The positive and negative aspects of each technique is briefed. An assessment of the techniques is also made for the researchers to choose the best one for their research.

Keywords: Viable source of energy, solar energy, photovoltics, energy, greenhouse gas

Abbreviations

AE	Autoencoder
ALSM	Attention-based Long-term and Short-term temporal neural network prediction Model
DFT	Discrete Fourier transform
ENS	Ensemble of Methods
GHG	Green House Gas
GTI	1-hr prediction of global irradiance directed on the inclined surface of the panel
GRNN	Generalized Regression Neural Network
GWO	Grey Wolf Optimization
IDBN	Improved Deep Belief Network
IPSO	Improved Particle Swarm Optimization
IRENA	International Renewable Energy Agency
ISOA	Improved Seagull Optimization Algorithm
IVMD	Improved Variational Mode Decomposition
MW	Mega Watt
NN	Neural Network
NREL	National Renewable Energy Laboratory
OVMD	Optimized Variational Mode Decomposition
PV	Photo Voltaic

PVGF	Photovoltaic Power Generation Forecasting
QRA	Quantile Regression Averaging
RES	Renewable Energy Sources
RRF	RReliefF
SPF	Solar Power Feature
SPV	Solar Photo Voltaic
SUNSET	Stanford University Neural Network for Solar Electricity Trend
SVRGC	Support Vector Regression with Gaussian Kernel
t	sampling time
WOA	Whale Optimization Algorithm
σ	activation function
$\phi_k(t)$	instantaneous phase
$\hat{Q}(q, \hat{p}_t)$	quantile value of qth condition
^oC	Degree Celsius
$a_1, a_2, \cdots, a_u$	AR coefficients
$b_1, b_2, \cdots, b_w$	MA coefficients
$\hat{P}_{GB}$	produced power generated hourly
c_n	PV plant dependent parameters
$A_k(t)$	instantaneous amplitude
$w_{q,t}$	weight vector
β_i	ith hidden nodes' output weight
$h(x)$	output mapping of ELMs' hidden layer
a_i, b_i	ith hidden node parameters
f_t	forget gate
i_t	input gate
$\hat{C}_t$	cell state
o_t	output gate
W_f	weight of the forget gate
W_c	weight of the cell state
b_f	bias of forget gate

b_i	bias of input gate
b_c	bias of output gate
x_t	input at time t
h_{t-1}	hidden layer output at time t-1

26.1 Introduction

In these recent times, out of all the various energies available, electrical energy holds a major stake in our lives every day. According to the data published in [1] almost 940 million people, which accounts for 13% of the world population, do not have access to electricity yet. Industrial development, globalization, worldwide integration, modernization, accelerated urbanization, escalation of people's standard of living and the ever-increasing population have all contributed to the surge in electricity demand. This climbing power requirement is leading towards a hike in electrical energy generation and distribution. It has become a fundamental aspect to a magnitude that a nation's economic expansion is calculated from its capacity to generate per capita energy, as reported in [1]. There is a huge noticeable difference in the per capita energy consumption between the rich and low-income countries, so much that the latter group consumes less than 100 times that of the former group.

In the past, fossil fuels were the primary sources for the generation of electricity and its massive consumption has triggered environmental challenges along with the concern over their diminishing reserves. In 2019, about 64% of the energy consumed was extracted from fossil fuels. Based on a study conducted in 2015 by [2], the known reserves of fossil fuels such as coal will last for 114 years, natural gas for 52.8 years and oil for 50.7 years. The use of such non-renewable energy sources causes environmental problems like GHG emission, ozone depletion, acid rain, global warming, air pollution, oil spill, etc. The Earth has a weather cycle but the one we are witnessing now is scary and new. The extreme meteorologic conditions leaving devastation and destruction behind are raging heat waves, furious wildfires, devastating droughts, torrential hurricanes, crippling ice storms, intense flooding and high winds. In 2015, the Paris Agreement, an international accord [3], was signed by 196 nations to combat climate change. The aim is to curb global warming to less than 2°C or rather lower than 1.5°C with regard to pre-industrial levels.

With all the above facts in mind, the usage of non-conventional energy sources such as hydro, solar, wind, tidal, biomass and geothermal heat is brought to the attention for eco-friendly power production. The energy generated from RESs has the following advantages: low GHG release, zero carbon emission, green, clean, replenishes naturally, does not pollute, cheaper form of electricity in the long run, etc. The remote locations where the electricity grid has no reach can be provided with electricity by creating off-grid power systems. One such arrangement of supplying power is the SPV system as its rate of penetration in the power market is high and also because it provides economic, ecological and technical benefits. The lifetime of PV systems is greater, maintenance cost is low, the setup is robust and the investment cost is procured within a quantified duration. The number of people with access to off-grid solar electricity worldwide has gone up from 85 million in 2016 to 105 million in 2019 as stated in the Energy Progress Report 2021 [4] released by IRENA. In the last decade, the worldwide SPV installed capacity has increased from 40,334 MW in 2010 to 7,09,674 MW in 2020 as illustrated in Figure 26.1.

Environmental and meteorological parameters play an important role in the generation of PV power. Parameters such as humidity, irradiance, cloud cover, temperature, dust particles, rain, wind speed etc., obstruct

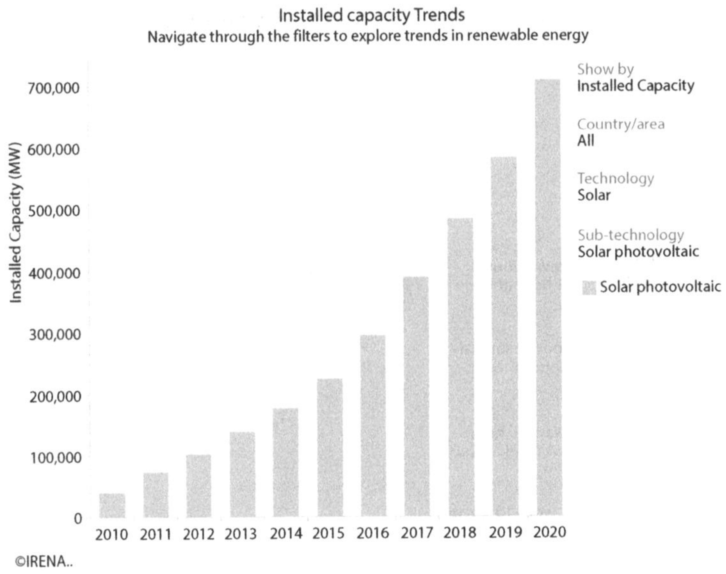

Figure 26.1 Worldwide SPV installed capacity trends [5].

in the continuous power generation with the presence of fluctuations. Accurate, precise and efficient PVGF is essential to guarantee stable and reliable power flow. Previously for predicting the generated power, mathematical techniques were used. But they usually provide little accuracy and do not give proper results for non-linear data. Nowadays, machine learning techniques are employed which deliver the predicted output with great results.

26.2 Durations of Forecasting

The duration for which prediction is required determines the type of forecasting. As per [6] the solar power forecasting can be classified into two ways as denoted in Figure 26.2. One type of classification is based on forecasting time horizon. This time scale is divided into four categories:

- Very-short term and ultra-short-term forecasting – to forecast a few minutes to 1 hour ahead
- Short-term forecasting – to forecast 1 hour to several hours ahead
- Medium-term forecasting – to forecast several hours to 1 week ahead
- Long-term forecasting – 1 week to 1 year or more ahead.

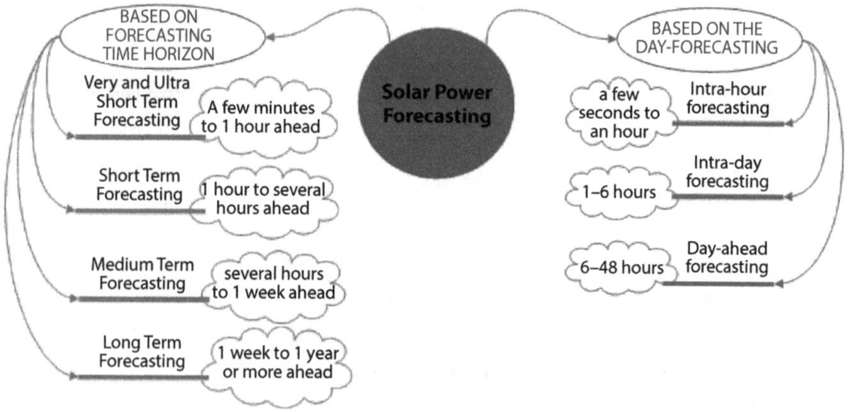

Figure 26.2 Classification of solar power forecasting.

The other type of classification is based on the day-forecasting and its time scale is divided into three categories:

- Intra-hour forecasting – to forecast a few seconds to an hour ahead,
- Intra-day forecasting – to forecast 1 hour to 6 hours ahead,
- Day-ahead forecasting – 6 hours to 48 hours ahead.

26.3 Forecasting Techniques

There are numerous power forecasting techniques available nowadays. They are majorly classified into two categories: Machine learning (together with deep learning) and Statistical method, as demonstrated in Figure 26.3. Many academicians, scientists, and scholars have studied and compared the power forecasting techniques. Some of them are listed in the following sections. Table 26.1 gives a fair idea of the type of forecast and the method undertaken for a set of collected data for a time period.

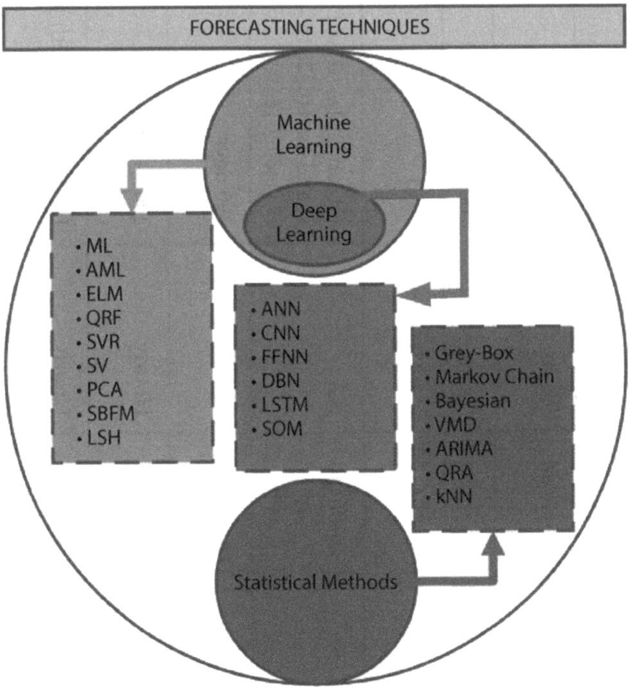

Figure 26.3 Forecasting techniques.

Table 26.1 Recent publications on power forecasting.

Ref. no.	Type of forecast	Sampling time	Data period	Forecast method	Location
[7]	1 Day-ahead	30 min	2016 - 2017	AML	Hokkaido, Japan
[8]	1 Day-ahead	5 min	MFU - 51 days. KHH - 278 days.	LSTM	MFU, Thailand and KHH, Taiwan
[9]	1 Day-ahead	5 min	2006, 2017 - 2020	Deep learning	NREL and Southeast China
[10]	1 Day-ahead	Hourly	2014 - 2017	ALSM	DKASC, Alice Springs, Australia
[11]	1 Day-ahead	NA	1st Jan. 2014 - 31st Dec. 2014 and 1st Jan. 2016 - 31st Dec. 2016	WOA-LSSVM	China
[12]	1 Day-ahead	3 hours	990 days	LSTM, Bayesian Optimization	Germany
[13]	1 Day-ahead	Hourly	2005 - 2015	Logistic Model Trees	Spanish Meteorological Agency
[14]	1 Day-ahead	Hourly	1st Apr. 2012 - 30th June 2014	SVRGC	Australia
[15]	1 Day-ahead	15 min	1st Jan. 2011 - 31st Dec. 2011	AE-LSTM	Catania, Sicily, Italy

(*Continued*)

Table 26.1 Recent publications on power forecasting. (*Continued*)

Ref. no.	Type of forecast	Sampling time	Data period	Forecast method	Location
[16]	1 Day-ahead	15 min	14th Sep. 2017 - 13th Sep. 2018	LSTM-QRA	China
[17]	1 Day-ahead	Hourly average	210 days (UCY) and 170 days (SNL)	ANN, k-means clustering	UCY, Cyprus and Sandia National Lab., Albuquerque, USA
[18]	1 Day-ahead	NA	5th Jan. 2017 - 4th Jan. 2018	LSTM	Fuhai, Taiwan
[19]	1 Day-ahead	1 min	2016	GT-DBN	Taiwan
[20]	1 Day-ahead	5 min	Sept. 2016 - Nov. 2018	hSBFM	State University of New York-Binghamton University
[21]	1 Day-ahead	5 min	2014 - 2017	Deep neural network	DKASC, Alice Springs, Australia
[22]	1 Day-ahead	5 min	2017	IVMD, ARIMA, IDBN	DKASC, Alice Springs, Australia
[23]	1 Day-ahead	1 min, 5 min, 30 min	1 year	Neural network	University of Queensland, Australia
[24]	1 Day-ahead	Hourly	1st Apr. 2012 - 31st May 2013	Cluster Analysis, Ensemble Model	Australia

(*Continued*)

Table 26.1 Recent publications on power forecasting. (*Continued*)

Ref. no.	Type of forecast	Sampling time	Data period	Forecast method	Location
[25]	1 Day-ahead	Hourly	11th Nov. 2008 - 24th Dec. 2018	CNN	DKASC, Alice Springs, Australia
[26]	2 Day-ahead	15 min	1st Nov. 2016 - 28th Oct. 2017	LSTM	Shandong province, China
[27]	1 Day-ahead	Hourly	1st Jan. 2015 - 31st Dec. 2015	DFT-PCA-Elman	China
[28]	1 Day-ahead	15 min	1st Apr. 2016 - 30th Apr. 2018	Deep learning	China
[29]	1 Day-ahead	30 min	2006	SOM-LSSVM	Australia
[30]	1 Day-ahead	Hourly	June 2014 - Dec. 2016	ENS	Italy
[31]	Intra-day	Bihourly	1st July 2010 - 16th June 2012	Markov switching	NA
[32]	Short-term	15 min	1st Jan. 2018 - 31st Dec. 2018	PCA-GWO-GRNN	Jiangsu province, China
[33]	Short-term	Hourly	2012 - 2018	LSTM NN	Desoto solar farm, Arcadia, Florida
[34]	Short-term	5 min	1st Mar. 2015 - 1st Mar. 2016	ISOA-DBN	DKASC, Alice Springs, Australia
[35]	Short-term	NA	NA	LSTM	Jeju island, Korea

(*Continued*)

Table 26.1 Recent publications on power forecasting. (*Continued*)

Ref. no.	Type of forecast	Sampling time	Data period	Forecast method	Location
[36]	Short-term	NA	21st May 2018 - 4th July 2018	OVMD-IPSO-LSTM	Changzhou, China
[37]	Short-term 24 hours	NA	2015	Deep neural network	Taiwan
[38]	Short-term	2 min	1st Mar. 2017 - 1st Mar. 2018	Deep learning	SUNSET
[39]	Short-term	Hourly	1 year	UC-M3	NREL, Colorado
[40]	Short-term	15 min	2014 - 2015	WGANGP-CNN	SURFRAD station, Desert Rock
[41]	Short-term	NA	1st Jan. 2013 - 31st Dec. 2014	BM, MM, QM	NREL, USA
[42]	Very short-term	10 min	2015 - 2017	FFNN	Basque Country and Vitoria-Gasteiz, Spain
[43]	Very short-term	Hourly	1 year	Multiscale LSTM-based deep learning	South Korea (Sejong, Yeongam, Yeonseong)
c	Very short-term	15 min	1st Jan. 2017 - 31st Dec. 2018	SPF-RRF-NN	Netherlands

(*Continued*)

Table 26.1 Recent publications on power forecasting. *(Continued)*

Ref. no.	Type of forecast	Sampling time	Data period	Forecast method	Location
[44]	Ultra-short-term	NA	2015, 2017	CNN and LSTM	University of Massachusetts, Amherst
[45]	Very short-term	1 min	July 2018 and Feb. 2019	Fuzzy	Swansea University, Bay Campus, UK
[46]	Ultra-short-term	15 min	June 2014 - Dec. 2014	LSH	Jilin Province
[47]	Very short-term	1 min	2012 - 2013	ELM	St Lucia, University of Queensland and NTU, Singapore

26.4 Statistical Methods

The collection of historical data (meteorological and electrical) and the capability to derive the required information to predict time series is the principal characteristic of statistical methods. They implement mathematical calculations for deriving the information from the recorded data, which is usually a large dataset. An important aspect of this method is that the quality of the recorded data dictates the precision and accurateness of the predicted output.

26.4.1 Grey-Box Model (GB)

The GB model in [30] is extensively employed as the initial testing point for the PV prediction issues. This model projects a correct relationship between the input (irradiance, temperature) and output (produced energy) variables. The relationship between the variables is given in (26.1).

$$\hat{P}_{GB}(i) = c_1.GTI(1) + c_2.GTI(i)^2 + c_3.GTI(i).T(i) \tag{26.1}$$

26.4.2 Grey Theory (GT)

The colours black and white represent the totally unfamiliar and totally familiar areas of a defined system. Between black and white, another colour grey represents part of the data which is undoubtedly understood and another part which is uncertain. This is explained in [19] by researchers by implementing the grey theory in multiple areas of research.

26.4.3 Markov Chain Model (MM)

The MM model is used to inspect the time-series information as reported in [31, 41]. A discrete Markov process oversees the existence of switching mechanism and its time series development. Markov chain model for forecasting 24-hour ahead and intraday PV irradiation is still under research.

26.4.4 Bayesian Optimization

This optimization process contemplates over the earlier instances for choosing the right hyper-parameters to estimate the next instance. Choosing parameters in such a way makes way for evaluating the next best constraints as in [12, 41]. It comprises three portions: search for sampled data, apply objective function and substitution.

26.4.5 Linear Pool Ensemble (LPE)

The ensemble design enhances the overall conduct of all the decisions taken by the various models. This method aids in reducing the unwanted factors, hence boosting the accuracy and stability of the entire algorithm. LPE is an advantageous technique of implementing the amalgamation of cumulative density function because of its benefit over accumulation of other predictors as reported in [41].

26.4.6 Variational Mode Decomposition (VMD)

From [22, 36] it is evident that this technique can transform the generated PV power to multiple nodes having limited bandwidth. The node $u_k(t)$ at k is given in (26.2):

$$u_k(t) = A_k(t) \cos(\phi_k(t)) \tag{26.2}$$

26.4.7 Autoregressive Integrated Moving Average (ARIMA)

ARIMA is a common time series analysis method and one that is extensively applied to linear forecasting and modelling as described in [9, 22]. The ARIMA equation is given in (26.3).

$$p_t = a_1 p_{t-1} + a_2 p_{t-2} + \cdots a_u p_{t-u} + \varepsilon_t + b_1 \varepsilon_{t-1} + \cdots + b_w \varepsilon_{t-w} \tag{26.3}$$

26.4.8 Quantile Regression Averaging (QRA)

The probabilistic and deterministic prediction models seldom get associated with each other. The QRA associates both the models of forecasting as applied in [16, 41]. The QRA equation is expressed in (26.4).

$$\hat{Q}(q, \hat{p}_t) = \hat{p}_t w_{q,t} \tag{26.4}$$

26.4.9 Logistic Model Trees

Logistic model trees in [13] are one where linear regression is substituted by logistic regression.

26.4.10 k-Nearest Neighbours (kNN)

The concept of kNN is it computes the smallest distance between the test data and the earlier recording as displayed in [30, 39]. It does so by making comparisons between the two values. The smallest distance is calculated by the Euclidean distance equation given by (26.5).

$$d(p,q) = \sqrt{\sum_{i=1}^{n} (q_i - p_i)^2} \tag{26.5}$$

For example, for a given observation it will find out the closest neighbour from all the instances available and give the predicted results.

26.5 Machine Learning Techniques

26.5.1 Machine Learning (ML)

Machine learning algorithms have an advantage over the clear-cut statistical algorithms in a way as presented in that they are able to simplify the most difficult complications. ML algorithms concentrate on the real statistics than on the complexities of the environment.

26.5.2 Automatic Machine Learning (AML)

AML boosts the work rate by carrying out the tiresome repetitive errands such as preparation of data, feature-selection, -extraction, parameter regulation. It also makes the best out of the model performance. In [7, 45] genetic programming systematizes the monotonous part of ML. By running the AML algorithm successively for a longer period, it produces the most appropriate output.

26.5.3 Extreme Learning Machine (ELM)

ELMs comprise three layers: the input layer, the hidden layer and the output layer. It requires a small amount of data for training as demonstrated in [47] and also the algorithm is simpler than standard ANNs. With 'K' hidden nodes, the calculation of output function is given by (26.6).

$$f_K(x) = \sum_{i=1}^{K} \beta_i h_i(x) \tag{26.6}$$

The matrix form of the hidden layer is,

$$H = \begin{bmatrix} h(x_1) \\ \vdots \\ h(x_N) \end{bmatrix} = \begin{bmatrix} G(a_1,b_1,x_1) & \cdots & G(a_K,b_K,x_1) \\ \vdots & \ddots & \vdots \\ G(a_1,b_1,x_N) & \cdots & G(a_K,b_K,x_N) \end{bmatrix} \; and \; T = \begin{bmatrix} t_1 \\ \vdots \\ t_N \end{bmatrix}$$

26.5.4 Quantile Random Forest (QRF)

QRF follows the decision tree approach but it is essentially a collection of classification models. The combination of various models intensifies the performance of power forecast. Rather than having a single decision tree, this combination tree approach increases the output efficiency. The same quantity of input variables as used in kNN for [30] is applied here and analyzed.

26.5.5 Support Vector Regression (SVR)

For the output to turn out linearly distinguishable and allowing extraction of patterns from an advanced spaced dimension, a Kernel function with X_i input patterns is implemented in SVR. The purpose is to transform the output pattern into an optimization problem by locating the finest fit. Out of the multiple interpretations of SVR, the one used in [30] is given in (26.7).

$$\nu \in \text{SVR}, \tag{26.7}$$

where $\nu = \in (0, 1]$ signifies an upper and lower bounded portion of SVs respectively.

26.5.6 Least-Square Support Vector Machine (LSSVM)

To simplify the disparity restraint present in the SVM, another LSSVM algorithm is proposed in [11, 29]. This modification increases the problem-solving speed and simplifies linear equations by comprehending the objective function.

26.5.7 Principal Component Analysis (PCA)

The dimensions of the actual dataset are decreased by PCA. It does so by employing an orthogonal transformation hence eliminating the superfluous information as explained in [27, 32]. The eigen vectors of the principal components are placed in decreasing order of their eigen values.

26.5.8 Hierarchical Similarity-Based Forecasting Model (hSBFM)

For the precise prediction of next days' PV power, with a 5-minute sampling time, the study in [20] is carried out. SBFMs are employed on the available data for the preparation of next days' forecast. The data is fed to a hierarchical SBFM whose accuracy of prediction naturally increases.

26.5.9 Local Sensitive Hashing Algorithm (LSH)

An issue that is commonly faced is coming across huge dimensional information. With this also come the problem of not finding its nearest neighbour. Processing of such data is time consuming. The solution to this issue is discussed in [46] by creating a distinct hash function. Similar kinds of data are hashed to a particular hash function and then it is mapped. This way the nearest neighbour can be found out easily. The hash functions are grouped as shown in (26.8).

$$H = \{h: S \rightarrow U\} \tag{26.8}$$

26.6 Deep Learning (DL)

Deep learning or deep neural network is a sub-group of machine learning. It is a rapidly growing area according to the research done in [21, 28, 37, 38, 43]. DL follows the unsupervised ML method to analyse and forecast the data.

26.6.1 Artificial Neural Network (ANN)

ANN is the most commonly used ML technique for PV power prediction as shown in [17, 23, 24, 30, 44]. The function of ANN is same as that of the human brain where numerous neurons form an interconnection with the neural network. It contains input, hidden and output layer sections where the supervisory functions take place.

26.6.2 Feed Forward Neural Network (FFNN)

FFNN presented in [14, 42] is a regression model for the nonlinear data. The flow of data is from the input to hidden to the output layer, since it is a feed forward network. The number of hidden and output layers decide on the type of network. Log-sigmoid and tan-sigmoid are some activation functions of the hidden layer and are usually nonlinear. This characteristic of nonlinearity helps in understanding the relationship between the training patterns of input and output vectors. The three major stages of FFNN are training, authentication and testing. The training phase begins with the logging in of the network biases and weights. The training phase halts when the maximum set iteration is attained.

26.6.3 Convolutional Neural Network (CNN)

The parameter distribution characteristic of CNN lessens the quantity of parameters to be adjusted compared with the other available network models. This boosts the scalability and productivity of the model according to the study in [10, 25, 40, 45]. The process of convolution is mainly employed in the Euclidean space to process the information. But this technique is not common in PVGF applications since the raw data received is in one-dimension. It will have to be converted to two-dimensional values. After analysis, the final result obtained is to be converted back to one-dimension.

26.6.4 Elman-Based Neural Network (ENN)

This type of neural network is a multiple layered network. The feedback layer takes care of the dynamic and nonlinear processes. Hence the performance gets enhanced as reported in [27].

26.6.5 Deep Belief Network (DBN)

DBN works on the principle of FFNN and it is one of the techniques that is applied to DL problems regularly according to the study of [19]. DBN is capable of making predictions using the asymmetrical PV output as shown in [22, 34]. The main characteristic of DBN is that it has a robust nonlinear plotting capability.

26.6.6 Long Short-Term Memory (LSTM)

LSTM consists of interconnected components, each carrying data about the earlier state according to [8]. It consists of cells that decide on performing functions like storing, reading, writing or deleting the information as discussed in [10, 12, 18]. These functions are based on the incoming signals and their weights that carry instructions for performing any action. These weights are an integrated part of LSTM. From [26, 33, 35, 36, 43, 45] it is clear that LSTM contains four layers namely, the forget gate, input gate, output gate and hidden gate. The gates are represented as given in equations (26.9) to (26.13).

$$f_t = \sigma\,(W_f\,[h_{t-1}, x_t] + b_f) \tag{26.9}$$

$$i_t = \sigma\,(W_i\,[h_{t-1}, x_t] + b_i) \tag{26.10}$$

$$\hat{C}_t = tanh(W_c[h_{t-1}, x_t] + b_c) \tag{26.11}$$

$$o_t = \sigma\,(W_o\,[h_{t-1}, x_t] + b_o) \tag{26.12}$$

$$h_t = o_t + \tanh\,(C_t) \tag{26.13}$$

26.6.7 Autoencoder Long Short-Term Memory (AE-LSTM)

AEs have the ability of learning even though it is categorized under unsupervised learning. They can learn the forecasting data models and use it to encode the data in the incoming layer. Later it can decode as well as discussed in [15].

26.6.8 Self-Organizing Maps (SOM)

SOM is built in two layers: the input layer and the output map as studied in [29]. The output map is a two-dimensional mesh which usually consists of a set of nodes. The SOM is capable of simulating and mapping the brains' functions.

26.7 Evaluation Index and Metrics

The papers reviewed are evaluated based on indices like RMSE, RMSPE, MAE, MAPE, MAD, MSE, MRE, MAE, SDE, R^2, sMAPE, nRMSE, nMAE, nMBE, rMAE, rRMSE. The respective formulae are mentioned in Table 26.2. Evaluation metrics of each work reviewed is listed in Table 26.3.

Table 26.2 Evaluation indies.

S. no.	Evaluation index	Formula
1	RMSE (Root Mean Square Error)	$\sqrt{\frac{1}{N}\sum_{i=1}^{N}(y_j - t_j)^2}$
2	RMSPE (Root Mean Square Percentage Error)	$\frac{1}{M}\sqrt{\sum_{m=1}^{M}\left(\frac{x_m - x_{m,f}}{S}\right)^2} * 100\%$
3	MAE (Mean Absolute Error)	$\frac{1}{N}\sum_{i=1}^{N}\lvert y_j - t_j \rvert$
4	MAPE (Mean Absolute Percentage Error)	$\frac{1}{N}\sum_{i=1}^{N}\frac{\lvert y_j - t_j \rvert}{y_j} * 100\%$
5	MAD (Median Absolute Deviation)	$\frac{1}{N}\sum_{t=1}^{N}\left\lvert\frac{S_i - Y_i}{Y_{total}}\right\rvert * 100\%$
6	MSE (Mean Squared Error)	$\frac{1}{n}\sum_{i=1}^{n}(y_i - \hat{y}_i)^2$
7	MRE (Mean Relative Error)	$\frac{1}{N}\frac{\sum_{i=1}^{N}\lvert x_i - \hat{x}_i \rvert}{\overline{x}}$
8	MADE (Mean Absolute Derivative Error)	$\frac{1}{n}\sum_{i=1}^{n}\left\lvert\frac{\Delta y_i - \Delta_i'}{\Delta t}\right\rvert$
9	SDE (Standard Deviation of Errors)	$\sqrt{\frac{1}{K-k}\sum_{T=k}^{K}(e_{T+6} - \overline{e}_{T+6})^2}$

(Continued)

Table 26.2 Evaluation indies. (*Continued*)

S. no.	Evaluation index	Formula
10	R^2 (Coefficient of Determination)	$R^2(y,\hat{y})=1-\frac{\sum_{i=0}^{n-1}(y_i-\hat{y}_i)^2}{\sum_{i=0}^{n-1}(y_i-\bar{y}_i)^2}$
11	sMAPE (Symmetric MAPE)	$\frac{1}{n}\sum_{i=1}^{n}\frac{\lvert y_i-y_i'\rvert}{\lvert y_i\rvert+\lvert y_i'\rvert}*100\%$
12	nRMSE (normalized RMSE)	$\frac{\sqrt{\sum_{i=1}^{N}(y_j-t_i)^2}}{\sqrt{\sum_{i=1}^{N}y_i^2}}$
13	nMAE (normalised MAE)	$\frac{1}{N}\sum_{i=1}^{N}\left\lvert\frac{\hat{p}_i-p_i}{p_{cap}}\right\rvert*100\%$
14	nMBE (normalised Mean Bias Error)	$\frac{1}{N}\sum_{i=1}^{N}(y_i-\hat{y}_i)*100\%$
15	rMAE (relative MAE)	$\frac{\frac{1}{N}\sum_{i=1}^{N}\lvert y_j-t_j\rvert}{y_j}$
16	rRMSE (relative RMSE)	$\frac{1}{p_{cap}}\sqrt{\frac{\sum_{i=1}^{N}(\hat{p}_i-p_i)^2}{N}}*100\%$

Table 26.3 Evaluation metrics.

Ref. no.	Evaluation metrics
[7]	RMSE – 7.77% MAE – 4.27% R2 – 94.80%
[8]	MFU data RMSE – 0.828
	KHH data RMSE – 1.653
[9]	nRMSE – 6.83% nMAE – 4.12%
[10]	R2 – 97.50% nRMSE – 6.34% nMAE – 4.20%
[11]	RMSE – 2.55% MAE – 2%
[12]	RMSE – 0.064 MAE – 0.032 R2 – 0.892
[13]	Automatic data rMAE – 12.69% rRMSE – 19.775%
	Expert data rMAE – 12.89% rRMSE – 19.835%
[15]	PM data nRMSE – 12.83% nMAE – 6.83%
	AE-LSTM data nRMSE – 8.39% nMAE – 4.56%
[16]	RMSE – 46.87 pu MAE – 27.526 pu

(*Continued*)

Table 26.3 Evaluation metrics. (*Continued*)

Ref. no.	Evaluation metrics
[17]	UCY data MAPE - 4.70% nRMSE - 6.11%
	SNL data MAPE - 6.27% nRMSE - 7.35%
[18]	RMSE - 0.184 MAE - 0.133 R2 - 0.711
[19]	RMSPE - 4.699 MAPE - 3.76
[20]	MAE - 618.3W MRE - 7.60% nRMSE - 10.60%
[21]	LSTM data RMSE - 1.465 MAE - 0.565 MAPE - 0.089
	CNN data RMSE - 0.971 MAE - 0.478 MAPE - 0.083
	CLSTM data RMSE - 0.886 MAE - 0.405 MAPE - 0.08
[22]	RMSE - 13.683 MAE - 9.625 MAPE - 2.723%
[23]	MAPE - 8.73%
[24]	nRMSE - 8.8 nMAE - 4.06 nMBE - 1.01

(*Continued*)

Table 26.3 Evaluation metrics. (*Continued*)

Ref. no.	Evaluation metrics
[25]	ResNet data MAE – 0.152 kW
	DenseNet data 0.180 kW
[26]	RMSE – 9.07% MAD – 2.213%
[27]	RMSE – 127.29 MAE – 74.33 R2 – 0.83
[28]	Station 1 data RMSE – 1.135 MAE – 0.621 MSE – 1.287 R2 – 0.879
	Station 2 data RMSE – 1.111 MAE – 0.613 MSE – 1.235 R2 – 0.843
[29]	RMSE – 72.5 MAE – 58.59 MAPE – 0.66%
[30]	MAE – 118 kW nRMSE – 7.16 nMAE – 3.07 nMBE – 0.356
[31]	RMSE – 103.4 MAE – 86.69 MSE – 4909.07 SDE – 103.29
[32]	RMSE – 122.32 kW nMAE – 2.55%

(*Continued*)

Table 26.3 Evaluation metrics. (*Continued*)

Ref. no.	Evaluation metrics
[33]	RMSE – 1.238 MW MAE – 0.63 MW MAPE – 36.285% MRE – 2.25%
[34]	RMSE – 0.057 MW MAPE – 1.20%
[36]	MAPE – 8.49% MRE – 0.0042 R2 – 0.9578
[37]	RMSE – 163.151 MAE – 109.485
[38]	RMSE – 2.51 kW
[39]	nRMSE – 7.94 nMAE – 4.79
[42]	RMSE – 37.84W/m2 MAE – 17.71 W/m2 MAPE – 9.67%
[43]	Sejong data RMSE – 14.142 MADE – 170.332 sMAPE – 49.262
	Yeongam data RMSE – 9.871 MADE – 173.381 sMAPE – 34.125
	Yeonseong data RMSE – 13.478 MADE – 133.508 sMAPE – 42.008
[48]	RMSE – 21.97 MW MAE – 4.5 MW MRE – 0.196%

(*Continued*)

Table 26.3 Evaluation metrics. (*Continued*)

Ref. no.	Evaluation metrics
[44]	RMSE – 0.071 MAE – 0.04 MSE – 0.005
[45]	RMSE – 1.28 kW MAE – 0.65 kW nMAE – 3.06%
[46]	RMSE – 8.567% MRE – 0.078
[47]	RMSE – 12.802% MAE – 7.863%

26.8 Conclusions

After going through the papers reviewed in this work and a few additional previously reviewed works [6, 49–54] we come to the following conclusions. The statistical methods are simple, flexible, accurate and completely capable in representing the linear time series model. They have an intuitive and interpolating characteristic. The model only needs the previous data to make the prediction. Addition of new data to the pool does not hamper with the forecasting accuracy. But while working with non-linear data, the forecasting accuracy reduces. Also, the technique becomes computationally and theoretically infeasible and expensive. If the data set is huge then the model does not work. The data will need to undergo standardization and normalization before analysis. The model is hypersensitive to outliers, missing and noisy information. There is no training time assigned to this method as it does not learn anything in the training duration. The machine learning algorithms have wide applications in all the fields. Its chief characteristics are speediness in learning, easy execution, simple, economic computation and precision in interpreting outcomes. The algorithms can effortlessly recognize patterns and trends, human interference is not desired to perform any actions. Also, the algorithms show constant development with the ability to handle multidimensional and variations of information. The disadvantages of machine learning are that it involves assigning time for algorithms to learn. In doing so, enormous resources are used up for performing the functions. But they are still vulnerable to errors and debugging them takes a while, relatively. The advantages of deep learning algorithms are its accuracy, immense fidelity, cancel noise at the time of data extraction and quite a long-term memory. It also has the skills to remove redundant costs, data labelling and feature engineering and gives the highest possible results even after dealing with unstructured information sets. The main disadvantages of deep learning are that huge quantity of data is required for the analysis. The data needs to flow continuously into the pool. It is an enormous resource-consuming technique. In some cases, the algorithm is not able to justify the conclusions it reaches. The deep learning in its fundamental core is a black box, which means the inside of the system is still unknown.

References

1. H. Ritchie, "Access to Energy - Our World in Data," OurWorldInData.org., 2019. https://ourworldindata.org/energy-access#access-to-electricity%0Ahttps://ourworldindata.org/energy-access (accessed Sep. 29, 2021).
2. "How long before we run out of fossil fuels? - Our World in Data." https://ourworldindata.org/how-long-before-we-run-out-of-fossil-fuels (accessed Sep. 29, 2021).
3. "The Paris Agreement | UNFCCC." https://unfccc.int/process-and-meetings/the-paris-agreement/the-paris-agreement (accessed Sep. 29, 2021).
4. "Tracking SDG 7: The Energy Progress Report," 2021.
5. "Solar energy." https://www.irena.org/solar (accessed Sep. 30, 2021).
6. R. Ahmed, V. Sreeram, Y. Mishra, and M. D. Arif, "A review and evaluation of the state-of-the-art in PV solar power forecasting: Techniques and optimization," *Renew. Sustain. Energy Rev.*, vol. 124, p. 109792, 2020, doi: https://doi.org/10.1016/j.rser.2020.109792.
7. W. Zhao *et al.*, "A point prediction method based automatic machine learning for day-ahead power output of multi-region photovoltaic plants," *Energy*, vol. 223, p. 120026, 2021, doi: https://doi.org/10.1016/j.energy.2021.120026.
8. C.-H. Liu, J.-C. Gu, and M.-T. Yang, "A Simplified LSTM Neural Networks for One Day-Ahead Solar Power Forecasting," *IEEE Access*, vol. 9, pp. 17174–17195, 2021, doi: 10.1109/ACCESS.2021.3053638.
9. Y. Qu, J. Xu, Y. Sun, and D. Liu, "A temporal distributed hybrid deep learning model for day-ahead distributed PV power forecasting," *Appl. Energy*, vol. 304, p. 117704, 2021, doi: https://doi.org/10.1016/j.apenergy.2021.117704.
10. J. Qu, Z. Qian, and Y. Pei, "Day-ahead hourly photovoltaic power forecasting using attention-based CNN-LSTM neural network embedded with multiple relevant and target variables prediction pattern," *Energy*, vol. 232, p. 120996, 2021, doi: https://doi.org/10.1016/j.energy.2021.120996.
11. B. Gu, H. Shen, X. Lei, H. Hu, and X. Liu, "Forecasting and uncertainty analysis of day-ahead photovoltaic power using a novel forecasting method,"

Appl. Energy, vol. 299, p. 117291, 2021, doi: https://doi.org/10.1016/j.apenergy.2021.117291.

12. M. Aslam, S.-J. Lee, S.-H. Khang, and S. Hong, "Two-Stage Attention Over LSTM with Bayesian Optimization for Day-Ahead Solar Power Forecasting," *IEEE Access*, vol. 9, pp. 107387–107398, 2021, doi: 10.1109/ACCESS.2021.3100105.
13. J. del Campo-Ávila, A. Takilalte, A. Bifet, and L. Mora-López, "Binding data mining and expert knowledge for one-day-ahead prediction of hourly global solar radiation," *Expert Syst. Appl.*, vol. 167, p. 114147, 2021, doi: https://doi.org/10.1016/j.eswa.2020.114147.
14. C. L. Dewangan, S. N. Singh, and S. Chakrabarti, "Combining forecasts of day-ahead solar power," *Energy*, vol. 202, p. 117743, 2020, doi: https://doi.org/10.1016/j.energy.2020.117743.
15. Y. Zhang, C. Qin, A. K. Srivastava, C. Jin, and R. K. Sharma, "Data-Driven Day-Ahead PV Estimation Using Autoencoder-LSTM and Persistence Model," *IEEE Trans. Ind. Appl.*, vol. 56, no. 6, pp. 7185–7192, 2020, doi: 10.1109/TIA.2020.3025742.
16. F. Mei *et al.*, "Day-Ahead Nonparametric Probabilistic Forecasting of Photovoltaic Power Generation Based on the LSTM-QRA Ensemble Model," *IEEE Access*, vol. 8, pp. 166138–166149, 2020, doi: 10.1109/ACCESS.2020.3021581.
17. S. Theocharides, G. Makrides, A. Livera, M. Theristis, P. Kaimakis, and G. E. Georghiou, "Day-ahead photovoltaic power production forecasting methodology based on machine learning and statistical post-processing," *Appl. Energy*, vol. 268, p. 115023, 2020, doi: https://doi.org/10.1016/j.apenergy.2020.115023.
18. Y. Hong, J. J. F. Martinez, and A. C. Fajardo, "Day-Ahead Solar Irradiation Forecasting Utilizing Gramian Angular Field and Convolutional Long Short-Term Memory," *IEEE Access*, vol. 8, pp. 18741–18753, 2020, doi: 10.1109/ACCESS.2020.2967900.
19. G. W. Chang and H. Lu, "Integrating Gray Data Preprocessor and Deep Belief Network for Day-Ahead PV Power Output Forecast," *IEEE Trans. Sustain. Energy*, vol. 11, no. 1, pp. 185–194, 2020, doi: 10.1109/TSTE.2018.2888548.
20. H. Sangrody, N. Zhou, and Z. Zhang, "Similarity-Based Models for Day-Ahead Solar PV Generation Forecasting," *IEEE Access*, vol. 8, pp. 104469–104478, 2020, doi: 10.1109/ACCESS.2020.2999903.
21. K. Wang, X. Qi, and H. Liu, "A comparison of day-ahead photovoltaic power forecasting models based on deep learning neural network," *Appl. Energy*, vol. 251, p. 113315, 2019, doi: https://doi.org/10.1016/j.apenergy.2019.113315.
22. J. Zhang, Z. Tan, and Y. Wei, "An adaptive hybrid model for day-ahead photovoltaic output power prediction," *J. Clean. Prod.*, vol. 244, p. 118858, 2020, doi: https://doi.org/10.1016/j.jclepro.2019.118858.
23. M. Q. Raza, N. Mithulananthan, J. Li, K. Y. Lee, and H. B. Gooi, "An Ensemble Framework for Day-Ahead Forecast of PV Output Power in Smart

Grids," *IEEE Trans. Ind. Informatics*, vol. 15, no. 8, pp. 4624–4634, 2019, doi: 10.1109/TII.2018.2882598.

24. C. Pan and J. Tan, "Day-Ahead Hourly Forecasting of Solar Generation Based on Cluster Analysis and Ensemble Model," *IEEE Access*, vol. 7, pp. 112921–112930, 2019, doi: 10.1109/ACCESS.2019.2935273.
25. H. Zang, L. Cheng, T. Ding, K. W. Cheung, Z. Wei, and G. Sun, "Day-ahead photovoltaic power forecasting approach based on deep convolutional neural networks and meta learning," *Int. J. Electr. Power Energy Syst.*, vol. 118, p. 105790, 2020, doi: https://doi.org/10.1016/j.ijepes.2019.105790.
26. M. Gao, J. Li, F. Hong, and D. Long, "Day-ahead power forecasting in a large-scale photovoltaic plant based on weather classification using LSTM," *Energy*, vol. 187, p. 115838, 2019, doi: https://doi.org/10.1016/j.energy.2019.07.168.
27. H. Lan, C. Zhang, Y.-Y. Hong, Y. He, and S. Wen, "Day-ahead spatiotemporal solar irradiation forecasting using frequency-based hybrid principal component analysis and neural network," *Appl. Energy*, vol. 247, pp. 389–402, 2019, doi: https://doi.org/10.1016/j.apenergy.2019.04.056.
28. T. Yang, B. Li, and Q. Xun, "LSTM-Attention-Embedding Model-Based Day-Ahead Prediction of Photovoltaic Power Output Using Bayesian Optimization," *IEEE Access*, vol. 7, pp. 171471–171484, 2019, doi: 10.1109/ACCESS.2019.2954290.
29. J. Huang, Y. Liang, H. Bian, and X. Wang, "Using Cluster Analysis and Least Square Support Vector Machine to Predicting Power Demand for the Next-Day," *IEEE Access*, vol. 7, pp. 82681–82692, 2019, doi: 10.1109/ACCESS.2019.2922777.
30. L. Gigoni *et al.*, "Day-Ahead Hourly Forecasting of Power Generation From Photovoltaic Plants," *IEEE Trans. Sustain. Energy*, vol. 9, no. 2, pp. 831–842, 2018, doi: 10.1109/TSTE.2017.2762435.
31. Y. Jiang, H. Long, Z. Zhang, and Z. Song, "Day-Ahead Prediction of Bihourly Solar Radiance With a Markov Switch Approach," *IEEE Trans. Sustain. Energy*, vol. 8, no. 4, pp. 1536–1547, 2017, doi: 10.1109/TSTE.2017.2694551.
32. L. Ge, Y. Xian, J. Yan, B. Wang, and Z. Wang, "A Hybrid Model for Short-term PV Output Forecasting Based on PCA-GWO-GRNN," *J. Mod. Power Syst. Clean Energy*, vol. 8, no. 6, pp. 1268–1275, 2020, doi: 10.35833/MPCE.2020.000004.
33. M. S. Hossain and H. Mahmood, "Short-Term Photovoltaic Power Forecasting Using an LSTM Neural Network and Synthetic Weather Forecast," *IEEE Access*, vol. 8, pp. 172524–172533, 2020, doi: 10.1109/ACCESS.2020.3024901.
34. W. Hu, X. Zhang, L. Zhu, and Z. Li, "Short-Term Photovoltaic Power Prediction Based on Similar Days and Improved SOA-DBN Model," *IEEE Access*, vol. 9, pp. 1958–1971, 2021, doi: 10.1109/ACCESS.2020.3046754.
35. S. Lee, H. Jin, L. F. Vecchietti, J. Hong, and D. Har, "Short-Term Predictive Power Management of PV-Powered Nanogrids," *IEEE Access*, vol. 8, pp. 147839–147857, 2020, doi: 10.1109/ACCESS.2020.3015243.

36. L. Wang, Y. Liu, T. Li, X. Xie, and C. Chang, "Short-Term PV Power Prediction Based on Optimized VMD and LSTM," *IEEE Access*, vol. 8, pp. 165849–165862, 2020, doi: 10.1109/ACCESS.2020.3022246.
37. C. Huang and P. Kuo, "Multiple-Input Deep Convolutional Neural Network Model for Short-Term Photovoltaic Power Forecasting," *IEEE Access*, vol. 7, pp. 74822–74834, 2019, doi: 10.1109/ACCESS.2019.2921238.
38. Y. Sun, V. Venugopal, and A. R. Brandt, "Short-term solar power forecast with deep learning: Exploring optimal input and output configuration," *Sol. Energy*, vol. 188, pp. 730–741, 2019, doi: https://doi.org/10.1016/j.solener.2019.06.041.
39. C. Feng, M. Cui, B.-M. Hodge, S. Lu, H. F. Hamann, and J. Zhang, "Unsupervised Clustering-Based Short-Term Solar Forecasting," *IEEE Trans. Sustain. Energy*, vol. 10, no. 4, pp. 2174–2185, 2019, doi: 10.1109/TSTE.2018.2881531.
40. F. Wang *et al.*, "Generative adversarial networks and convolutional neural networks based weather classification model for day ahead short-term photovoltaic power forecasting," *Energy Convers. Manag.*, vol. 181, pp. 443–462, 2019, doi: https://doi.org/10.1016/j.enconman.2018.11.074.
41. A. Bracale, G. Carpinelli, and P. De Falco, "A Probabilistic Competitive Ensemble Method for Short-Term Photovoltaic Power Forecasting," *IEEE Trans. Sustain. Energy*, vol. 8, no. 2, pp. 551–560, 2017, doi: 10.1109/TSTE.2016.2610523.
42. F. Rodríguez, F. Martín, L. Fontán, and A. Galarza, "Ensemble of machine learning and spatiotemporal parameters to forecast very short-term solar irradiation to compute photovoltaic generators' output power," *Energy*, vol. 229, p. 120647, 2021, doi: https://doi.org/10.1016/j.energy.2021.120647.
43. D. Kim, D. Kwon, L. Park, J. Kim, and S. Cho, "Multiscale LSTM-Based Deep Learning for Very-Short-Term Photovoltaic Power Generation Forecasting in Smart City Energy Management," *IEEE Syst. J.*, vol. 15, no. 1, pp. 346–354, 2021, doi: 10.1109/JSYST.2020.3007184.
44. Y. Ju, J. Li, and G. Sun, "Ultra-Short-Term Photovoltaic Power Prediction Based on Self-Attention Mechanism and Multi-Task Learning," *IEEE Access*, vol. 8, pp. 44821–44829, 2020, doi: 10.1109/ACCESS.2020.2978635.
45. M. Monfared, M. Fazeli, R. Lewis, and J. Searle, "Fuzzy Predictor with Additive Learning for Very Short-Term PV Power Generation," *IEEE Access*, vol. 7, pp. 91183–91192, 2019, doi: 10.1109/ACCESS.2019.2927804.
46. M. Yang and X. Huang, "Ultra-Short-Term Prediction of Photovoltaic Power Based on Periodic Extraction of PV Energy and LSH Algorithm," *IEEE Access*, vol. 6, pp. 51200–51205, 2018, doi: 10.1109/ACCESS.2018.2868478.
47. F. Golestaneh, P. Pinson, and H. B. Gooi, "Very Short-Term Nonparametric Probabilistic Forecasting of Renewable Energy Generation—With Application to Solar Energy," *IEEE Trans. Power Syst.*, vol. 31, no. 5, pp. 3850–3863, 2016, doi: 10.1109/TPWRS.2015.2502423.
48. A. Rafati, M. Joorabian, E. Mashhour, and H. R. Shaker, "High dimensional very short-term solar power forecasting based on a data-driven heuristic

method," *Energy*, vol. 219, p. 119647, 2021, doi: https://doi.org/10.1016/j.energy.2020.119647.

49. T. Hong, P. Pinson, Y. Wang, R. Weron, D. Yang, and H. Zareipour, "Energy Forecasting: A Review and Outlook," *IEEE Open Access J. Power Energy*, vol. 7, pp. 376–388, 2020, doi: 10.1109/OAJPE.2020.3029979.
50. M. Guermoui, F. Melgani, K. Gairaa, and M. L. Mekhalfi, "A comprehensive review of hybrid models for solar radiation forecasting," *J. Clean. Prod.*, vol. 258, p. 120357, 2020, doi: https://doi.org/10.1016/j.jclepro.2020.120357.
51. M. N. Akhter, S. Mekhilef, H. Mokhlis, and N. Mohamed Shah, "Review on forecasting of photovoltaic power generation based on machine learning and metaheuristic techniques," *IET Renew. Power Gener.*, vol. 13, no. 7, pp. 1009–1023, May 2019, doi: https://doi.org/10.1049/iet-rpg.2018.5649.
52. D. Su, E. Batzelis, and B. Pal, "Machine Learning Algorithms in Forecasting of Photovoltaic Power Generation," in *2019 International Conference on Smart Energy Systems and Technologies (SEST)*, 2019, pp. 1–6, doi: 10.1109/SEST.2019.8849106.
53. D. W. van der Meer, J. Widén, and J. Munkhammar, "Review on probabilistic forecasting of photovoltaic power production and electricity consumption," *Renew. Sustain. Energy Rev.*, vol. 81, pp. 1484–1512, 2018, doi: https://doi.org/10.1016/j.rser.2017.05.212.
54. J. Antonanzas, N. Osorio, R. Escobar, R. Urraca, F. J. Martinez-de-Pison, and F. Antonanzas-Torres, "Review of photovoltaic power forecasting," *Sol. Energy*, vol. 136, pp. 78–111, 2016, doi: https://doi.org/10.1016/j.solener.2016.06.069.

27

A Review on Optimum Location and Sizing of DGs in Radial Distribution System

P. Tejaswi[1] and O.V. Gnana Swathika[2]*

[1]School of Electrical Engineering, Vellore Institute of Technology, Chennai, India
[2]Centre for Smart Grid Technologies, School of Electrical Engineering, Vellore Institute of Technology, Chennai, India

Abstract
This chapter presents a review on optimum placement and sizing of DGs in the radial distribution. This will enable the minimization of real power loss, enhance VSI and VP. The optimum placement and sizing of DG in turn increases load ability, reliability and security of environmental friendly power system. In this chapter, a hybrid method which is a combination of Analytical and PSO method is incorporated to attain optimum location and sizing of DG with minimum real power loss, increased VSI and VP. The location and size of multiple DG units are determined for loss minimization and MOF for an IEEE 33-bus RDS. The hybrid method is implemented on an IEEE 33-bus RDS using MATLAB with varying load conditions.

***Keywords*:** Particle swarm optimization, multiple DG locations, enhancement of voltage profile, distributed generator, power loss reduction

Abbreviations

DG	Distributed generator
DS	Distribution system
DES	Distributed energy sources
RDS	Radial distribution system

**Corresponding author*: gnanaswathika.ov@vit.ac.in

Milind Shrinivas Dangate, W.S. Sampath, O.V. Gnana Swathika and P. Sanjeevikumar (eds.) Integrated Green Energy Solutions Volume 2, (103–132) © 2023 Scrivener Publishing LLC

DSM	Demand side management
DLF	Distribution load flow
DR	Demand response
ANM	Active network management
VP	Voltage profile
VSI	Voltage stability index
RNN	Recurrent neural network
PBEST	Present best value
GBEST	Global best value
PSO	Particle swarm optimization
MSA	Moth Swarm algorithm
AHP	Analytic hierarchy process
RES	Renewable energy sources
GA	Genetic algorithm
GWO	Grey Wolf Optimizer
EMA	Exchange market algorithm
MOWOA	Multi-objective whale optimization algorithm
WOA	Wolf Optimisation algorithm
TLBO	Teaching Learning Based Optimization method
DNEP	Distribution network expansion planning
BOA	Bat optimization algorithm
NSGA	Non-dominated sorting genetic algorithm
MPOP	Multi-period optimal power
EVCS	Electric Vehicle Charging Stations
FSM	Fuzzy satisfying method
ELPSO	Embedded local particle-swarm optimisation
THD	Total harmonic distortion
MFO	Moth Flame optimization
BC-BV	Branch current - Bus voltage matrix
BI-BC	Bus injection - Branch current matrix
MOF	Multi-Objective function

27.1 Introduction

The demand for RES is gradually increasing to provide continuous power supply to customers. Problems like low voltage regulation, power losses, lack of transmission capacities and increased environmental issues are overcome by DG. Power loss reduction is a critical entity for the distribution utilities to maintain losses at low levels. For planning and operation of DG, loss reduction is the crucial factor. The optimum location and sizing of DGs enables planning of distribution network considering the constraints of DG capacity.

In [1], a novel heuristic technique for capacity of DG investment planning is proposed which obtains the optimal realizable DG capacity investment plan. Multi-objective nonlinear mixed integer programming is presented [2] to enhance the performance of DG units location in the RDS. [3] presents a novel algorithm to obtain the optimal sizes, sites and optimal payment incentives that compensates cost of investment in the distribution network. An analytical method is proposed in [4] for DG placements to reduce the customer blackout costs, enhance service reliability, and save the power cost. Multi-objective PSO tool is discussed in [5] to obtain DG sizes and locations that decrease power loss and enhance the voltage stability, grid reliability and VP. The Harmony-Search technique is presented in [6] for optimal placement, size of DGs. The Fuzzy-Set theory method is utilized to find the resolution from the acquired Pareto solutions optimally. This algorithm's aim is to minimize the voltage deviation, real power loss, pollution and investment with maintenance cost. Evolutionary programming is proposed in [7], which normalizes the count of DGs and its location which may cause uncertainty in the individual capacity of DG.

The location and size of the DG units are found optimally using GA via a realistic case study in which investment and operation costs are included in the problem in [8]. A new multi-level stochastic distributed generation investment planning design is proposed in [9] for minimizing the emission, net present value of costs based on losses, cost of unreserved energy, operation and maintenance. Mixed-integer nonlinear programming method is proposed in [10] for the best planning of the routes of medium voltage feeders and the previous and recent sub-transmission substations. The decision-making dynamic investment issue occurring in a DS is solved by a new approach called mixed-integer trilevel program relying on Benders decomposition in [11]. Fault current Limiter allocation is explained in [12] as an optimization issue including multiple objective functions. PSO method is utilized to decrease the boost in fault current levels because of penetration of DG, total cost (size) of necessary limiters and voltage sag simultaneously. [13] presents challenges, trends and latest developments in planning of Multi-objectives of distributed energy resources to increase their advantages, such as decline of network energy losses, decrease of carbon emissions, and to decrease the adverse effects, which affects the network quality and sterilization, maximize operation and investment costs.

[14] presents an effective model for DNEP to determine the size, location and installation year of DG optimally in DS. The efficacy of the DNEP approach is compared with the combination of a Modified Differential

Evolution and Binary Enhanced PSO method to find power flow optimally and determine the optimal expansion strategy.

Systematic methodology is proposed in [15] for quick recovery of voltage at the DG bus. This approach decreases the rating and the number of STATCOMs required during the fine tuning of control parameters. At selected locations the rating of STATCOM is tested to be sufficient at certain dispatch levels of DG units. Based on location and sizing of DGs in DS planning and operation are solved in [54] utilizing new analytical and fuzzy logic methods. To demonstrate the effectiveness of the presented methodology, an exact performance analysis is done on 12-bus, 33-bus and 69-bus RDS.

[79] presents an extensive literature review of different methods technically for integration of DG and its fundamentals are discussed in the DS. Non-heuristic, analytical, meta-heuristic and hybrid optimization methods are compared based on its application to various test systems, considering the advantages, objective function, and disadvantages. Improved switch-exchange method along with GA are utilized to obtain loss minimization in reconfigured network. These algorithms are aided with Graph theory tool to obtain topologies in the radial network and in turn enhances power factor, THD minimization and control of VP in the entire day [80]. To obtain optimal allocation of capacitors, optimal power factor of DGs in the 119 node DS, a novel heuristic method is employed [81]. The performance of the novel heuristic method is compared with Improved Analytical, PSO, Hybrid Analytical PSO, Quasi Oppositional Teaching Learning based optimization, Comprehensive TLBO and TLBO methods.

[82] presents an MSA technique for coordination of optimal allocation of thyristor-controlled series capacitor placement and steady-state shedding of loads in the IEEE 30-node distribution test system. This in turn decreases the amount of power loss and load shedding, enhances voltage stability and VP in the test system. The outcomes of the MSA algorithm are correlated among other known optimization methods such as PSO, GWO, TLBO, MFO and WOA.

[83] proposes a multi-objective optimal location and sizing of DGs which is tested on 33-node distribution network under uncertainty related to D-S evidence theory and affine arithmetic. [84] presents chaos map theory and Sine Cosine Algorithm to attain the DGs placement and sizing optimally in the DS. CSCA is implemented on IEEE radial 33 and 69-nodes distribution feeders to solve the optimal multiple DGs allocation with high convergence rate and minimum power loss.

To assist the microgrid portfolio and DSM, a new model is proposed that combines the microgrid modules and DSM methods in a two-layer configuration. To decrease the electricity demand related to the zone temperature set-point, DSM is employed in the first layer. Utilizing the optimal load profile attained in the first layer, entire operation and investment costs of a microgrid are optimized in the second layer [85]. Artificial Bee Colony algorithm is employed in the 14-node and 57-node DS to attain optimal placement and sizing of the DGs with decrease in power losses [86]. PSO method and GA are implemented on standard IEEE 33-node and 69-node DS for optimal placement and sizing of DG, with an aim of reducing the losses, improving the reliability and VP. In [87], GA is applied in DS for identifying the optimal location of DGs such that power losses are reduced in the system. But this method requires extensive computations and hence is slow in convergence. An analytical method is presented in [88] for optimal DG location in the radial and meshed system with constant DG size. [89] presents an Improved Gravitational-Search Algorithm for optimal location and sizing of all DG types for minimizing the losses and voltage variations.

The integrated execution of GA and RNN technique are used for location of fuel cell optimally to decrease power losses and increase the bus VP. In GA first stage, a Fuel cell is placed optimally in the DS. Its execution is evaluated by comparing various techniques such as PSO, GA and other hybrid PSO techniques [90].

The optimum location and sizing of DGs is tested on IEEE 33-node and 69-node DSs simultaneously utilizing Symbiotic Organisms Search algorithm to enhance reliability, VP and decrease the power loss [91]. Multi-objective optimisation methods such as NSGA-II and FSM is presented in [92] for placement of droop related DGs in islanded microgrid system to reduce voltage fluctuations and frequency from the nominal values. The suggested analysis is implemented on a standard IEEE 33-node DS.

A novel approach is presented to assure voltage stability during load variations for DG allocation and sizing optimally. By utilizing voltage stability margin index, sensitive buses are attained. DG sizing is solved optimally by utilizing MATLAB curve fitting. For different scenarios of load, the determined DG locations and size are tested on the IEEE 33-bus and 69-bus DSs utilizing PSAT MATLAB tool in [93]. The advances in DG technology, overview of DG, various optimization techniques, challenges and key problems are presented in [94]. Various models and methods are discussed for the solution of the optimal DG placement problem. Also optimization techniques are proposed considering the load models, DG variables, number of DGs and objectives to be satisfied are discussed in [95]. In the DS, maintenance and reliability prevents the perforation of DES [96–98]. For the best allocation of the DG and deregulation of power industry, optimisation techniques are employed in [99]. Taking into account the short-circuit level and system losses, exhaustive-search algorithm is utilized in a meshed system to determine DG location and sizing in an optimal way of a DG unit [100]. An improved analytical method is proposed for location of multiple DG units where both real and reactive power are injected by the DG.

27.1.1 DG Planning Based on Multi-Objective Optimization Techniques

[16] presents a multi-objective performance index-based GA in the DS to obtain sitting and sizing of DG units. The suggested method is implemented on 16-node and 37-node DSs with various load models. [17] presents a relative study of reactive and real power consumption and loss at the substation. Installation of DG Units for various load type models aids in MVA support. AC Optimal Power Flow is used in [18] for DG optimisation and planning to maximize energy loss minimisation and DG sizing. To evaluate the potential benefits, control approaches like advanced smart grid control are included in this planning tool. [19] presents a novel state-reduction algorithm to obtain the states less in number which are necessary to represent the behaviour of speed of wind and solar irradiance in DG planning issues and reliability analysis.

DG planning in coordination with Distribution network expansion planning reduces planning cost and it is employed along with PSO algorithm on the nine bus and Kianpars-Ahvaz test systems [20]. Covariance Matrix Adaptation Evolutionary approach is employed in [21] to maximize the income of the DS and to obtain the planning scheme of the DGs optimally.

Non-dominated-Sorting GA is an important methodology which handles multiple objectives for obtaining location and size of multi-type and multi-source DG which is implemented on a 37-bus DS. The objective involves DG operation and maintenance, costs for DG investment, and quantization for enhancement of reliability and VP with loss reduction [22].

To optimize the entire cost spent on power consumption and reliability of the distribution network, an approach with a set of solutions for DG placement is proposed in [23] for the DS operator. A linear programming and the simplex method are used in [24] to optimally identify settings of relay for every possible DG planning scenarios in the future and implemented on meshed IEEE 13-node and the 14-node RDS. A novel integrated model is proposed in [25] for resolving the DS planning issue to reduce entire payments towards compensation of system losses during the period of planning, DG's operating and investment costs, and various costs according to the available alternative scenarios.

Contingencies, load fluctuations, dispatch and control operations are simulated precisely. Planning tools are required to obtain the capacity limits, the depth of analysis and related costs [26]. Integration of optimal power flow and GA are designed in [27] to resolve the issue of expansion planning in a multistage DS. The feeders installation and reinforcement of substations and DG units are considered. The combination of Successive-elimination technique and multistage planning is tested in [28] on a meshed, generic and U.K. DS which focuses on rise in demand and security of the system related to investment.

[29] presents a comprehensive search technique to assess particular DG location, size of utility DG applications. The effect of increasing DG size on the VP of the feeders at different locations, short circuit levels and loading are known by using this technique. For optimal DG planning, AHP method is utilized in [30] to choose the best plans related to the renewable fraction, emission rate and fuel consumption. Firstly, ANM strategies are examined to know the impact of configuration of multiple DGs. MPOP flow method is presented secondly for estimating the maximum size of DG in [31]. Modified Honey-Bee Mating Optimization with Chaotic Local Search optimizes objective functions simultaneously such as emission cost, overall cost and VP by determining the schemes of sizing, timing, DG technologies and location optimally in a long duration planning [32].

DG planning model is proposed in [33] for coordinating demand flexibility, in which the behaviour of customers in DG expansion plan and DR programmes are co-optimised for the best utilisation of renewable generation and the maximum social welfare. A critical review is done on challenges ahead for distributed generation planning, techniques and optimization to produce the best control strategies for planning in [34]. Fuzzy decision-making techniques and multi-objective PSO are implemented on IEEE 33-bus RDS in [35] to attain the location, size, and technology of DG units optimally as well as handle the economic, environmental and technical problems. The Placement and size of RES, EVCS, distribution network expansion schemes, and Battery Energy Storage System are obtained by applying Multi-Objective Natural Aggregation Algorithm in [36]. The proposed method reduces total cost due to investment and reliability of the DS and maximizes the charging ability of EVCSs. The efficiency of the suggested technique is shown from the case studies implemented on coupled 25-bus traffic system and

54-bus DS. [37] presents extensive review on numerous methods applied to planning of DG and various methods of the DG placement. [38] presents two major sub-problems like DG Size and location of DG optimally. This paper discusses the techniques of DG deployment and their impact on the present research works.

27.1.2 Optimal Placement and Sizing of DG Based on Multi-Objective Optimization Techniques

A Bi-layer hierarchical control is presented in [39], in which the first layer manages the microgrid voltage and frequency. The second layer realizes the unbalance compensation of the sensitive load bus. Enhancement of the voltage unbalance factor at the Sensitive load bus will lead to boost in Voltage unbalance factor at DG terminals. [40] presents a Crow Search Algorithm and Analytical method implemented on an IEEE standard 69-bus test system to attain minimum power losses in the RDS. The impact of time varying loads, load modelling and highly voltage sensitive loads are considered in the DG placement.

To minimize line losses, improve VP and reliability, a combination of PSO with clonal selection principle of immune system is proposed in [41]. The suggested method is implemented on IEEE 14-bus loop and 33-bus RDS. [42] presents a stochastic multi-objective DR model to decrease the cost to distribution company and increase the profit to DG owner. Mixed-integer non-linear programming issue deals with electricity price, variations in generation of wind power and demand. On IEEE 33-node RDS, the suggested model is implemented in General Algebraic Modelling System *environment.*

EMA is presented in [43] for solving multiple DG allocation and sizing issue in the RDS. To evaluate this algorithm, EMA is applied on the test systems, including 33-node, 69-node DS to minimize power losses to the maximum extent. [44] presents a new technique for the placement and sizing of DGs to ensure voltage stability along with load variations and minimizing overall power losses in the electrical grid. It is implemented on the 33-node and 69-node DSs utilizing PSAT-MATLAB toolbox. [45] presents an efficient multi-objective Shuffled bat algorithm to determine the placement and size of DG optimally with various load models in the DS. The suggested method is implemented on 33-node RDS. The ideal sizes and placement of DGs are obtained to lower the cost, power losses and deviations in voltage. ELPSO method with the suitable particle formulation is tested on IEEE 30-node and IEEE 33-node test systems to locate the DG units optimally. The proposed method considers the effect of energy loss minimization and reverse power flow [46, 47]. To minimize the network loss, an iterative analytical method is proposed in [48] to find the location and size of DG units optimally in a IEEE 33-node, 69-node and 49-node RDS. A multi-objective BOA is suggested in [49] to reduce entire active power losses and maximize VSI. BOA and Fuzzy-Based technique are compared and implemented on IEEE test system. [50] presents an PSO algorithm to evaluate the effectiveness of a RDS by integrating RES of a microgrid. The PSO algorithm is tested on 2-bus, 9-bus and 34-bus RDS. [51] presents a method to determine variations in annual energy losses, when various concentration levels of DG and penetration of DG are linked to DS. The effect on losses for various technologies of DGs like wind power, combined heat and power, fuel cells and photovoltaic are analysed. In DG technologies, the wind power shows the least behaviour in loss reduction.

A MOWOA method is presented in [52] in the DS to combine different DG units optimally, to decrease the power losses, deviations in voltage, and increase the VSI. MOWOA and Multi-objective PSO are compared to check its efficiency. FSM and multiple objective NSGA-II method are proposed in [53] to solve the objectives line loading index, active power loss index and voltage deviation index. The proposed methods are tested on IEEE 33-node RDS. DG utilization is significant in [55] for generation of power securely and for loss reduction. The new challenges to power systems are voltage regulation, optimal location, power quality problems and settings of protective devices. The important point to be studied is particular DG technologies related to RES like solar and wind, owing to their uncertain generation of power.

Multi-objective differential evolution optimization algorithm is correlated with Multi-objective PSO algorithm to evaluate the different operating indices of the system like reactive power loss, active power loss and changes in voltage. The effectiveness of DGs placement is shown in [56] by implementing proposed algorithm on IEEE 69-node and 85-node RDS. To obtain location, rating of DG and capacitor units optimally, a perfect methodical approach is implemented on 130-bus Indian DS and standard IEEE 69-bus DS to minimize overall cost [57].

A new multi-objective opposition related chaotic differential evolution method is implemented on IEEE 33-node, 69-node RDSs to minimize yearly economic loss and power loss and improvement of VP [58]. For solving optimal DG placement and optimal RES, Unified PSO method is presented in [59]. In renewable energy resources region, the suggested technique is employed on 33-bus and 118-bus systems. Various types of load models like industrial, constant, commercial, residential and mixed loads are taken into account in [60] to apply PSO algorithm on IEEE 33-bus RDS. Local PSO-variant algorithm is suggested in [61] for energy loss minimization for variations of load, various compositions of load. The discussed method is employed in a IEEE-33 bus system.

A mixed-integer conic programming approach helps in reduction of costs and power losses in RDSs. The suggested approach is employed in [62] to obtain the DG sizing, placement and hourly generation on IEEE 33-bus RDS. [63] presents multi-objective PSO algorithm to resolve uncertainties in RES, minimizing power loss, enhancing power quality and reliability. The suggested method is implemented on a PV/win/hydrogen tank/fuel cell microgrid system. The sizing and location of the above sources optimally decreases the proportion of rise in the overall microgrid's cost taking into account the growth in load. Technically, the issues produced in secondary networks are synchronization of DGs to the network, inrush currents generated by the energization of network transformers, operation of network protectors, limits on generation resources, circulating currents among DGs, dynamic constraints, DG capacity and operational limits. A look-ahead load restoration framework is implemented on IEEE 342-bus low voltage test system [64].

The results in [65] are compared to obtain power allocation of DG at various power levels: the initial is assisted with the average wind speed; the latter is computed related to the generator capacity factor and the maximum generator power. To reduce losses in the DSs the methodology is applied on 34-bus and 70-bus system. Hybrid PSO is effective in maximizing system load ability, improving voltage quality and decreasing power losses in the RDS. The hybrid PSO method is implemented on 16-node, 33-node and 69-node RDS [66].

[67] presents a multi-objective index based particle swarm to handle technical issues in the protective devices like the VP, reactive and active power losses of the system, the MVA consumption by the grid, the loading of the line and short-circuit-level parameters. This method is implemented on the IEEE 38-node radial system and 30-node meshed system.

Population-based intelligent search techniques are broadly used in [68] for location of DG units optimally in the transmission systems to enhance reliability of the system. It is achieved by introducing a placement index. The DG units and DR programs are tested with a unit commitment issue and certain operation parameters are examined to improve the effectiveness of suggested method. [69] presents a Backtracking Search optimization algorithm related to the change of the conventional big bang- big crunch technique for location and sizing of voltage-controlled DGs optimally.

The suggested method is executed on balanced and unbalanced distribution feeders.

For capacitor allocation, a 2/3 rule is implemented for location of DGs. But this method cannot be used to meshed DS. For decreasing the entire power losses in the DSs, a novel method is utilized to obtain the best location and DG size. DG's ability to inject active power is studied in [70–72]. Sitting and sizing of DG units is done optimally by using GWO-based approach to minimize reactive and active energy losses in the DS. The numerical outcomes of PSO algorithm, GWO technique and Gravitational-search algorithm are compared in [73] in terms of reactive, active energy loss, convergence, VP characteristics are implemented on 15-bus and 33-bus RDSs. Invasive-weed optimization method is utilized in [74] to obtain sizing of the DGs optimally. On various load models, the suggested method is implemented for IEEE 33-bus, 69-node radial distribution systems. Local PSO algorithm resolves the optimal DG placement problem in [75] by considering the effect of power flow in reverse direction and is implemented in IEEE test systems.

Optimal sizing as well as placement of various sequences of capacitor with DG units is done in [76] by cuckoo-search algorithm to enhance the VP, minimize overall cost, real power loss, and provide reliability benefits. Power loss index and flower pollination algorithm are proposed in [77] for allocation and sizing of DGs. Power loss minimization is taken as an objective function in the suggested technique which is tested on IEEE 33-bus and 69-bus test systems. For a different distribution system, binary particle swarm optimization and shuffled frog leap algorithms are used to reduce losses, improve cost saving and VPs [78].

A comprehensive literature review is done to get information regarding DG planning, location and sizing of DG units optimally related to multi-objective optimization techniques with an objective of decreasing real power losses, enhancing the bus VP and VSI. In the DS, the location and sizing of DG is accomplished optimally considering full load and critical loads utilizing PSO method. The PSO algorithm is validated on IEEE 33-node RDS. Optimal Placement and sizing of DG is done on Unbalanced RDS with the application of faults in [103]. Distribution load flow solution, Multiple DG allocation and size, PSO algorithm and multi-objective function are discussed in section 27.2. The results of size and location of DG for an IEEE 33-bus system considering full load and minimum load are discussed in section 27.3. The conclusion is discussed in section 27.4.

27.2 Proposed Location and Sizing of DGs in RDS Using Analytical and PSO Methods

27.2.1 Methodology

The objective of minimizing loss, improving VP and increasing voltage stability is achieved by using analytical expressions incorporating PSO.

The placement of DG units are obtained optimally for the bus with minimal functional value of F is located. The location and Sizing of DG are resolved similar to the method followed in obtaining best location and size of DG where the aim is to reduce loss only. But instead of changing the DG size in small steps, random sizing of DG is carried out by using PSO algorithm as described above. The optimal DG size is the size for which the objective function F is minimal.

A. Flow chart

27.2.1.1 Distribution Load Flow Solution

In order to solve DLF solution, the BC-BV and BI-BC are developed. Multiplication of the matrices BI-BC and BC-BV gives the solution of load flow analysis. The load flow method is proposed in [101].

Algorithm (Figure 27.1):

Step 1: For a distribution network, input the line and load data.
Step 2: Form the BI-BC and BC-BV matrices.
Step 3: Obtain DLF matrix by multiplying the matrices BI-BC and BC-BV.
Step 4: Initialize iteration count r = 0.
Step 5: Obtain the values of current injection at the r-th iteration using (27.1).

$$I_l^r = \mathrm{conj}\left(\frac{S_l}{V_l^r}\right) \tag{27.1}$$

Step 6: Compute the change in voltage using (27.2).

$$[\Delta^{Vr+1}] = [DLF]\,[I^r] \tag{27.2}$$

Step 7: Update the voltage values using (27.3).

$$[V^{r+1}] = [V^0]\,[\Delta V^{r+1}] \tag{27.3}$$

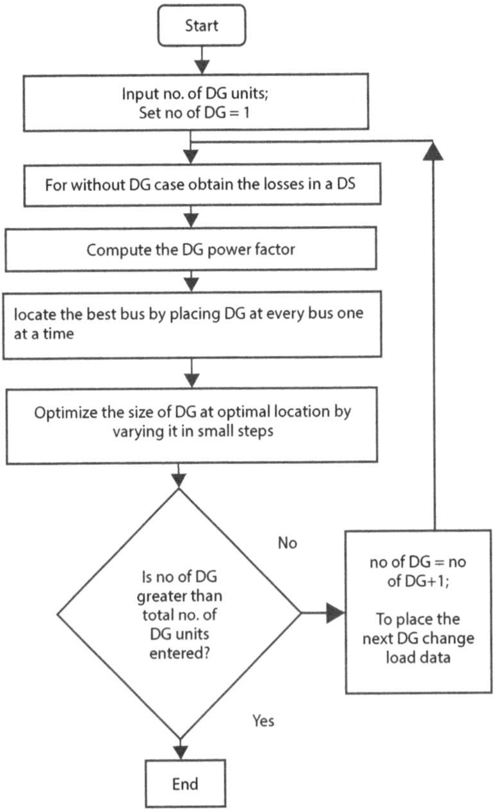

Figure 27.1 Flow chart for the optimization algorithm.

Step 8: Repeat steps 5 to 7 until iteration r is equal to maximum iteration.

Step 9: If iteration = maximum iteration, compute branch currents using (27.4).

$$i_{br} = BI\text{-}BC \times I. \tag{27.4}$$

Step 10: The power losses are computed for every branch using (27.5-27.6).

$$\text{Real Power Loss} = |i_{br}|^2 \text{ x branch resistance} \tag{27.5}$$

$$\text{Reactive Power Loss} = |i_{br}|^2 \text{ x branch reactance} \tag{27.6}$$

27.2.1.2 Multiple DG Allocation and DG Size

In this paper, to achieve power loss reduction in a DS, an analytical method is presented to allocate multiple DG units.

B. Expression for Size of DG
At each and every bus, size of DG is given using (27.7-27.8).

$$P_{DG_l} = \frac{a_{ll}(P_{D_l} + uQ_{D_l}) - G_l - uH_l}{u^2 a_{ll} + a_{ll}} \tag{27.7}$$

$$Q_{DG_l} = uP_{DG_l} \tag{27.8}$$

in which,

$$u = (\text{polarity}) \tan(\cos^{-1})(PF_{DG})) \tag{27.9}$$

If DG injects Reactive power, polarity is positive
If DG consumes Reactive power, polarity is negative

$$G_l = \sum_{\substack{m=1 \\ m \neq l}}^{N} (a_{lm}P_m - b_{lm}Q_m) H_l = \sum_{\substack{m=1 \\ m \neq l}}^{N} (a_{lm}Q_m + b_{lm}P_m) \tag{27.10}$$

Where

$$a_{lm} = \frac{r_{lm}}{V_l V_m} \cos(\theta_l - \theta_m); b_{lm} = \frac{r_{lm}}{V_l V_m} \sin(\theta_l - \theta_m) \tag{27.11}$$

C. Types of DG
There are four types of DG depending on the injection of real and reactive powers.

Type 1: Injection of both P and Q by the DG.
The DG's Power factor lies in between 0 and 1. By using (27.7) and (27.8) the size of DG at every bus is obtained optimally to achieve minimal loss.

Type 2: Injection of P but consumption of Q by the DG.

The sizing of DG is done optimally by using (27.7) and (27.8), similar to Type1 DG

Type 3: Injection of P by the DG.

The DG injects the real power only and the power factor of the DG is unity. The size of DG at every bus is obtained optimally to attain minimal power loss by using (27.12)

$$P_{DG_l} = P_{D_l} - \frac{1}{a_{ll}} \sum_{\substack{m=1 \\ m \neq l}}^{N} (a_{lm} P_m - b_{lm} Q_m) \tag{27.12}$$

Type 4: Injection of Q by the DG.

The power factor of the DG is zero, when the reactive power is injected by the DG. The size of the DG at each and every bus is obtained optimally for minimal power loss by using (27.13)

$$Q_{DG_l} = Q_{D_l} - \frac{1}{a_{ll}} \sum_{\substack{m=1 \\ m \neq l}}^{N} (a_{lm} Q_m + b_{lm} P_m) \tag{27.13}$$

D. Calculation of combined Power Factor

The power factor of the system is obtained by using (27.14)

$$PF_D = \frac{P_D}{\sqrt{P_D^2 + Q_D^2}} \tag{27.14}$$

Where

P_D and Q_D are the combined loads given by (27.15)

$$P_D = \sum_{l=1}^{N} P_{Dl}; Q_D = \sum_{l=1}^{N} Q_{Dl} \tag{27.15}$$

E. Algorithm to find optimal location and size

Step 1: Initialize the total count of DG units to be integrated in the RDS.

Step 2: When DG is not integrated in the RDS, obtain the power losses utilizing the load flow solution described in section 27.2.1.1.
Step 3: DG's Power factor is computed utilizing (27.14).
Step 4: Get the values of P_{DG} and Q_{DG} by using (27.7) and (27.8).
Step 5: By placing the DG at every bus one at a time, compute the power loss utilizing the BIBC and BCBV load flow method.
Step 6: The optimal location is given by the bus which provides the minimal loss.
Step 7: Optimize the P_{DG} value at optimal location by varying it in small steps.
Step 8: The best size of DG provides the minimal power loss.
Step 9: To integrate the next DG, load data is updated by locating the DG with best size attained in previous step.
Step 10: Repeat steps 2 to 9 until all DG units are integrated.

27.2.1.3 PSO Algorithm

PSO method is utilized for sizing of DG optimally. PSO is a technique proposed by Kennedy and Eberhart by observing the nature of birds going together in a crowd. In this, every individual or particle shifts its location depending on its own knowledge named PBEST and on the knowledge of neighbouring particle named GBEST. It produces arbitrary values of a given parameter such as DG size, location, velocity, etc.

A. Basic algorithm:

Step 1: The population of velocities and size of DGs are initialised arbitrarily.
Step 2: The number of iterations are initialised with k = unity.
Step 3: At every bus the objective function is computed.
Step 4: The objective function attained for each bus gives the values of PBEST and among the values of PBEST, GBEST is the minimal value for first iteration.
Step 5: Iteration count is incremented by changing the values of velocity and size of DG's using (27.16) and (27.17).

$$v_{jd}^{k+1} = \omega v_{jd}^{k} + c_1 \text{rand x}\left(PBEST_{jd} - s_{jd}^{k}\right) + c_2 \text{rand x}\left(GBEST_{jd} - s_{jd}^{k}\right); \tag{27.16}$$

$$s_{jd}^{k+1} = s_{jd}^{k} + v_{jd}^{k+1}; \tag{27.17}$$

Where

$$\omega = \omega_{max} - \left(\frac{\omega_{max} - \omega_{min}}{\text{maxk}}\right) \text{x } k; \tag{27.18}$$

$$\omega_{max} = 0.9; \omega_{min} = 0.4$$

Step 6: At a specific bus, the objective value acquired is correlated with the earlier iteration. If it is lower than the value acquired in the earlier iteration, the current value is fixed as present PBEST. The objective value with minimum is considered as the GBEST (out of all the present values of PBEST).

Step 7: If number of iterations are reached maximum then go to next step, else go to Step 3.

Step 8: GBEST provides the size of DGs optimally.

27.2.2 Multi-Objective Function

F_1, F_2 and F_3 are the three objective functions which are combined into one where F_1 represents the minimising loss, F_2 represents the improving VP and F3 represents the improving VSI. The three parameters F_1, F_2 and F_3 are in p.u.

A. Objective Function

Multi-objective function expression is given by (27.19)

$$F = \min(F_1 + k_1 F_2 + k_2 F_3); \tag{27.19}$$

F_1 is the loss in the DS computed using the load flow solution described above.

F_2 is the VP given by (27.20)

$$F_2 = \sum_{n=2}^{N} (|V_n - V_{rated}| - (V_n - V_{rated})); \qquad (27.20)$$

F_3 is the VSI of the network given by (27.21)

$$F_3 = \frac{1}{\min\{SI_2,\ SI_3, \ldots\ldots, SI_N\}}; \qquad (27.21)$$

Where

$$SI(n) = |V_m|^4 - 4[P_n(n)R_n + Q_n(n)X_n]|V_m|^2 - 4[P_n(n)R_n + Q_n(n)X_n]^2; \qquad (27.22)$$

27.3 Result

The results of the locations and sizes of DG is done optimally for loss-minimization objective and MOF for IEEE 33-Bus RDS and is tabulated in Table 27.1 and Table 27.2, respectively. Tables 27.1 and 27.2 show the enhancement in VP and reduction in DG size using MOF. The size of DG and location of DG is mentioned for location of one, two and three DG units. Table 27.1 and Table 27.2, respectively. Tables 27.1 and 27.2 show the enhancement in VP and reduction in DG size using MOF. The size of DG and location of DG is mentioned for location of one, two and three DG units. From Table 27.1 and Table 27.2, it is evident that there is a high power loss reduction in the IEEE 33-bus RDS when three DG units are integrated. The data of IEEE 33-bus system is available in [102].

Figure 27.2 shows the comparison of the VPs of 33-bus system without DG, with loss minimization objective, with multi-objective function at 0.85 p.f.

Table 27.1 PQ injection in IEEE 33-bus system considering full load.

No. of distributed generator units	Objective	Location of DG	Size of DG (KVA)	F1(KW)	F2	F3
0		-	-	211.00	3.62	1.50
1	Loss minimization objective	6	3103.00	68.17	1.1580	1.1857
	Multi-objective Function	6	325.760	68.49	1.0532	1.1731
2	Loss minimization objective	6, 15	2749, 505.80	52.02	0.3543	1.0893
	Multi-objective Function	6, 16	282.22, 48.80	54.61	0.3035	1.0796
3	Loss minimization objective	6, 15, 25	2248.3, 427.12, 765.15	37.849	0.2807	1.0875
	Multi-objective Function	6, 16, 32	235.30, 46.05, 52.37	38.67	0.1599	1.0580

Table 27.2 PQ injection in IEEE 33-bus system considering minimum/critical load.

No. of distributed generator units	Objective	Location of DG	Size of DG (KVA)	F1(KW)	F2	F3
0		-	-	20.44	1.17	1.17
1	Loss minimization objective	13	545.41	4.4568	0.2729	1.0388
	Multi-objective Function	13	60.15	0.894	0.2048	1.0351
2	Loss minimization objective	13,27	544.31,343.63	2.4923	0.0619	1.0133
	Multi-objective Function	13,28	52.64,29.07	0.7871	0.0382	1.0133
3	Loss minimization objective	13,27,20	436.525,343.069,212.854	1.98	0.0401	1.0111
	Multi-objective Function	13,28,20	45.063,29.064,23.70	0.6809	0.0411	1.0061

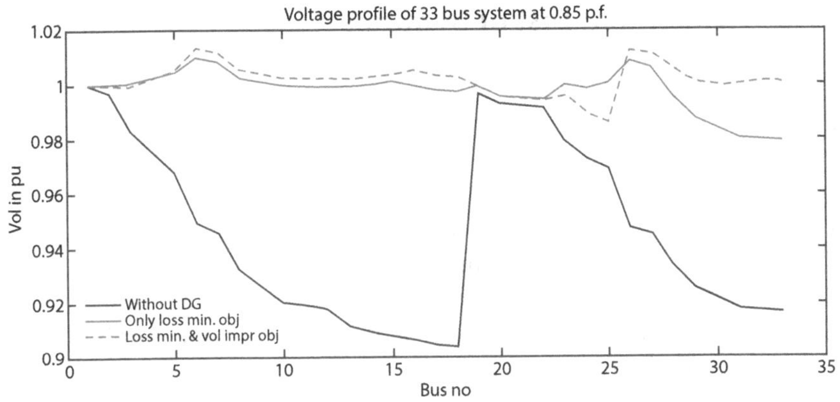

Figure 27.2 VP of 33-Bus RDS without and with DG placement at 0.85 p.f.

27.4 Conclusion

To show the effectiveness of DG placement and sizing, various algorithms are surveyed to overcome the problems of loss minimization, VP, voltage stability, cost, reliability and flexible operation. In this paper, the optimal locations are obtained using analytical expressions and the size of DG is finely adjusted considering the MOF, utilizing PSO algorithm. The power Loss, VP and VSI are compared with the optimal sizes of DG which are obtained considering only loss minimization. For IEEE 33-Bus RDS the location as well as sizing is done optimally considering full load and critical loads (Table 27.3).

Table 27.3 Power loss reduction of DG units for IEEE 33-Bus.

Number of DG units	Power factor	Percentage of power loss reduction (%)
1	Unity	47.3
2		54.4
3		59.2
1	0.85 lagging	67.6
2		75.3
3		80.9

27.5 Appendix: List of Symbols

i_{br}: Branch current;
I: Bus current injection;
P_D: Active power demand;
Q_D: Reactive power demand;
$V_l \angle \theta_l$: l^{th} bus voltage magnitude and angle;
$Z_{lm} = r_{lm} + jx_{lm}$: lm^{th} element of impedance matrix;
P_l and P_m: Real power is injected at the lth and m^{th} nodes or buses;
Q_l and Q_m: Reactive Power is injected at the lth and m^{th} nodes or buses;
N: Total count of nodes;
P_{DG} and Q_{DG}: Size of DG
r: iteration count;
k: Iteration count;
n: Receiving end bus number;
m: Sending end bus number;
c_1, c_2 (learning factors): 2;
V_{rated}: Rated voltage in 1 p.u.;
k_1: penalty constant ($k_1 = 0.6$);
k_2: penalty constant ($k_2 = 0.35$);
P: net real power flow;
Q: net reactive power flow.

References

1. W. El-Khattam, K. Bhattacharya,Y. Hegazy, M.M.A. Salama, "Optimal investment planning for distributed generation in a competitive electricity market", *IEEE Transactions on Power Systems*,19(3), 1674-1684 (2004).
2. Aboelsood Zidan, Mostafa F.Shaaban, Ehab F.El-Saadany, "Long-term multi-objective distribution network planning by DG allocation and feeders' reconfiguration", *Electric Power Systems Research*, 105, 95-104 (2013).
3. H. A. Hejazi, Ali R. Araghi, Behrooz Vahidi, S. H. Hosseinian, M. Abedi, Hamed Mohsenian-Rad, "Independent distributed generation planning to profit both utility and DG investors", *IEEE Transactions on Power Systems* 28(2), 1170-1178 (2013).
4. Lei Han; Renjun Zhou; Xuehua Deng, "An analytical method for DG placements considering reliability improvements", *2009 IEEE Power & Energy Society General Meeting Conference* (2009).
5. Amir Ameli; Mohammadreza Farrokhifard; Alireza Shahsavari; Amir Ahmadifar; Heidar-Ali Shayanfar, "Multi-objective DG planning considering

operational and economic viewpoints", *13th International Conference on Environment and Electrical Engineering (EEEIC)* (2013).

6. S. Kayalvizhi; D. M. Vinod Kumar, "Dispatchable DG planning in distribution networks considering costs", *International Conference on Recent Developments in Control, Automation and Power Engineering (RDCAPE) (2015).*
7. Majid Bavafa, "A new method of Evolutionary programming in DG planning", *International Conference on Energy, Automation and Signal (2011).*
9. Sérgio F. Santos; Desta Zahlay Fitiwi; Abebe Worke Bizuayehu; Miadreza Shafie-khah; Miguel Asensio; Javie, "Novel Multi-Stage Stochastic DG Investment Planning with Recourse", *IEEE Transactions on Sustainable Energy*, 8(1), 164-178, (2017).
10. MehdiJalali, KazemZare, Mehrdad TarafdarHagh, "A multi-stage MINLP-based model for sub-transmission system expansion planning considering the placement of DG units", *International Journal of Electrical Power & Energy Systems*, 63, 8-16, (2014).
11. Gregorio Muñoz-Delgado; Javier Contreras; José M. Arroyo, "Distribution System Expansion Planning Considering Non-Utility-Owned DG and an Independent Distribution System Operator", *IEEE Transactions on Power Systems*, 34 (4), 2588-2597 (2019).
12. A.Elmitwally, E.Gouda S. Eladawy, "Restoring recloser-fuse coordination by optimal fault current limiters planning in DG-integrated distribution systems", *International Journal of Electrical Power & Energy Systems*, 77, 9-18, (2016).
13. Arturo Alarcon-Rodriguez, Graham Ault, Stuart Galloway, "Multi-objective planning of distributed energy resources: A review of the state-of-the-art", *Renewable and Sustainable Energy Reviews*, 14(5), 1353-1366 (2010).
14. Masoud Ahmadigorji, Nima Amjady, "A new evolutionary solution method for dynamic expansion planning of DG-integrated primary distribution networks", *Energy Conversion and Management*, 82, 61-70, (2014).
15. T. Aziz; M. J. Hossain; T. K. Saha; N. Mithulananthan, "VAR Planning With Tuning of STATCOM in a DG Integrated Industrial System", *IEEE Transactions on Power Delivery*, 28(2), 875-885 (2013).
16. Deependra Singh; Devender Singh; K. S. Verma, "Multiobjective Optimization for DG Planning With Load Models", *IEEE Transactions on Power Systems*, 24(1) 427-436, (2009).
17. Devender Singh; R. K. Misra; Deependra Singh, "Effect of Load Models in Distributed Generation Planning", *IEEE Transactions on Power Systems*, 22(4) 2204-2212, (2007).
18. Luis F. Ochoa; Gareth P. Harrison, "Using AC Optimal Power Flow for DG planning and optimisation", *IEEE PES Conference: 25-29 (July 2010).*
19. Majed A. Alotaibi[ab], M.M.A. Salama[b], "An efficient probabilistic-chronological matching modeling for DG planning and reliability assessment in power distribution systems", *Renewable Energy*, 99, 158-169, (2016).

20. Reza Hemmati, Rahmat-Allah Hooshmand, Nabi Taheri, "Distribution network expansion planning and DG placement in the presence of uncertainties", *International Journal of Electrical Power & Energy Systems*, 73, 665-673, (2015).
21. Mahmood Sadeghi, Mohsen Kalantar, "Multi types DG expansion dynamic planning in distribution system under stochastic conditions using Covariance Matrix Adaptation Evolutionary Strategy and Monte-Carlo simulation", *Energy Conversion and Management*, 87, 455-471 (2014).
22. LimeiZhang, WeiTang, Yongfu Liu, TaoLv, "Multiobjective optimization and decision-making for DG planning considering benefits between distribution company and DGs owner", *International Journal of Electrical Power & Energy Systems*, 73, 465-474 (2015).
23. Neelakanteshwar RaoBattu, A.R.Abhyankar, NilanjanSenroy, "DG planning with amalgamation of economic and reliability considerations", *International Journal of Electrical Power & Energy Systems* , 73, 273-282 (2015).
24. Łukasz Huchel; Hatem H. Zeineldin, "Planning the Coordination of Directional Overcurrent Relays for Distribution Systems Considering DG", *IEEE Transactions on Smart Grid*, 7(3) (2016).
25. W. El-Khattam; Y.G. Hegazy; M.M.A. Salama, "An integrated distributed generation optimization model for distribution system planning", *IEEE Transactions on Power Systems*, 20(2) 1158-1165 (2005).
26. R.C. Dugan; T.E. McDermott; G.J. Ball, "Planning for distributed generation", *IEEE Industry Applications Magazine*, 7(2), 80-88 (2001).
27. H. Falaghi, C. Singh, M.-R. Haghifam, M. Ramezan, "DG integrated multistage distribution system expansion planning", *International Journal of Electrical Power & Energy Systems*, 33(8), 1489-1497 (2011).
28. David T.-C. Wang; Luis F. Ochoa; Gareth P. Harrison, "DG Impact on Investment Deferral: Network Planning and Security of Supply", *IEEE Transactions on Power Systems*, 25(2), 1134-1141 (2010).
29. Limuel Khin L. Estorque; Michael Angelo A. Pedrasa, "Utility-scale DG planning using location-specific hosting capacity analysis", *IEEE Innovative Smart Grid Technologies - Asia (ISGT-Asia) Conference: 28 Nov.-1 Dec. 2016*, 984-989 (2016).
30. M. Sahraei-Ardakani; M. Peydayesh; A. Rahimi-Kian, "Multi attribute optimal DG planning under uncertainty using AHP method", *2008 IEEE Power and Energy Society General Meeting - Conference*: 20-24 (2008).
31. Sultan S. Al Kaabi; H. H. Zeineldin; Vinod Khadkikar, "Planning Active Distribution Networks Considering Multi-DG Configurations", *IEEE Transactions on Power Systems*, 29(2) 785-793, (2014).
32. MajidEsmi Jahromi, MehdiEhsan, AbbasFattahi Meyabadi, "A dynamic fuzzy interactive approach for DG expansion planning", *International Journal of Electrical Power & Energy Systems*, 43(1), 1094-1105, (2012).
33. Can Dang; Xifan Wang; Xiuli Wang; Furong Li; Baorong Zhou, "DG planning incorporating demand flexibility to promote renewable integration", *IET Generation, Transmission & Distribution*, 12(20), 4419–4425 (2018).

34. Andrew Keane; Luis F. Ochoa; Carmen L. T. Borges; Graham W. Ault; Arturo D. Alarcon-Rodriguez; Robert A, "State-of-the-Art Techniques and Challenges Ahead for Distributed Generation Planning and Optimization", *IEEE Transactions on Power Systems*, 28(2),1493-1502 (2013).
35. Amir Ameli; Mohammad-Reza Farrokhifard; Ehsan Davari-nejad; Hashem Oraee; Mahmoud-Reza Haghifam, "Profit-Based DG Planning Considering Environmental and Operational Issues: A Multiobjective Approach", *IEEE Systems Journal*, 11(4), 1959-1970 (2017).
36. ShuWang[a], FengjiLuo[b], Zhao YangDong[a], GianlucaRanzi[b], "Joint planning of active distribution networks considering renewable power uncertainty", *International Journal of Electrical Power & Energy Systems*, 110, 696-704 (2019).
37. Tarannum Bahar,Omveer Singh, Vinod Yadav, "Optimal Planning Strategies of DG in Distribution Systems", in *Applications of Computing, Automation and Wireless Systems in Electrical Engineering*, Springer, 553, 333-345 (2019).
38. Sanjay Jain, Shilpa Kalambe, Ganga Agnihotri, Anuprita Mishra, "Distributed generation deployment: State-of-the-art of distribution system planning in sustainable era", *Renewable and Sustainable Energy Reviews*, 77, 363-385 (2017).
39. Mohammad Hadi Andishgar, Mehdi Gholipour, Rahmat-allah Hooshmand, "Improved secondary control for optimal unbalance compensation in islanded microgrids with parallel DGs", *Electrical Power and Energy Systems*, 105535, 116 (2020).
40. Mareddy Padma Lalitha, Oruganti Hemakesavulu, "Effect of Load Model and Load Level on DG Placement by Crow Search Algorithm", *Emerging Trends in Electrical, Communications, and Information Technologies*, 569, 185-198 (2019).
41. Vikas Singh Bhadoria, Nidhi Singh Pal & Vivek Shrivastava, "Artificial immune system based approach for size and location optimization of distributed generation in distribution system", *International Journal of System Assurance Engineering and Management*, 339-349, (2019).
42. Ehsankianmehra Saman Nikkhahb Abbas Rabiee, "Multi-objective stochastic model for joint optimal allocation of DG units and network reconfiguration from DG owner's and DisCo's perspectives", *Renewable Energy*, 132, 471-485, (2019).
43. Mohammadreza Daneshvar, "Exchange Market Algorithm for Multiple DG Placement and Sizing in a Radial Distribution System", *Journal of Energy Management and Technology*, 2(1), 54-65 (2018).
44. Sirine Essallah, Adel Khedher, Adel Bouallegue, "Integration of distributed generation in electrical grid: Optimal placement and sizing under different load conditions", *Computers & Electrical Engineering*, 79, 106461 (2019).
45. Chandrasekhar Yammani, Sydulu Maheswarapu, Sailaja Kumari Matam, "A Multi-objective Shuffled Bat algorithm for optimal placement and sizing

of multi distributed generations with different load models", *International Journal of Electrical Power & Energy Systems*, 79,120-131 (2016).

46. KallisthenisI.Sgouras; AggelosS. Bouhouras; Paschalis A. Gkaidatzis; Dimitrios I. Doukas; Dimitris P. Labridis, "Impact of reverse power flow on the optimal distributed generation placement problem", *IET*, 11 (18), 4626-4632 (2017).
47. PaschalisA.Gkaidatzis[a]Aggelos S. Bouhouras[ab] Dimitrios I.Doukas[a]Kallisthenis I.Sgouras[a], "Load variations impact on optimal DG placement problem concerning energy loss reduction", *Electric Power Systems Research* 152, 36-47, (2017).
48. Nematollahi, A. Dadkhah, O. Asgari Gashteroodkhani, and B.Vahidi, "Optimal sizing and siting of DGs for loss reduction using an iterative-analytical method", *Journal of Renewable and Sustainable Energy 8, 055301 (2016).*
49. Souhaib Remha, Saliha Chettih, Salem Arif, A, "Novel Multi-Objective Bat Algorithm for Optimal Placement and Sizing of Distributed Generation in Radial Distributed Systems", *Advances in Electrical Engineering* (2018).
50. Shalaka N. ChaphekarAnandkumar Nale Anjali A., Dharme Nitant Mate "Optimal Location and Sizing of Microgrid for Radial Distribution Systems", in *Intelligent Computing Techniques for Smart Energy Systems*, Springer, *105-113* (2019).
51. V.H.M. Quezada,J.R. Abbad,T.G.S. Roman, "Assessment of energy distribution losses for increasing penetration of distributed generation", *IEEE Transactions on Power Systems*, 21(2), 533-540 (2006).
52. Salah Kamel; Ali Selim; Francisco Jurado; Juan Yu; Kaigui Xie; Changqing Yu, "Multi-Objective Whale Optimization Algorithm for Optimal Integration of Multiple DGs into Distribution Systems" *IEEE Innovative Smart Grid Technologies - Asia (ISGT Asia)*, 1312-1317 (2019).
53. Ankit Uniyal[a] Ashwani Kumar, "Optimal Distributed Generation Placement with Multiple Objectives Considering Probabilistic Load", *Procedia Computer Science*, 125, 382-388, (2018).
54. S. Kumar Injeti, Dr. Navuri P Kumar, "Optimal Planning of Distributed Generation for Improved Voltage Stability and Loss Reduction", *International Journal of Computer Applications* (0975 – 8887), Volume 15(1), 40-46 (2011).
55. Seyed-Ehsan Razavi[a] Ehsa Rahimi[b] Mohammad Sadegh Javadi[b]Ali Esmaeel Nezhad[c], "Impact of distributed generation on protection and voltage regulation of distribution systems: A review, *Renewable and Sustainable Energy Reviews*, 105, 157-167 (2019).
56. Snigdha R. Behera & B. K. Panigrahi, "A multi objective approach for placement of multiple DGs in the radial distribution system", *International Journal of Machine Learning and Cybernetics*,10, 2027–2041 (2019).
57. Sarfaraz Nawaz, Shahbaz Ahmed Siddqui, "Cost–Benefit Analysis in Distribution System of Jaipur City After DG and Capacitor Allocation", *Ankush Tandon, Intelligent Computing Techniques for Smart Energy Systems*, 607, 351-358 (2019).

58. Sajjan Kumar , Kamal K. Mandal, Niladri Chakraborty, "Optimal DG placement by multi-objective opposition based chaotic differential evolution for techno-economic analysis", *Applied Soft Computing Journal* 78, 70–83 (2019).
59. Paschalis A. Gkaidatzis, "Efficient RES Penetration under Optimal Distributed Generation Placement Approach", *Energies* (2019).
60. Mahesh Kumar, Perumal Nallagownden, Irraivan Elamvazuthi, "Optimal placement and sizing of distributed generators for voltage-dependent load model in radial distribution system", *Renewable Energy Focus*, 19–20, 23-37 (2017).
61. Paschalis A.Gkaidatzisa Aggelos S.Bouhourasab Dimitrios I.Doukasa Kallisthenis I.Sgourasa, "Load variations impact on optimal DG placement problem concerning energy loss reduction", *Electric Power Systems Research*, 152, 36-47 (2017).
62. Mohammad Mousavi; Ali Mohammad Ranjbar; Amir Safdarian, "Optimal DG placement and sizing based on MICP in radial distribution networks", *Smart Grid Conference (SGC)* (2017).
63. Seyed Mehdi Hakimi1, Arezoo Hasankhani, Miadreza Shafie-khah , João P.S. Catalão, "Optimal sizing and siting of smart microgrid components under high renewables penetration considering demand response", *IET Renewable Power Generation*, 13(10), 1809-1822 (2019).
64. Yin Xu, Chen-Ching Liu, Zhiwen Wang, Kefei Mo,Kevin P. Schneider, Francis K. Tuffner, and Dan T. Ton, "DGs for Service Restoration to Critical Loads in a Secondary Network", IEEE Transactions on Smart Grid, 10(1), 435-447 (2019).
65. William M. da Rosa*, Julio C. Teixeira, Edmarcio A. Belati, "New method for optimal allocation of distribution generation aimed at active losses reduction", *Renewable Energy*, 123, 334-341 (2018).
66. M.M. Aman a,*, G.B. Jasmon, A.H.A. Bakar, H. Mokhlis, "A new approach for optimum simultaneous multi-DG distributed generation Units placement and sizing based on maximization of system loadability using HPSO", *Energy*, 66, 202-215 (2014).
67. A.M.El-Zonkoly, "Optimal placement of multi-distributed generation units including different load models using particle swarm optimisation", *IET Generation, Transmission & Distribution*, 5(7) 760-771 (2011).
68. HadiChahkandi Nejad, aSaeedTavakoli,bNoradinGhadimi,cSamanKorjani, SayyadNojavan, "Reliability based optimal allocation of distributed generations in transmission systems under demand response program", *Electric Power Systems Research*, 176, (2019), 105952.
69. D. Sai Krishna Kanth; N. Sree Ramula Reddy, "Optimal placement & sizing of DG's using backtracking search algorithm in IEEE 33-bus distribution system", *IEEE* (2017).
70. Caisheng Wang, M.H.Nehrir, "Analytical Approaches For Optimal Placement Of Distributed Generation Sources in Power System". *IEEE Transactions on Power Systems*, 19(4), 2068–76 (2004).

71. H. L. Willis, "Analytical Methods and Rules of thumb for Modelling DG-Distribution Interaction," in *Proceedings of the IEEE Power Engineering Society Summer Meet.*, 3, 1643-1644 (2000).
72. N. Acharya, P. Mahat, and N. Mithulananthan, "An analytical Approach for DG Allocation in Primary Distribution Network," *International. Journal of Electrical. Power Energy System*, 28(10), 669-678 (2006).
73. Umbrin Sultana, Azhar Khairuddin, A. S. Mokhtar, Sajid Hussain Qazi, Beenish Sultana, "An Optimization approach for minimizing energy losses of distribution systems based on distributed generation placement", *Jurnal Teknologi*, 79:4, 87-96 (2017).
74. D.Rama Prabha,T.Jayabarathi, "Optimal placement and sizing of multiple distributed generating units in distribution networks by invasive weed optimization algorithm", *Ain Shams Engineering Journal*, 7(2), 683-694 (2016).
75. Kallisthenis I. Sgouras 1; Aggelos S. Bouhouras 1, 2; Paschalis A. Gkaidatzis 1; Dimitrios I. Doukas 1; Dimitris P. Labridis, "Impact of reverse power flow on the optimal distributed generation placement problem", *IET*, 11(18) 4626-4632 (2017).
76. Suman Biswal; Arnab Ghosh; Sajjan Kumar; Niladri Chakraborty; Swapan Kumar Goswami, "Cuckoo Search Algorithm Based Cost Minimization by Optimal DG and Capacitor Integration in Radial Distribution Systems", *IEEE* (2018).
77. G Sabarinath; T.Gowri Manohar, "Optimal Placement and Sizing of Distributed Generation using Flower Pollination Algorithm for Power Loss Reduction Maximization in Distribution Networks", *2nd IEEE International Conference on Power Electronics, Intelligent Control and Energy Systems (ICPEICES)*, (2018).
78. Abdurrahman ShuaibuHassan. YanxiaSun, ZenghuiWang, "Multi-objective for optimal placement and sizing DG units in reducing loss of power and enhancing voltage profile using BPSO-SLFA", *Energy Reports*, 6, 1581-1589 (2020).
79. Sambaiah Sampangi Kola, "A Review on Optimal Allocation and Sizing Techniques for DG in Distribution Systems", *International Journal of Renewable Energy Research*, 8(3) (2018).
80. Faheem Ud Din, Ayaz Ahmad, Hameed Ullah, Aimal Khan, Tariq Umer, Shaohua Wan, "Efficient sizing and placement of distributed generators in cyber-physical power systems", *Journal of Systems Architecture*, 97, 197–207 (2019).
81. IA Bayat, A Bagheri, "Optimal active and reactive power allocation in distribution networks using a novel heuristic approach", *Applied Energy*, 233–234, 71-85 (2019).
82. Fatma Sayed, Salah Kamel, Juan Yu, Francisco Jurado, "Optimal Load Shedding of Power System Including Optimal TCSC Allocation Using Moth Swarm Algorithm", *Iranian Journal of Science and Technology, Transactions of Electrical Engineering* (2019).

83. Qianyu Zhaoa, Shouxiang Wanga,, Kai Wanga, Bibin Huangb, "Multi-objective optimal allocation of distributed generations under uncertainty based on D-S evidence theory and affine arithmetic", *Electrical Power and Energy Systems* 112, 70–82 (2019).
84. Ali Selim, Salah Kamel, and Francisco Jurado, "Efficient Optimization Technique for Multiple DG Allocation in Distribution Networks", *Applied Soft Computing Journal*, 86, 105938 (2019).
85. Seyyed Danial Nazemi, Khashayar Mahani, Ali Ghofrani, Mahraz Amini, Burcu E. Kose, Mohsen A. Jafari, "Techno-Economic Analysis and Optimization of a Microgrid Considering Demand-Side Management", *IEEE Texas Power and Energy Conference (TPEC)* (2020).
86. Rashmi Deshmukh; Amol Kalage,. "Optimal Placement and Sizing of Distributed Generator in Distribution System Using Artificial Bee Colony Algorithm", *IEEE Global Conference on Wireless Computing and Networking (GCWCN)* (2018).
87. M. S. Sujatha,V. Roja, T. Nageswara Prasad, "Multiple DG Placement and Sizing in Radial Distribution System Using Genetic Algorithm and Particle Swarm Optimization", in *Computational Intelligence and Big Data*, Springer, pp. 21-36 (2018).
88. Mithulananthan Nadarajah, Than Oo, and Le Van Phu. "Distributed Generator Placement In Power Distribution System Using Genetic Algorithm to Reduce Losses" *TIJSAT*, 9(3), 55–62 (2004).
89. M. Faisal Nadeem; T. N. Malik; Abdur Rehman Imtiaz; Ghulam Ali, "Simultaneous placement and sizing of multi-DG units in distribution system using IGSA", *International Conference on Computing, Mathematics and Engineering Technologies (iCoMET)* (2018).
90. T. C. Subramanyam, S. S. Tulasi Ram, J. B. V. Subrahmanyam, "Optimal Placement and Sizing of DG in a Distributed Generation Environment with Comparison of Different Techniques", in *Artificial Intelligence and Evolutionary Computations in Engineering Systems*, Springer, 2018, pp. 609-19.
91. M. Padma Lalitha; P. Suresh Babu; B. Adivesh, "SOS algorithm for DG placement for loss minimization considering reverse power flow in the distribution systems", *International Conference on Advanced Communication Control and Computing Technologies (ICACCCT)* (2016).
92. Ankit Uniyal, Saumendra Sarangi, "Optimal DG Allocation in a Microgrid Using Droop-Controlled Load Flow", Intelligent Computing Techniques for Smart Energy Systems, Springer, 745-752 (2019).
93. SirineEssallah,AdelKhedher,AdelBouallegue," Integration of distributed generation in electrical grid: Optimal placement and sizing under different load conditions", *Computers & Electrical Engineering*, 79, 106461 (2019).
94. Kumar Mahesh, Perumal Nallagowndena, and Irraivan Elamvazuthi, "Optimal Configuration of DG in Distribution System: An Overview", *MATEC Web of Conferences*, 38, (2016), *UTP-UMP Symposium on Energy Systems* (2015).

95. Pavlos S. Georgilakis, and Nikos D. Hatziargyriou, "Optimal Distributed Generation Placement in Power Distribution Networks: Models, Methods, and Future Research", *IEEE Transactions on Power Systems*, 28(3), 3420-3428 (2012).
96. T. Ackermann, G. Anderson, and L. Soder, "Distributed generation: A definition," *Elect. Power Syst. Res.*, 57, 195–204 (2001).
97. Distributed Energy Resources Guide, California Energy Commission, 2003. Online. Available: http://www.energy.ca.gov/.
98. "Wind power systems," in *Encyclopedia of Physical Science and Technology*, 3rd ed., New York: Academic (2002).
99. Thong, V.V., Driesen, J., Belmans, R. "Transmission system operation concerns with high penetration level of distributed generation". I, 867–871 (2007).
100. Elnashar, M.M., El-Shatshat, R., Salama, M.A. "Optimum siting and sizing of a large distributed generators in a mesh connected system', *Int. J. Electr. Power Syst. Res.*, 80, 690–697 (2010).
101. Jen-Hao Teng, "A Direct Approach for Distribution System Load Flow Solutions", *IEEE Transactions on Power Delivery*, 18(3) (2003).
102. Vasiliki Vita, "Development of a Decision-Making Algorithm for the Optimum Size and Placement of Distributed Generation Units in Distribution Networks", *Energies*, 10, 1433 (2017).
103. P.Tejaswi, G.Sujatha, "Fault Analysis in Unbalanced radial distribution system without and with Distributed Generator", *International Journal of Research*, 8, 1109-1114 (2019).

28

High Step Up Non-Isolated DC-DC Converter Using Active-Passive Inductor Cells

Kanimozhi, G.[1]*, Amritha, G.[2] and O.V. Gnana Swathika[1]

[1]*Centre for Smart Grid Technologies, School of Electrical Engineering, Vellore Institute of Technology, Chennai, India*
[2]*School of Electrical Engineering, Vellore Institute of Technology, Chennai, India*

Abstract
The implementation and analysis of a non-isolated DC-DC converter with high gain employing APICs (Active-Passive Inductor Cells) is presented. The function of the converter is established on the parallel charging and series discharging arrangement of inductors. Since the converter is designed using APICs the topology is extendable by increasing APICs. The topology's key benefit is that it allows for high gain with a much shorter duty cycle and reduced voltage stress on the power devices. The main parameters like voltage gain, voltage stress across all power devices, minimum output voltage ripple and efficicay are calculated. The results are compared in both software and hardware implementations. A 200 W prototype is designed and simulated using Simulink (MATLAB) and a gain of 13 is achieved for an input voltage of 30V. The simulated results are realized with that of experimental results to provide an output voltage of 400 V.

Keywords: DC-DC converter, voltage stress, output voltage ripple, continuous conduction mode, active passive inductor cells

28.1 Introduction

Electricity has become one of the major necessities globally. The sources of electricity become the main factor of concern in this scenario due to the depletion of natural resources. The use of coal, petroleum, oil, etc., has to

**Corresponding author*: kanimozhi.g@vit.ac.in

Milind Shrinivas Dangate, W.S. Sampath, O.V. Gnana Swathika and P. Sanjeevikumar (eds.) Integrated Green Energy Solutions Volume 2, (133–150) © 2023 Scrivener Publishing LLC

be narrowed down since they are non-renewable sources of energy and also their usage leads to the emission of carbon dioxide and other gases leading to pollution. Drawbacks in the production of electricity from conventional energy sources can be replaced slowly by natural processes. The cost of obtaining these forms of energy is increasing at a faster rate due to increase in demand, thus affecting the economy. The majority of the renewable resources [1] deliver output voltages at lower levels. These levels cannot be utilized for commercial applications. To step up these voltage levels to meet the required levels step-up converters are used [2–4]. The efficiency of the overall system (refer Figure 28.1) for the conversion of renewable energy sources into electricity is affected by the DC-DC converter. High gain converters are not limited to this application. It is also used in various other fields like automobile applications, grid systems, energy back-up for UPS, etc. As it acts as a connection between available low-voltage sources and output loads that run at extremely high voltages, a high efficiency and high step-up converter is necessary in renewable energy applications.

It is possible to attain high voltage improvement with traditional boost converters, but this would necessitate extremely high duty cycles. Due to the extreme variations in a duty cycle that directly influence the output voltage, the high gain of a conventional boost converter is restricted in real-time scenarios due to losses caused by intrinsic resistance, resulting in accurate and high-cost drive circuitry for the switch. Low input current ripple and low voltage across the component are the two most important considerations while designing a high gain converter [5–8].

In spite of the advantages of renewable energy sources, there are certain drawbacks that limit the utilization of these, one of the main reasons being low output voltage. DC choppers with high step-up are used to overcome this restriction. High gain for a reasonable duty cycle, high reliability, low voltage stress across devices, and lower design costs are the key concerns with a high gain dc-dc converter [9–11]. The energy in the capacitor is stored in parallel

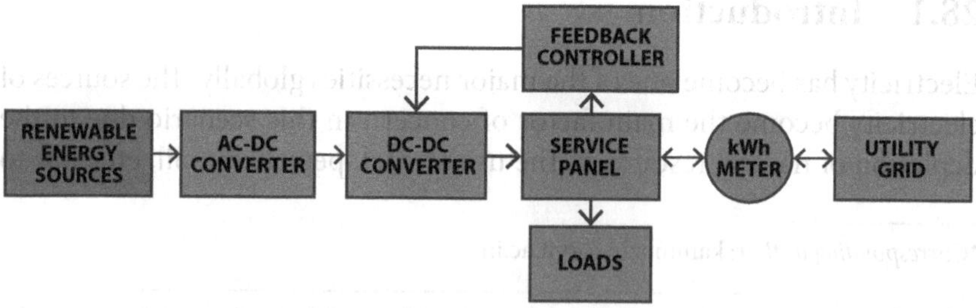

Figure 28.1 Block diagram for grid interface with renewable energy sources.

and released in series in high step-up converter topologies.To accomplish high output gain, the topology is made extendable by using APICs [12–17].

High voltage gain DC-DC converters eliminate the limitation of low output voltage from renewable energy sources by extending the framework by increasing the number of APICs. The merits of the given structure includes low voltage stress and low current on the switches at lower duty cycle. The inductors are of small sizes and the filter size is also small. The proposed structure can be analyzed in different operating modes. The equations are calculated in the selected mode of operation and the components are designed according to the design equations. The losses, as well as efficiency, can be calculated.

28.2 Proposed Converter

Figure 28.2 shows the circuit schematic for the suggested n-cell converter construction. The converter is designed to achieve high gain under lower duty cycles. In this converter, the energy in the inductors is stored in parallel and released in series. The voltage stress across the switches should be kept as low as possible by the converter. The converter operates in two operational modes: CCM (continuous conduction mode) and DCM (discontinuous conduction mode). Full inductor supply mode (FISM) and Partial inductor supply mode (PISM) are two types of CCM [18–23]. All the inductors are of the same inductance value, and all the elements are assumed to be ideal. The load current and the minimum current passing through the inductors are compared to evaluate FISM and PISM operations. In FISM, the inductors' minimum current is greater than the output current. In PISM, the inductors' least value of current is lower than the

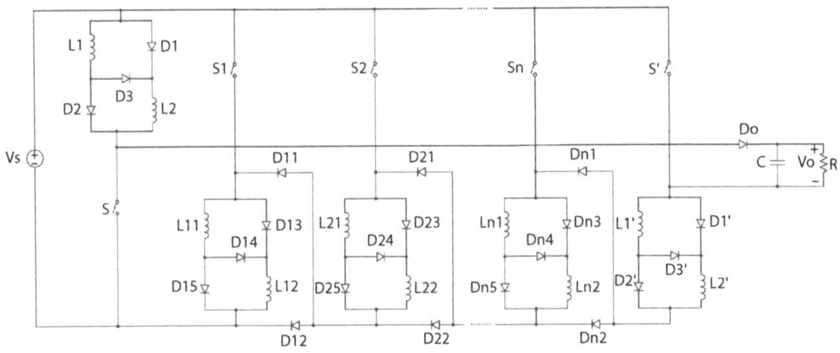

Figure 28.2 Converter circuit for n cells.

load current. The FISM-CCM mode of operation is used to examine the converter.

The proposed converter uses three switches (MOSFETs), six inductors and twelve diodes. On the input end are the diodes D1, D2, D3 and the diode Do is on the output side of the structure. All the switches S, S1, S' are turned on simultaneously and all the switches are provided with the same duty cycle. The output capacitor C serves as a filter, lowering the output voltage ripple. The Do, output diode is switched off during Ton of switches.

28.2.1 Features of the Suggested Converter

The following are the features of the high gain dc-dc converters (Figure 28.3).

- To expand the topology, APICs are used.
- The converter achieves high gain under lesser values of the duty cycle.
- The converter is set to operate in the FISM-CCM mode.
- The energy stored in the inductors in parallel and released in series in the converter.
- The reliability of the system has improved, and the power devices subjected to voltage stress has decreased.
- Considering filter size, the output voltage ripples are reduced.

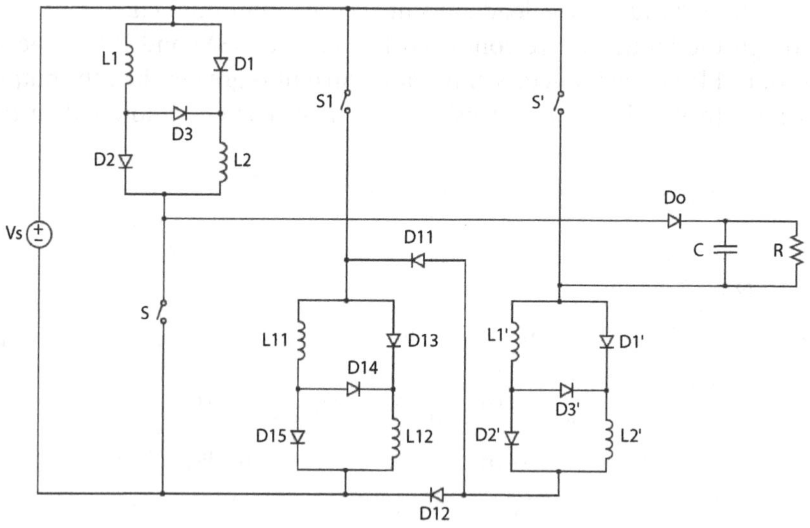

Figure 28.3 Proposed converter ciruit for n=1 cell.

28.3 Modes of Operation

a) Converter evaluation in CCM

T_{on} time: All the power switches are switched ON at the same time. The diodes are switched on D_{11}, D_{21},… D_{N1} and D_{12}, D_{22},.. D_{N2}, switched OFF, respectively.

The inductor's voltage is expressed as:

$$v_L = V_i \tag{28.1}$$

$$\text{Current flowing through inductors, } i_L = \frac{V_i}{L}t + I_{LV} \tag{28.2}$$

In the case of (28.2), the inductors are charged during T_{on} period, and the inductor current is increased until it reaches its determined value at $t = DT$. The maximum current through the inductor can be calculated by adding $t = DT$ to (28.2):

$$I_{LP} = \frac{V_i DT}{L} + I_{LV} \tag{28.3}$$

The capacitor current equals $-I_0$ and delivers load current throughout this time period. The capacitor's stored energy is released at the completion of this period and voltage is reduced to V_{CV}.

Toff time: The power semiconductor devices are turned off at the same time and diodes D_{12},D_{22},……D_{n2} and D_{11},D_{21},…….D_{n1}, Do are turned off and on accordingly. The voltage between inductor terminals are given by:

$$v_L = \frac{V_i - V_0}{2n+4} \tag{28.4}$$

Where, n is the number of APICs. Therefore the new inductor current is:

$$i_L = \frac{V_i - V_0}{(2n+4)L}t + I_{LV} \tag{28.5}$$

The inductors stored energy is released during this time frame, and their current is I_{LV} at the end. Substituting $t = (1 - D)T$ to equation (28.5) yields least value of current through inductors:

$$I_{LP} = \frac{(V_i - V_0)(1-D)}{(2n+4)L} + I_{LP} \tag{28.6}$$

$$i_C = \frac{(V_i - V_0)}{(2n+4)L} + I_{LP} - I_0 \tag{28.7}$$

As equation (28.7) is used, the capacitors and inductors currents are lowered. The inductor provides load current while still charging the capacitor. The voltage across capacitor is increased from V_{CV} *to* V_{CP} during Toff. In PISM-CCM, the Toff duration is divided into (t_1, t_2) and (t_2, t_3) time interval.

Time period (t_1, t_2): Considering (28.6), as i_L decreases, i_C decreases and attains zero at time period t_2.

Time period (t_2, t_3): Considering (28.7), i_L decreases to the value of $I_{LV} - I_0$ at t_3. During this time frame, both the inductor and the capacitor supply the load current. By combining (28.1) and (28.4), the voltage gain can be calculated as given below:

$$M_{ccm} = \frac{V_0}{V_i} = \frac{1+(2n+3)D}{(1-D)} \tag{28.8}$$

b. Evaluation of the converter in DCM

In DCM, the converter's analysis is as follows: There are four time intervals in this mode (Figure 28.4).

Ton-Time interval: The proposed converter's analysis during Ton is identical to CCM. The maximum inductor current is obtained by applying $I_{LV} = 0$ and $t = DT$ as given below:

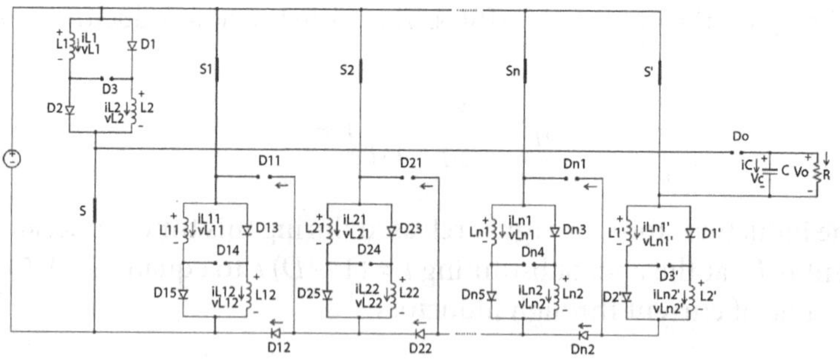

Figure 28.4 Operational mode of proposed converter for Ton.

$$i_L = \frac{V_i}{L} DT \tag{28.9}$$

Toff-time period: This time span is divided into three parts: (t1,t2), (t2,t2a), and (t3,t2b) (t2a,t3). The energy diffusion procedure throughout these time periods is given as follows (Figure 28.5).

Time period (t1,t2): The inductors and capacitor current can be acquired from the equation (28.5) and (28.6) during this time period. During this time frame, the energy transfer procedure is similar to CCM.

Time period - (t2,t2a): The energy transfer mechanism is identical to the CCM during the time span (t1,t2), with the exception that the inductor's current is nil at $t = t2a$. The capacitor's current (IC) equals the output current Io at this period. The voltage across capacitor drops as capacitor energy gets released. The maximum inductor current I_{LP} on assuming $t1 = 0$ as the new period reference is expressed as,

$$I_{LP} = \frac{(Vo - Vi)D'T}{(2n+4)L} \tag{28.10}$$

$$D' = \frac{(2n+4)ViD}{(Vo - Vi)} \tag{28.11}$$

Time interval - (t2a,t3): At this moment, the capacitor current (IC) equals $-Io$. The sole source of output current (Io) is the capacitor current. As the capacitor discharges, the voltage of the capacitor decreases.

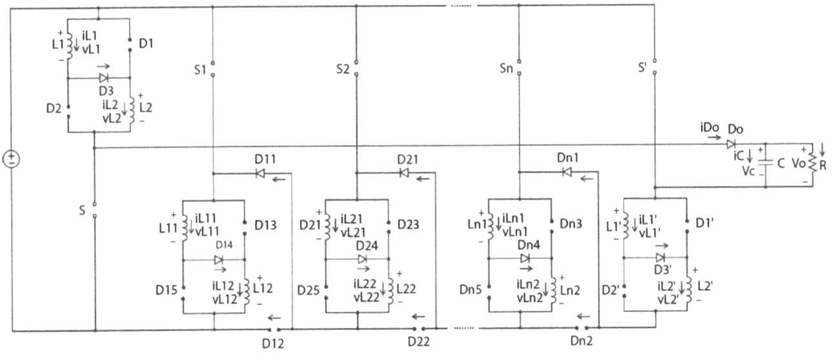

Figure 28.5 Operational modes of the converter: Toff period.

Voltage gain Calculation: The voltage gain is found by applying the current-second balance rule to the capacitor:

$$\frac{V_o}{V_i}=\frac{1}{2}+\sqrt{\frac{(n+2)RD^2}{2Lf}+\frac{1}{4}} \tag{28.12}$$

28.4 Design Considerations

The prototype is designed for 200W and operated at CCM. An output voltage of 400V is achieved for a supply voltage of 30V.

(i) Voltage gain: The voltage gain may be computed by combining equations (28.1) and (28.4) as follows:

$$M_{ccm}=\frac{V_0}{V_i}=\frac{1+(2n+3)D}{(1-D)} \tag{28.13}$$

Where V_0 is the output voltage in volts, V_i is the supply voltage in volts and D is the duty ratio.

(ii) Voltage stress analysis: The normalized stress across the switch S are as given;

$$S=\frac{n+1+M}{(n+2)M} \tag{28.14}$$

The voltage stress experienced by the power devices S' and S1 is given by

$$S=\frac{1+(n+1)M}{(n+2)M} \tag{28.15}$$

The voltage across output diode D0 is given as,

$$D0=\frac{1+M}{M} \tag{28.16}$$

(iii) OVR and the peak switch current: In the FISM, the output voltage ripple is calculated by averaging the capacitor current over the current interval Ton. The OVR in the operative mode of the converter is given by:

$$V_{PP} = \frac{Vo(Vo - Vi)}{fRC(Vo + (2n+3)Vi)} \tag{28.17}$$

The OVR is inversely proportional to Vi and R and is independent of inductance values (Figure 28.6). The peak current passing through the power devices is twice the peak current flowing via the inductors. It is written as

$$I_{SP} = \frac{2Vo[(2n+3)Vi]}{(n+2)RVi} \tag{28.18}$$

(iv) Design of inductors and capacitors: Between CCM and DCM, the critical inductance is obtained as

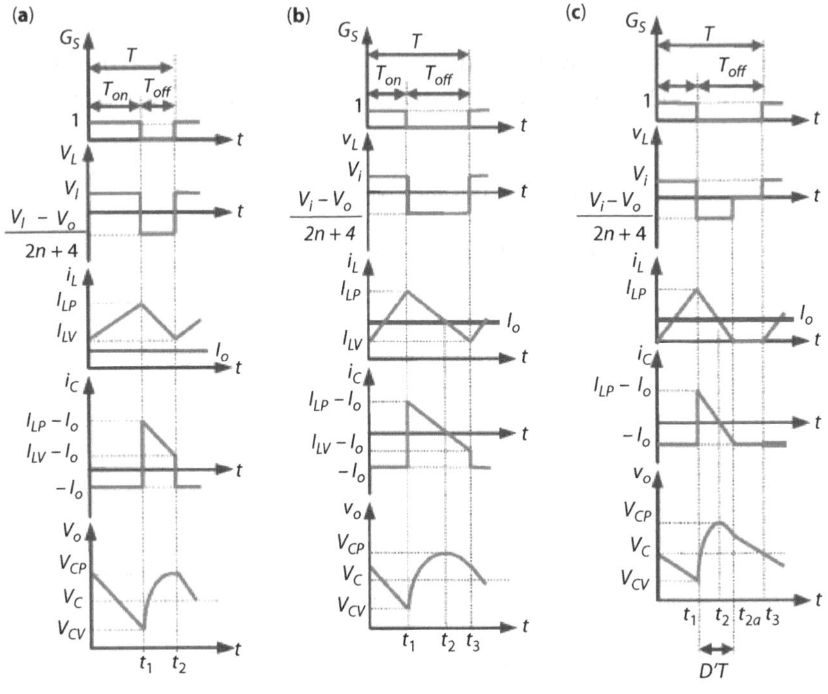

Figure 28.6 Theoretical waveform for (a) FISM-CCM (b) PISM-CCM (c) PISM-DCM.

$$L_C = \frac{(n+2)(V_o - V_i)V_i^2 R}{fV_o[(2n+3)V_i + V_o]^2} \tag{28.19}$$

Between FISM and PISM, the critical inductance is expressed as

$$L_k = \frac{(n+2)V_i^2 R}{fV_o[(2n+3)V_i + V_o]} \tag{28.20}$$

After calculating the critical values of inductance the proper inductance value is chosen by using Table 28.1. The capacitance value is calculated from the following equation:

$$C = \frac{DV_0}{f\Delta Vo} \tag{28.21}$$

28.5 Simulation

The analysis of this converter is done using Simulink (MATLAB). The converter (for n=1) is simulated with the design parameters as portrayed in Table 28.1.

Table 28.1 Design parameters.

Parameters	Values
V_0	388 V
I_0	0.538 A
D	68 %
V_{in}	30 V
I_{in}	7.25A
P_{in}	220W
P_0	209W
Efficiency	96.09%
Gain	13

28.5.1 Simulation for n=1

The converter is designed for n=1 at 30 V supply voltage, a load voltage of 388.5 V is acquired. Duty cycle of 68% is given to all the switches. The switches are operated at a switching frequency of 50kHz as portrayed in Figure 28.7(a).

Figure 28.7(b), Figure 28.7(c), and Figure 28.7(d) show the voltage across the power switches for n=1. The output voltage and current waveforms are portrayed in Figure 28.8(a) and Figure 28.8(b). The load voltage and load current values are 388.5 V and 0.53 A, respectively.

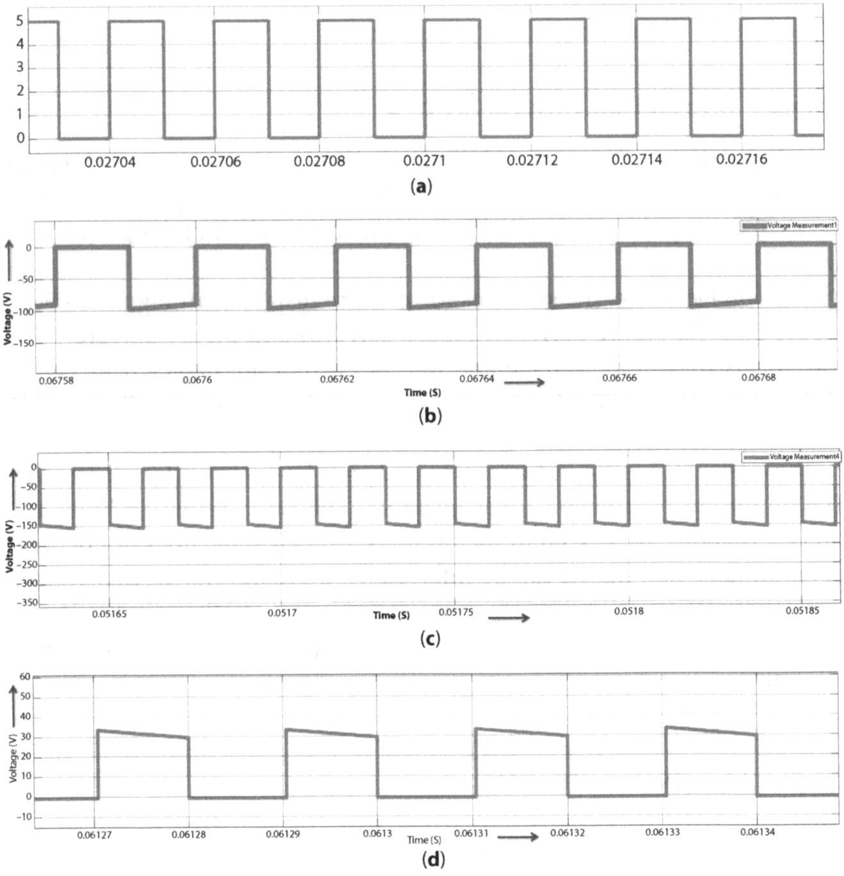

Figure 28.7 (a) Gating pulse for switches S, S1 and S'. (b) Voltage stress across switch S. (c) Voltage stress obtained across switch S'. (d) Voltage through diode D1.

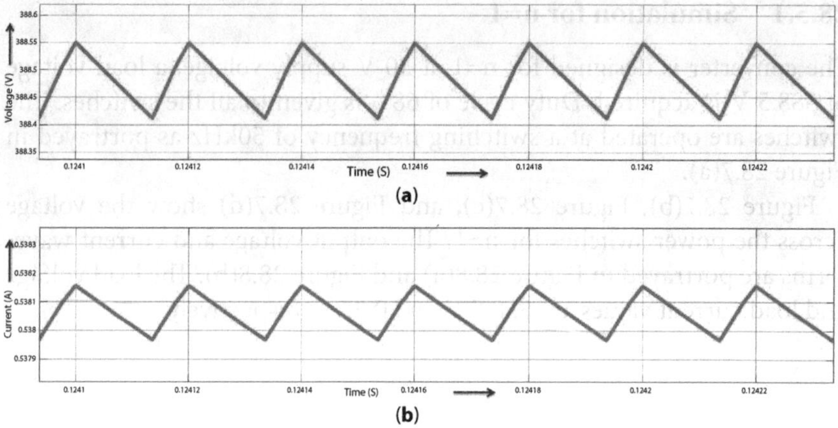

Figure 28.8 (a) Output voltage (b) output current.

28.5.2 Simulation Results for n=2

The simulation is done for 30V input voltage and an output voltage of 441.4V is obtained. Figure 28.9(a) and Figure 28.9(b) displays the waveforms of firing pulses with switching frequency as 50kHz and input current around 9A.

The voltage across the switches for n=2 are displayed in Figure 28.9(c), Figure 28.9(d), Figure 28.10(a), Figure 28.10(b) and Figure 28.10(c). The waveforms for output voltage and output current are depicted in Figure 28.10(d) and Figure 28.10(e) has the average value of output voltage as 441 V and output current as 0.6 A.

The critical values of inductances are $L_c = 0.127mH$ and $L_K = 0.193mH$. The important parameters are as given in Table 28.2. When L=300H, the MOVR is about 2V, as seen in the findings. MOVR value does not depend upon inductance value. It can be inferred from the waveforms that voltage stress across the output diode is more as compared to other diodes. When compared to alternative topologies, the voltage stress via the switches is minimised.

28.6 Hardware Results

The hardware set up is as presented in Figure 28.11(a) and Figure 28.11(b) displays the waveforms of the gating pulses given the switches with the duty cycle of 68%. The hardware components with the component specification, ratings and requirements are shown below in Table 28.2.

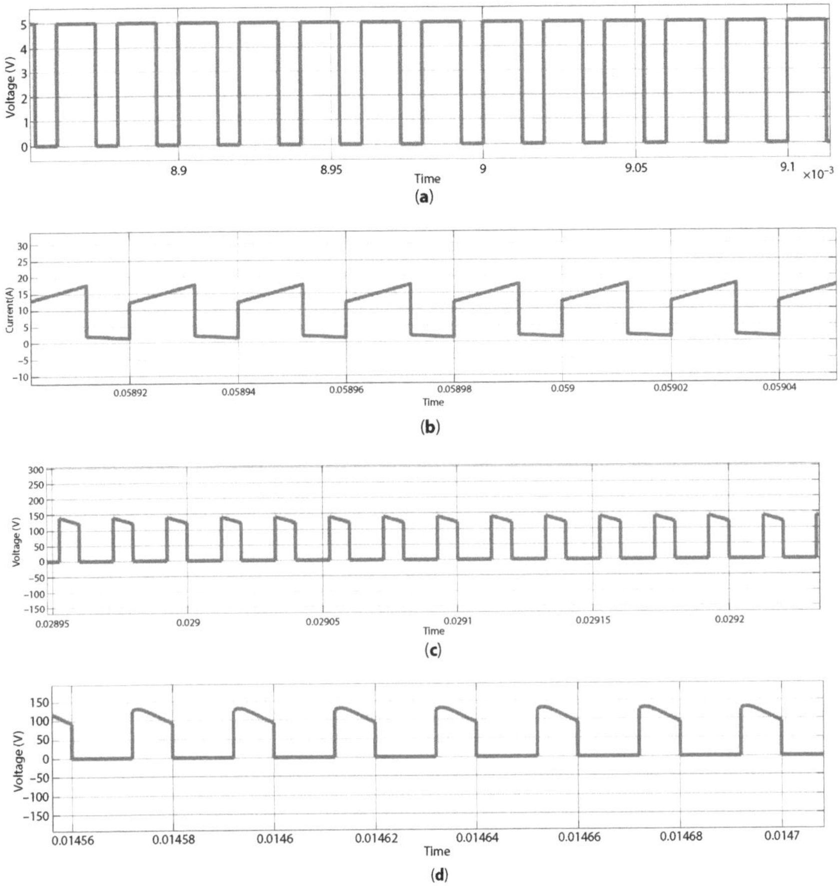

Figure 28.9 (a) Gating pulse for switches S,S1,S2,S'. (b) Input current. (c) Voltage across switch S. (d) Voltage across switch S1.

As given in Figure 28.11(c), an output of 209 V with a voltage gain of 17 is achieved for 12V input voltage. The output gain variation concerning the duty cycle is revealed in Figure 28.12. It can be perceived from the graph that voltage gain rises when the duty cycle increases. And higher values for the conventional boost converter if the duty cycle is needed to obtain higher voltage gain values. With a rise in n, the voltage gain increases by approximately 18% for a given duty cycle. Significantly high gain is obtained for lower values of duty cycle.

The efficacy of the converter is also improved in the presented topology. For both n=1 and n=2, Figure 28.13(a) shows the difference in efficiency as a function of duty cycle. It can be inferred from the graph that for n=1 the

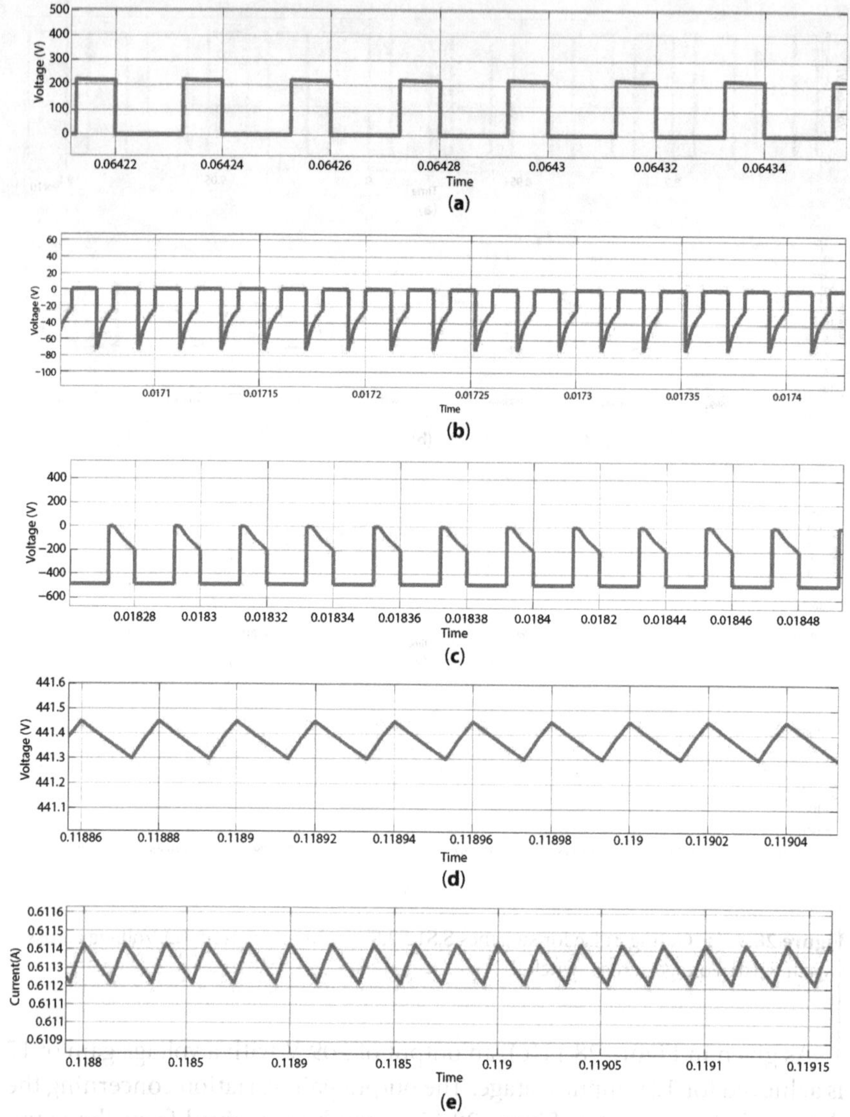

Figure 28.10 (a) Voltage across switch S2. (b) Voltage across diode D. (c). Voltage across ouput diode D0. (d) Output voltage. (e) Output current.

efficiency is 96.09% at 68% duty cycle. At a given duty cycle the efficiency is improved when the number of APIC's increases. Efficiency versus output power graphs is also given for both n=1 and n=2. With a higher n, the converter's performance improves as well.

Table 28.2 Hardware specifications.

Components	Specifications
Inductors(L1,L2,L11,L22,L1',L2')	0.3mH
Diodes(D1,D2,D3,D11,D12,D13, D14, D15, D1',D2',D3',D0)	SUF30J(400 V, 3A)
Switches(S,S1,S')	MOSFET IRF640(18A, 200V)
Capacitor(C)	68uF(400 V)
Microcontroller	PIC16f877A
Gate Driver circuit	TLP250

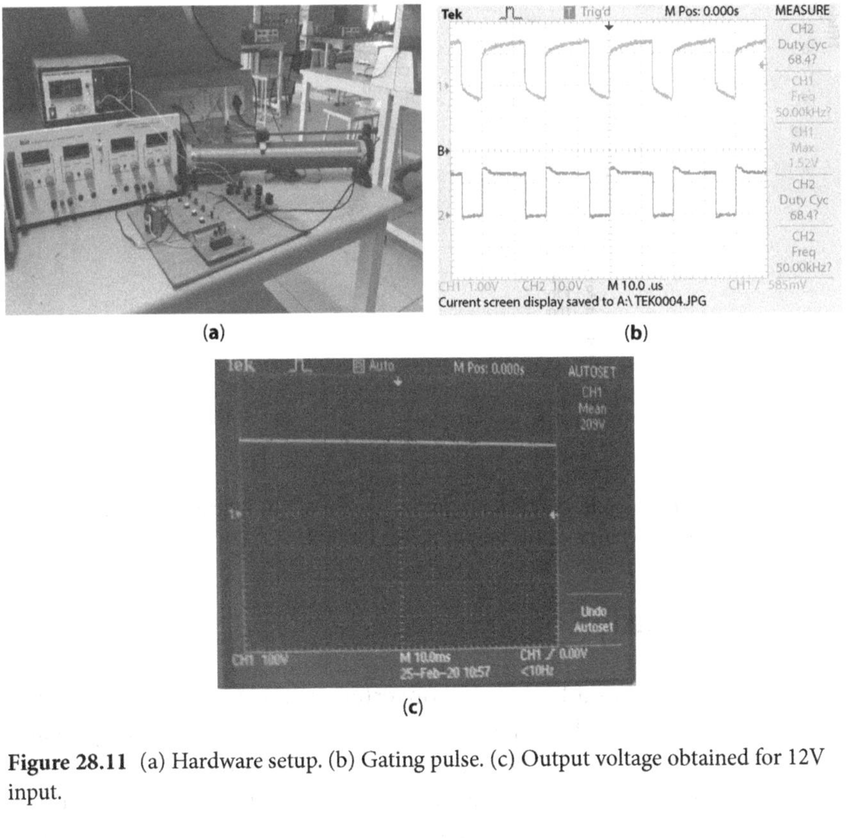

Figure 28.11 (a) Hardware setup. (b) Gating pulse. (c) Output voltage obtained for 12V input.

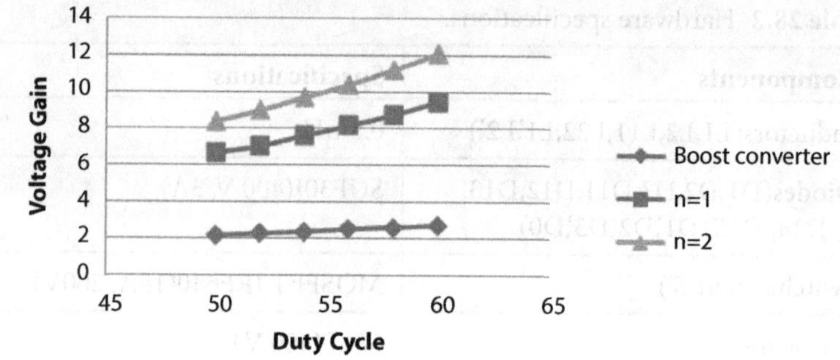

Figure 28.12 Duty cycle versus voltage gain for n=1, n=2 and conventional converter.

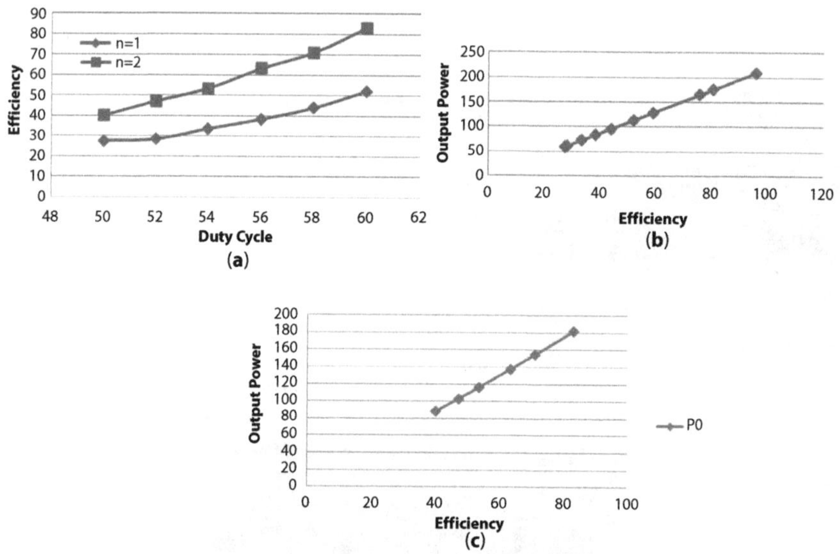

Figure 28.13 (a) Duty cycle vs. efficiency for n=1 and n=2. (b) Efficiency vs. output power for n=1. (c) Efficiency versus output power for n=2.

28.7 Conclusion

The chapter describes a high-gain converter architecture based on APICs that increases the number of APICs while decreasing the duty cycle and conduction losses. The circuit uses a small-value inductor that reduces the size of the magnetic core needed. The filter is smaller than in other topologies since this converter is designed to generate the smallest possible output voltage ripple, minimizing the converter's cost. In FISM-CCM, the OVR is the lowest. In comparison to other topologies, the total current and voltage stress across the switches are lower. The suggested converter is also adaptable, with the number of APICs modified according to the necessary gain. The validity of the proposed concepts is established by comparing the simulation and experimental findings to the theoretical results. The converter has extendability in realizing its usage for renewable energy applications.

References

1. R. J. Wai, C. Y. Lin, and C. C. Chu, "High step-up dc–dc converter for fuel cell generation system," in *Proc. IECON*, 2004, pp. 57–62. 23 May 2005.
2. Y.P. Siwkoti, F. Blaabjerg, and P.C. Loh, "High step-up trans inverse dc-dc converter for the distributed generation system," *IEEE Trans.Ind. Electron.*, vol. 63, no. 7, pp. 316-322, March 2016.
3. H. Ardi, A. Ajami, and M. Sabahi, "A novel high step-up dc-dc converter with continuous input current integrating coupled inductor for renewable energy applications," *IEEE Trans. Ind. Electron*, in press (2017).
4. K. B. Park, G. W. Moon, and M. J. Youn, "Nonisolated high step-up boost converter integrated with sepic converter," *IEEE Trans. Power Electron*, vol. 25, no. 9, pp. 6476–6482, Sep. 2010.
5. J. Ai and M. Li, "Ultra-large gain step-up coupled inductor dc-dc converter with asymmetric voltage multiplier network for a sustainable energy system," *IEEE Trans. Power Electron*, vol. 32, no. 9, pp. 6896–6903, Sept. 2017.
6. H. Wu, K, Sun, L. Chen, L. Zhu, and Y. Xing, "High step-up/stepdown soft-switching bidirectional dc-dc converter with coupled inductor and voltage matching control for energy storage systems," *IEEE Trans. Ind. Electron.*, vol. 63, no. 5, pp. 2892-2903, May 2016.
7. M. Forouzesh, Y. Shen, K. Yari, Y. P. Siwakoti, and F. Blaabjerg "High-efficiency high step-up dc-dc converter with dual coupled inductors for grid-connected photovoltaic Systems," *IEEE Trans. Ind.Electron*, in press (2017).
8. D. Gunasekaran, L. Qin, Y. Karki, Y. Lin, and F.Z. Peng, "A variable (n/m)X switched capacitor dc-dc converter," *IEEE Trans. Power Electron.*, vol. 32, no. 8, pp. 6219–6235, Aug. 2017.
9. S. Arslan, S. A. Ali Shah, J. J. Lee, and H. W. Kim, "An energy efficient charging technique for switched capacitor voltage converters with low duty ratio," *IEEE Trans. Circuit and system-II.*, in press (2017).
10. L. Hong-Chen and L. Fei, "Novel high step-up dc–dc Converter with active coupled-inductor network for a sustainable energy system," *IEEE Trans. Power Electron*, vol. 30, no. 12, pp. 6476–6482, Dec. 2015.

11. Jaisudha S., Sowmiya Srinivasan, Kanimozhi G,"Bidirectional Resonant DC-DC Converter for Microgrid Application", *International Journal of Power Electronics and Drive System (IJPEDS)*, Vol. 8, No. 4, December 2017, pp. 1548~1561, Oct 2017.
12. K. Likhitha, S Sathish Kumar, G. Kanimozhi, "Isolated DC-DC zero voltage switching converter for battery charging applications", *2016 Biennial International Conference on Power and Energy Systems: Towards Sustainable Energy (PESTSE)*, 10.1109/PESTSE.2016.7516437, 21 July 2016.
13. Kanimozhi. G, Umayal. C, Dhanasekar. S, "FPGA based Hybrid Resonant Switching DC/DC converter for Electric Vehicles", *International Journal of Innovative Technology and Exploring Engineering (IJITEE)* ISSN: 2278-3075, Vol. 8 no. 9, July 2019.
14. L. Hong-Chen and L. Fei, "Novel high step-up dc-dc converter with active coupled-inductor network for a sustainable energy system,"*IEEE Trans. Power Electron,* vol. 30, no. 12, pp. 6476–6482, Dec. 2015
15. L. Hong-Chen and L. Fei, "Novel high step-up dc-dc converter with an active coupled-inductor network for a sustainable energy system," *IEEE Trans. Power Electron,* vol. 30, no. 12, pp. 6476–6482, Dec. 2015.
16. Ebrahim Babaei, Hamed Maschinchi Maheri, Mehran Sabahi, Seyed Hossein, "Extendable Non Isolated High Gain DC-DC Converter Based on Active-passive Inductor Cells", *IEEE Trans.Ind. Electron,* Sept. 2017.
17. J. Ai and M. Li, "Ultra-large gain step-up coupled inductor dc-dc converter with asymmetric voltage multiplier network for a sustainable energy system," *IEEE Trans. Power Electron,* vol. 32, no. 9, pp. 6896–6903, Sept. 2017.
18. H. Wu, K, Sun, L. Chen, L. Zhu, and Y. Xing, "High step-up/stepdown soft-switching bidirectional dc-dc converter with coupled inductor and voltage matching control for energy storage systems," *IEEE Trans. Ind. Electron.*, vol. 63, no. 5, pp. 2892-2903, May 2016.
19. M. Forouzesh, Y. Shen, K. Yari, Y. P. Siwakoti, and F. Blaabjerg, "High-efficiency high step-up dc-dc converter with dual coupled inductors for grid-connected photovoltaic Systems," *IEEE Trans. Ind.Electron,* in press (2017).
20. G. Kanimozhi, O.V. Gnana Swathika, K. Logavani, A. Ambikapathy, "High gain DC-DC converter with extendable APIC's", *Materials Today: Proceedings* 2020, ISSN 2214-7853, https://doi.org/10.1016/j.matpr.2020.10.953.
21. S. Arslan, S. A. Ali Shah, J. J. Lee, and H. W. Kim, "An energy efficient charging technique for switched capacitor voltage converters with low duty ratio, *IEEE Trans. Circuit and system-II.,* in press (2017).
22. L. Schmitz, D. C. Martins, and R. F. Coelho, "Generalized high stepup dc-dc boost-based converter with gain cell," *IEEE Trans. Circuits and systems-I,* vol. 64, no. 2, pp. 480–493, Feb. 2017.
23. H. Mashinchi, E. Babaei, S.H. Hosseini, and M. Sabahi, "High step-updc-dc converter with minimum output voltage ripple," *IEEE Trans.Ind. Electron,* vol. 64, no. 5, pp. 3568–3575, May 2017.

29

A Non-Isolated Step-Up Quasi Z-Source Converter Using Coupled Inductor

Shashank, P.C.[1] **and Kanimozhi, G.**[2*]

[1]*School of Electrical Engineering, Vellore Institute of Technology, Chennai, India*
[2]*Centre for Smart Grid Technologies, School of Electrical Engineering, Vellore Institute of Technology, Chennai, India*

Abstract

A non-isolated high step-up quasi z-source dc-dc converter is examined and simulated in continuous conduction mode (CCM). The converter operation is based on continuous charging and discharging of the switched capacitors, coupled inductor, and switched inductor. The suggested converter has several advantages, including high and constant efficiency, high gain at lower duty cycles, and a larger range of operations based on the duty cycle. The main parameters like voltage gain, efficacy, voltage stress across power devices, minimum output voltage ripple are calculated. All the aforesaid parameters are analyzed through simulation. A comparison is also made between the suggested converter and a conventional converter on which an analysis was performed and the shortcomings have been rectified. A 400W high step-up quasi-z source dc-dc converter was designed and simulated in MATLAB/Simulink. For a 40V dc input and 50% duty cycle; a voltage gain of 9.5 and efficiency of 94% is achieved.

Keywords: Z-source converter, renewable energy, high gain, CCM, renewable energy sources

29.1 Introduction

Renewable energy sources like photovoltaic and fuel cells are manufactured using the currently available technology deliver output voltages at

**Corresponding author*: kanimozhi.g@vit.ac.in

Milind Shrinivas Dangate, W.S. Sampath, O.V. Gnana Swathika and P. Sanjeevikumar (eds.) Integrated Green Energy Solutions Volume 2, (151–166) © 2023 Scrivener Publishing LLC

lower levels. These levels cannot be utilized for commercial and industrial applications. To step up these voltage levels to meet the required levels step-up converters are used. The rectifier converts the AC to DC, while the boost converter increases the DC voltage. The overall system's efficiency for the conversion of renewable energy sources into electricity is affected by the DC-DC converter. High gain converters are also used in various other fields, like automobile applications, microgrid systems, energy backup for UPS, etc. In renewable energy application, high efficiency and high step-up converters are required since they act as a proper interface between available low voltage sources and the output loads, which in turn operate at very high voltages. In a real-time scenario, the conventional boost converter's high gain is limited by losses caused by intrinsic resistance, resulting in precise and switch driving circuitry with a high cost due to deviations in the duty cycle. It directly affects load voltage and reduces the converter dependability. Even though renewable energy sources are clean and durable, the fact remains that a typical PV panel output voltage range is low, ranging from 12V to 80V. In commercial and industrial applications [1–11], where the needed output voltage ranges from 200V to 400V, and even up to 600 V, this is not ideal.

Isolated and non-isolated DC-DC converters are the two main types of DC-DC converters [12–14]. Isolated converter topologies necessitate a high-frequency transformer usage, which adds more components to the circuit, increasing expense, complexity, and performance. Non-isolated DC-DC converters have a simpler structure and fewer components than isolated DC-DC converters. Non-isolated DC-DC converters, on the other hand, have a poor voltage gain, which limits its usage in the industrial sector.

Various approaches and topologies of non-isolated DC-DC converters have been presented to address the drawbacks and boost the voltage gain. The main approaches to improve voltage gain are cascaded and voltage-lift techniques, using switched-inductors (SL) [15], switched-capacitors (SC) [16], and multiplier cells [17–19].

Switched inductor (SL) [15] in the basic qZSC network limits the duty cycle operating range, voltage stress across the power devices and the passive components increases and the voltage gain achieved is not significant. A switched capacitor (SC) [16] implements a filter and without any use of additional active switches, increases the boost factor range. By including SC, the converter may achieve a larger voltage gain while imposing less voltage stress on the capacitors in the impedance network. On practical implementation, the enhancement of voltage gain is not substantial, and the converter was unsuccessful in achieving high voltage gains. Voltage lift technique in the qZSC achieves high gain in the modified converter qZSC with SC, qZSC with SL and qZSC with SC-SL combination. However the duty cycle is limited, the voltage stress is increased in the device and the converter is unable to accomplish high voltage gains practically.

This converter depicted in Figure 29.1 is designed to achieve high gain and higher efficiency at lower duty cycles and widen the range of operation based on the duty cycle of the switch. The converter operates in CCM based on continuous switching of diodes, continuous charging, and discharging of SC and SL. It is designed to maintain low voltage stress across the switching devices and the passive elements utilized in the converter. The suggested converter is constructed in such a way that higher gains can be attained at lower duty cycles while maintaining high efficiency, boosting the converter's dependability.

The high step-up qZSC that is being analyzed in this section can be utilized for renewable energy applications. This converter is designed to achieve high gain and higher efficiency at lower duty cycles and widen the range of operation based on the duty cycle of the switch. The converter operates in CCM based on continuous switching of diodes, continuous charging, and discharging of SC and SL.

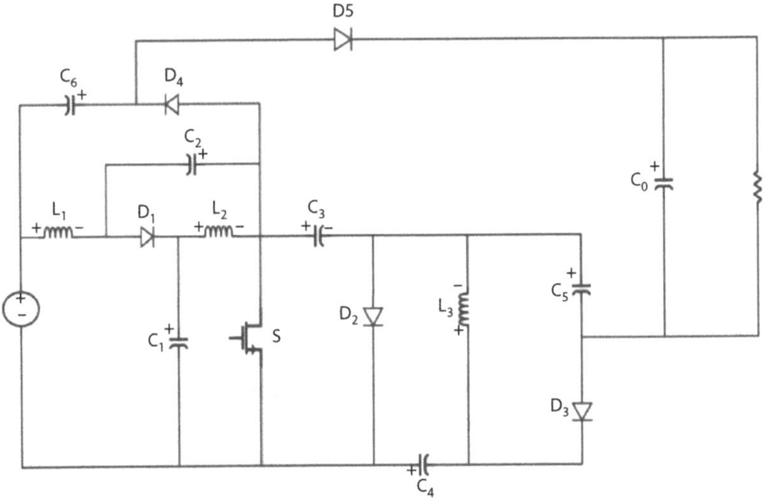

Figure 29.1 Modified Quasi ZSC.

29.2 Improved Quasi Z Source Converter with Coupled Inductor

The proposed qZSC uses a single IGBT switch S, four diodes D_1, D_2, D_3, D_4, one coupled inductor L_C, one switched inductor L_1, and four capacitors (C_1, C_2, C_3, C_4) to perform the role of switching diodes that makes the switched capacitors charge or discharge. Capacitors (C_1, C_2, C_3, C_4), are switched capacitors that charge and discharge based on the on/off state of the diodes. The switched inductor L_1 is essentially used to boost the voltage and it also charges and discharges based on the on/off state of the diodes. The output capacitor C_O performs the role of a low pass filter and reduces the ripple in the output voltage and output current waveforms.

To analyze the operation of the converter, all the switches and the passive components in the circuit are assumed to be ideal. The circuit diagram of the improved qZSC using coupled inductor is displayed in Figure 29.2.

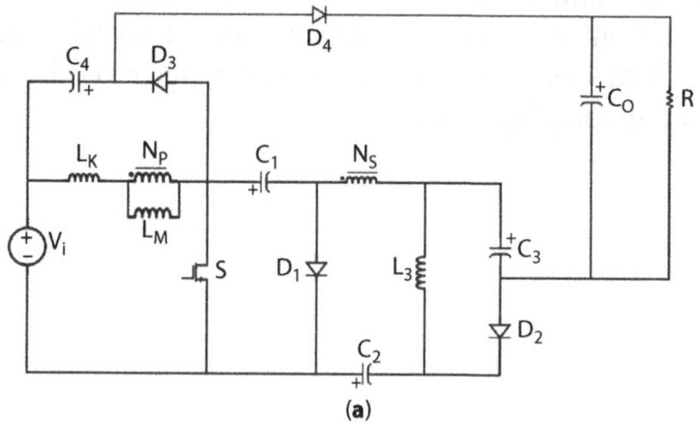

Figure 29.2 (a) Improved qZSC using coupled inductor.

29.3 Modes of Operation

The proposed qZSC consists of three operation modes as Mode-1 ($t_o < t < t_1$), Mode-2 ($t_1 < t < t_2$) and Mode-3 ($t_2 < t < T$).

For the forthcoming relationship, D signifies the duty cycle of the switching devices. Mathematically it is denoted as:

$$D = \frac{t_{ON}}{T} = \frac{t_1 - t_0}{T} \tag{29.1}$$

D' represents the duty cycle when circuit is in its second mode of operation. Mathematically,

$$D' = \frac{t_{ON}}{T} = \frac{t_2 - t_1}{T} \quad (29.2)$$

If the conduction period of the power device is denoted as $D'T$, then the non-conduction period is denoted as $(1 - D')T$. For the converter analysis, Mode 1 and Mode 2 operational modes are only considered as Mode 3 is similar to Mode 2.

a. MODE-1 (t_o<t<t_1)

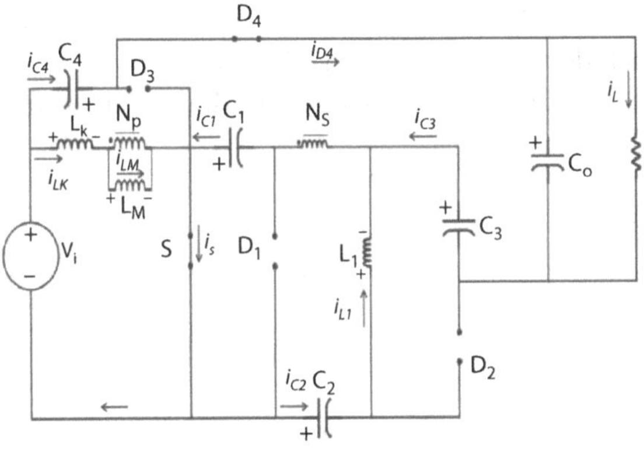

(b)

Figure 29.2 (b) Mode-1 circuit diagram.

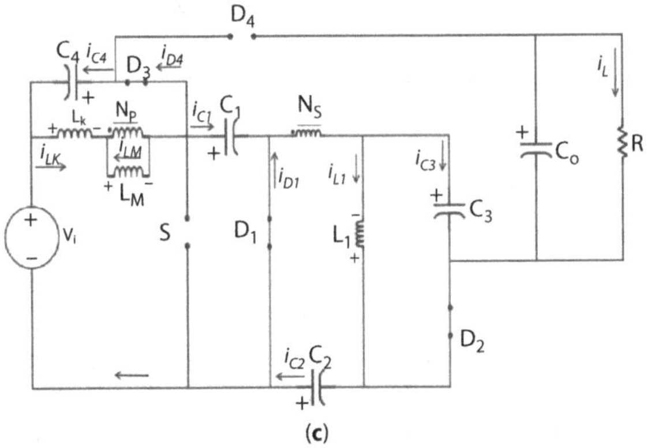

(c)

Figure 29.2 (c) Mode-2 circuit diagram.

Mode 1 begins as the switch S is switched ON. Diodes D_1, D_2, D_3, are reverse biased condition in this mode, while diode D_4 is forward biased. The direction of current flow and the on/off state of the switches are presented in Figure 29.2(b). L_K represents the leakage inductance of L_C which denotes an inductive component resulting from the imperfect magnetic linking between the primary winding and the secondary winding. Magnetizing inductance L_M of the coupled inductor L_C is charged by the DC input voltage source and the capacitor C_1. C_1 also charges the capacitor C_4 through the path provided by S. Capacitors C_3 charges the inductor L_1. The voltage required for the load is supplied by the DC voltage source and the capacitor C_4 through the path provided by D_4.

Based on the on/off state of the diodes, the current through the coupled inductor L_C, inductor L_1, and the capacitor C_2 voltage increases. The voltage across capacitors C_1, C_3, and C_4 decreases as they are getting discharged. This mode concludes when the switch S is turned off.

The KCL and KVL equations obtained in Mode-1 are as follows:

$$V_{NP} = V_i \tag{29.3}$$

$$V_{LM} = kV_i \tag{29.4}$$

$$V_{Lk} = V_i - V_{LM} = V_i(1 - k) \tag{29.5}$$

$$V_{Ls} = knV_i \tag{29.6}$$

$$V_i = V_o - V_{L1} - V_{C2} - V_{C3} - V_{C4} \tag{29.7}$$

$$V_{C1} = V_{C2} + V_{L1} - V_{NS} = V_{C2} + V_{L1} - knV_i \tag{29.8}$$

Here k denotes the coupling coefficient of L_C and n denotes the turns ratio in the coupled inductor.

b. MODE-2 (*$t_1<t<t_2$*): MODE 2 commences when the IGBT switch S is in its OFF state. In this Mode, D_1, D_2, D_3, diodes are forward biased and diode D_4 is reverse biased. The current flow and the on/off state of the switches are presented in Figure 29.2(c). Magnetizing inductance L_M of the coupled inductor L_C charges the capacitor C_1. L_C also charges the capacitor C_4 through the path provided by D_4 Capacitor C_4 and inductor L_1 charge the capacitor C_3 through the path provided by D_2. The voltage required for the load is also provided by C_2 and L_1 through the path provided by D_2.

Based on the switching state of the diodes, the current through the coupled inductor L_C, inductor L_1, and the capacitor C_2 voltage decreases as they are getting discharged. The voltage across capacitors C_1, C_3, C_4, increases as they are getting charged. This mode concludes when the circuit completes t_2 seconds of operation and the switch S is still OFF.

The KCL and KVL equations obtained in Mode-2 are as follows:

$$V_{C3} = -V_{L1} \tag{29.9}$$

$$V_{C2} = V_{NS} - V_{L1} = knV_i + V_{L1} \tag{29.10}$$

$$V_{C2} = knV_i + V_{C3} \tag{29.11}$$

$$V_{C4} = V_i + V_{C4} \tag{29.12}$$

$$V_{C1} = Vi + V_{C4} \tag{29.13}$$

$$V_{C1} = V_i - V_{LM} \tag{29.14}$$

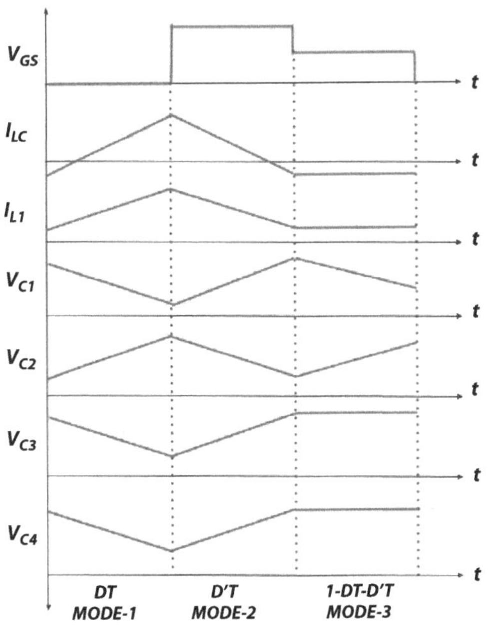

Figure 29.3 Theoretical waveforms of the proposed converter.

c. MODE-3 (*t_2<t<T*): Mode 3 commences when the converter crosses t_2 seconds of its operation and the switch S is still OFF. Mode-3 operation is alike to that of Mode-2, all the diodes are in the same switching state, and inductors and capacitors undergo the same operation. This mode concludes as the converter completes T seconds of operation and the switch is switched ON.

Based on the assumptions and analysis of the converters in the three modes of operation, the theoretical waveforms of all the passive elements in the circuit are provided in Figure 29.3.

29.4 Simulation Results

The simulation results for the proposed qZSC are obtained on simulating the circuit using MATLAB Simulink. For the seamless analysis of the converter, the ratings of the components present in the improved qZSC are specified in Table 29.1. The simulation circuit diagram for the high step-up qZSC is portrayed in Figure 29.4. A duty cycle of 50% is given to the switch with a switching frequency of 40kHz. Figure 29.5 presents the input voltage waveform of the circuit. A constant voltage of 40V DC is supplied to the circuit. The average current measured in the input side of the circuit is 11.18A.

Table 29.1 Converter specifications of proposed qZSC.

Parameters	Values
Input Voltage (V_{in})	40V
Capacitors (C_1, C_2, C_3, C_4, C_o)	100μF
Coupled Inductor (L_P, L_S, L_M)	4μH,5μH,3μH
Coupling Coefficient (k)	0.7
Inductor (L_1)	3mH
Switching Frequency (f_s)	40kHz
Duty Cycle (D)	50%
Load Resistance (R_O)	350Ω
Output Power (P_O)	400W

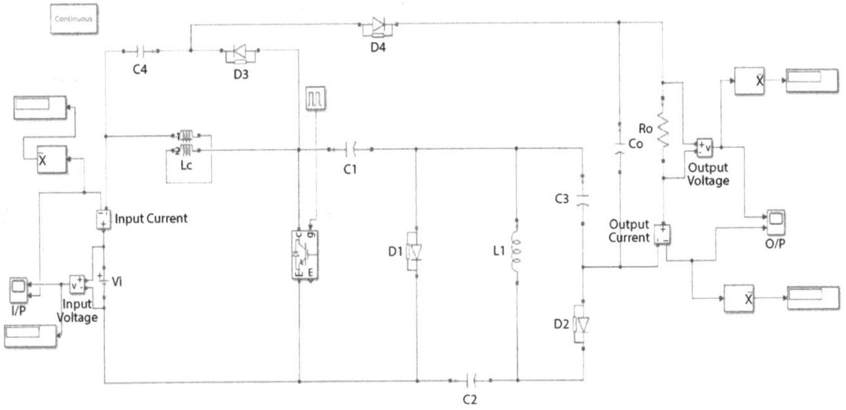

Figure 29.4 Improved qZSC simulation diagram.

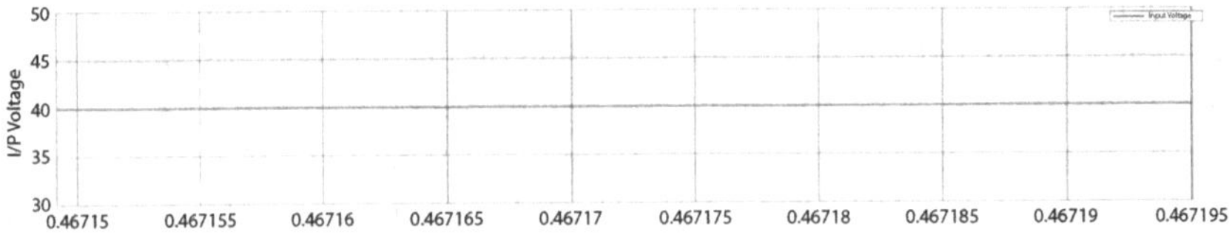

Figure 29.5 Input waveforms.

Figure 29.6 displays the output voltage measured across the load resistor R and the current waveform measured through R. A 350Ω resistor is utilized to measure the waveforms. The average output voltage measured is 383.4V and the average output current measured is 1.095A. The voltage gain obtained on simulating the circuit is 9.585 and an efficiency of 93.87% is achieved.

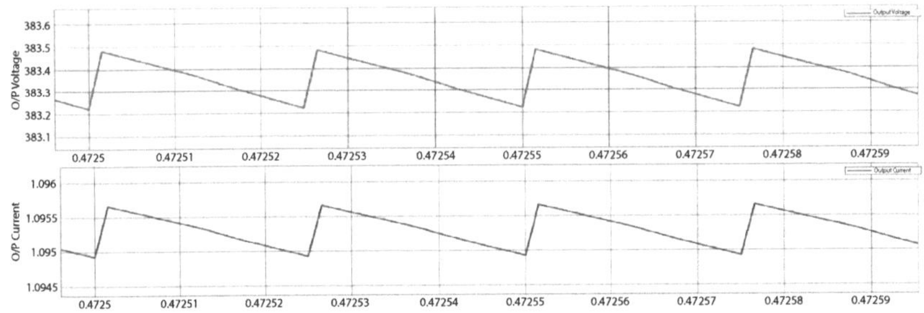

Figure 29.6 Output waveforms.

As viewed in Figure 29.7(a), the topmost graph represents the switching voltage. When V_S=0, the circuit operates in Mode-1. The current through the inductor L_C, increases from 0A to 34A, and current through L_1 increases from 1.08A to 1.2A. While V_S=140V, the circuit operates in Mode-2. The current through the inductor L_C decreases from 34A to -1.05A and current through L_1 decreases from 1.2A to 1.08A. The circuit operates in Mode-3 at V_S=40V. The current through the inductor L_C remains the same at -1.05A and the current through L_1 remains the same at 1.08A till the circuit operates in Mode-1.

As viewed in Figure 29.7(b), the first graph represents the switching voltage and the subsequent graphs represent the voltage across C_1, C_2, C_3. During Mode-1 operation, the voltage across C_1 decreases from 142V to 141.3V non-linearly, the voltage across C_2 increases linearly from 101.5V to 101.7V, and the voltage across C_3 decrease linearly from 101.6V to 101.3V. When V_S=140V, the circuit operates in Mode-2. The voltage across C_1 increases from 141.3V to 141.8V non-linearly, the voltage across C_2 decreases from 101.7V to 101.4V, and the voltage across C_3 increases from 101.3V to 101.6V. When V_S=40V, the circuit operates in Mode-3.

The voltage across C_1 decreases from 141.8V to 141.7V, voltage across C_2 increases from 101.4V to 101.5V, and voltage across C_3 is stable at 101.6V.

The topmost graph observed in Figure 29.7(c), represents the switching voltage and the subsequent graph represents the voltage across C_4. When

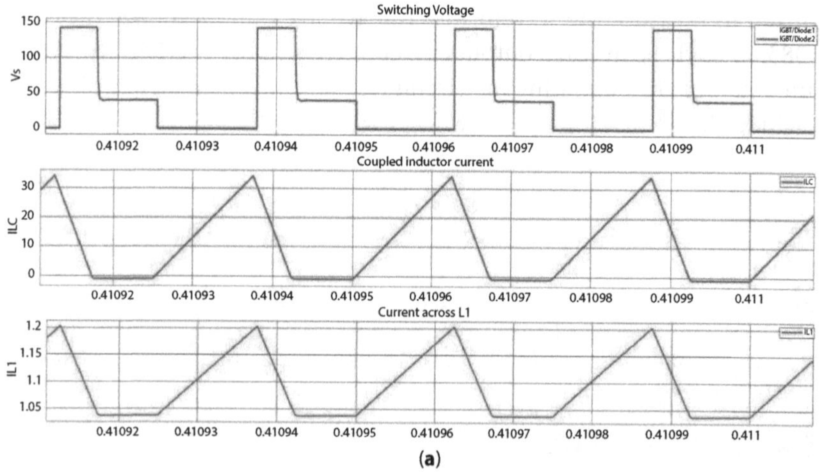

Figure 29.7 (a) Current through inductors L_C, L_1.

V_S=0 the circuit operates in Mode-1. The voltage across C_4 decreases from 101.8V to 101.5V. When V_S=140V, the circuit operates in Mode-2. The voltage across C_4 increases from 101.5V to 101.8V. When V_S=40V, the circuit operates in MODE-3. The voltage across C_4 remains at 101.8V during Mode-3.

Figure 29.7(d) represents the voltage stress across switch S and diodes D_1, D_2. During Mode-1, the switch S is ON and voltage measured across D_1, D_2 is -140V which implies that D_1, and D_2 are reverse biased. When V_S=140V, the circuit operates in Mode-2 and the switch S is OFF.

The voltage measured across D_1, D_2 is 0V which implies that D_1, D_2 is forward biased and the switches are ON. When V_S=40V, the circuit operates in Mode-3 and the switch is OFF. Voltage measured across D_1, D_2 is -100V which implies that D_1, D_2 is reverse biased and the switches are OFF.

Figure 29.8 represents the voltage stress across switch S and diodes D_3, D_4. When V_S = 0 the circuit operates in Mode-1 and the switch S is ON. Voltage measured across D_3 is -140V and across D_4 is 0V which implies that D_3 is reverse biased and the switch is OFF, D_4 is forward biased and the switch is ON. When V_S = 140V, the circuit operates in MODE-2 and the switch S is OFF.

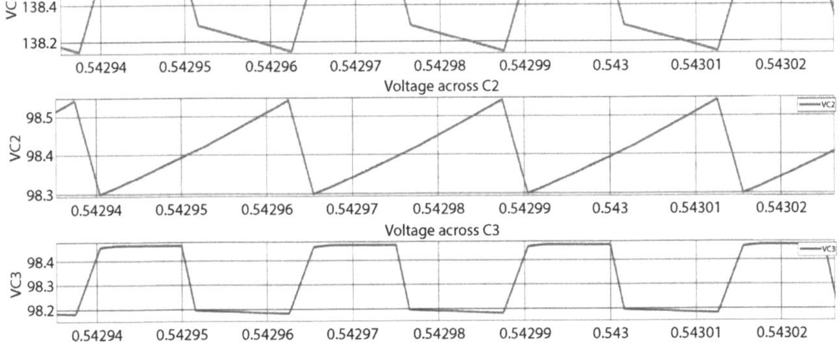

(b)

Figure 29.7 (b) Voltage across capacitors C_1, C_2, C_3.

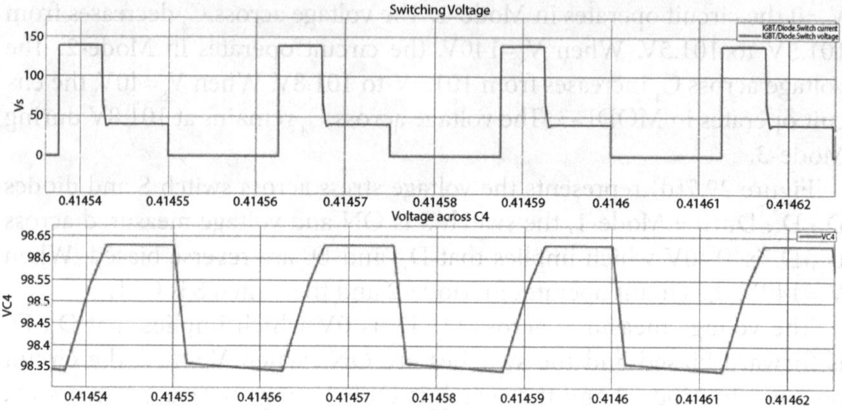

(c)

Figure 29.7 (c) Voltage across capacitors C_4.

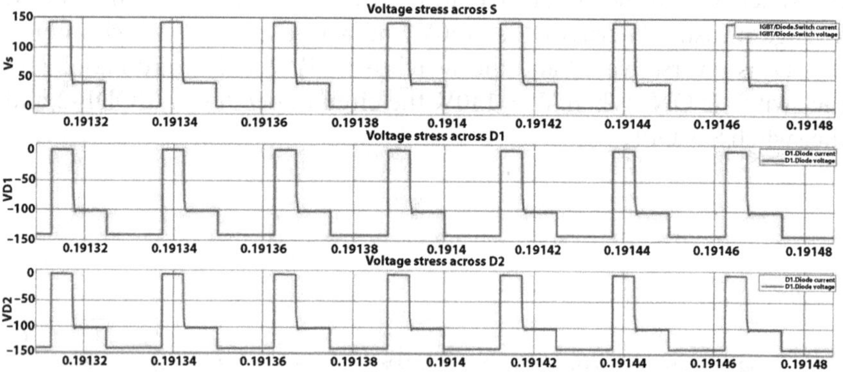

(d)

Figure 29.7 (d) Voltage stress across diodes D_1, D_2.

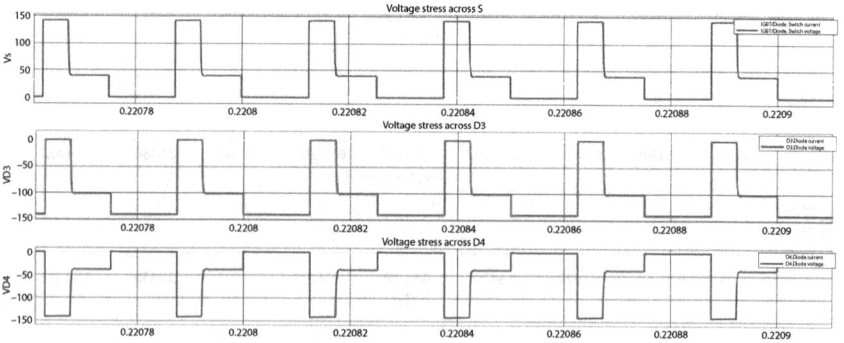

Figure 29.8 Voltage stress across diodes D_3, D_4.

The voltage measured across D_3 is 0V and across D_4 is -140V which implies that D_3 is forward biased and the switch is ON, D_4 is reverse biased and the switch is OFF. When V_S = 40V, the circuit operates in MODE-3 and the switch is OFF. Voltage measured across D_3, D_4 is -100V and -40V, respectively, which implies that D_3, D_4 is reverse biased and the switches are OFF.

29.5 Comparison

In the graph shown in Figure 29.9 (a), it can be noticed that on increasing the load power in the modified qZSC the efficiency of the converter ranges between 70%-75% peaking at 550W. It can also be seen that after 550W the efficiency of the modified qZSC drops to 70% which is considerably low and unreliable. For the proposed qZSC, it is observed that irrespective of the increase in the load power the efficiency of the converter is considerably consistent ranging from 93%-94% which is considerably high and based on current trends the converter can be considered reliable and efficient.

In Figure 29.9 (b), for the modified qZSC, it can be inferred that the gain rises steadily till the converter operates in a duty cycle of 35%. After the duty cycle crosses 35% the gain of the converter rises steeply to a point when the converter stops functioning at 50% duty cycle. The converter does not work beyond the duty cycle of 50% and thus it has a limited range of operation. For the proposed qZSC the gain rises steadily and the converter can operate at higher duty cycles also. It can be said that the proposed qZSC has a wider range of operations in terms of duty cycle.

In the graph shown in Figure 29.10, it can be viewed for the modified qZSC that at lower duty cycles between 5%-10%, the efficiency of the converter is approximately 90%. Subsequently, when the duty cycle is increased from 10% to 35% the efficiency of the converter drops gradually from 90% to 70%. For the proposed qZSC, irrespective of the duty cycle the efficiency remains stable in the range of 92% and 94%. Based on the observation of stable efficiency it can be claimed that the proposed qZSC is reliable.

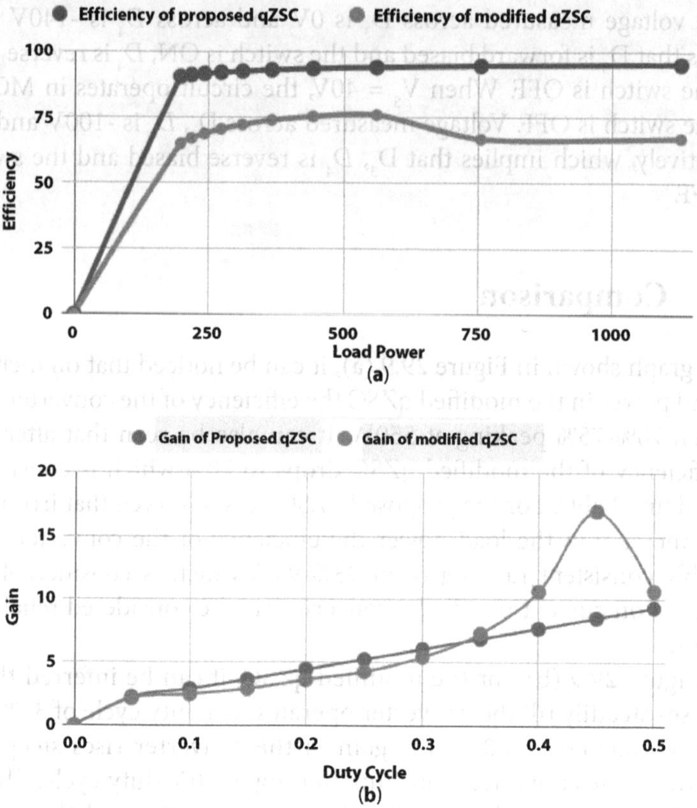

Figure 29.9 (a) Load power vs. efficiency graph. (b) Gain vs. duty cycle graph.

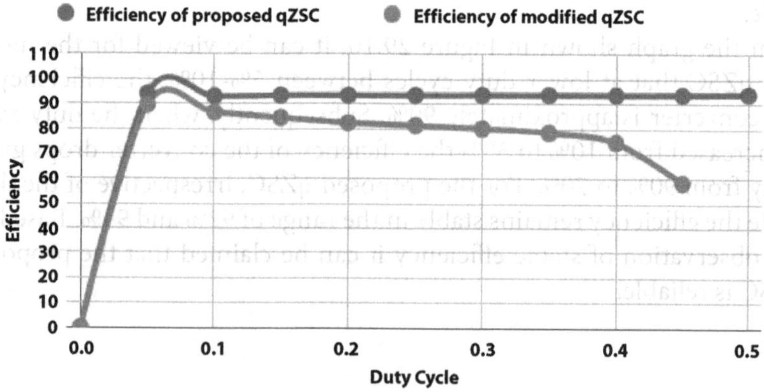

Figure 29.10 Efficiency vs. duty cycle graph.

29.6 Conclusion

A modified qZSC with a high step-up capability has been analyzed theoretically and through simulation. The modified qZSC is n-TCIB based with a combination of switching-capacitors and switching-inductors apart from QZS network. The converter could overcome the basic converter's voltage gain limitation while retaining its main benefits, such as continuous input current, low voltage stress on capacitors and keeping the duty cycle range unchanged. Additional stages to the modified qZSC to aid the capability to accomplish higher voltage gains. The circuit operating principle, analysis, and essential relationships are addressed. The limitations of the modified qZSC include high in-rush current, low efficiency at higher duty cycles, inconsistency due to input surges, lower range of operation with respect to the duty cycle, while retaining its main benefits, such as addition of extra stages, continuous input current, low voltage stress on capacitors and the diodes.

References

1. Xiaoquan Zhu, Bo Zhang, High step-up quasi-Z-source DC-DC converters with single switched capacitor branch. *Journal of Modern Power Systems and Clean Energy*, Vol. 5, Issue 4, pp: 537-547, July 2017.
2. M. Forouzesh, Y. P. Siwakoti, S. A. Gorji, F. Blaabjerg, and B. Lehman, Step-Up DC-DC Converters: A Comprehensive Review of Voltage-Boosting Techniques, Topologies, and Applications. *IEEE Trans. Power Elec*, 32(12) 9143-9178, 2017.
3. Fang Zheng Peng, Z-Source Inverter. *Industry Applications Conference, 37th IAS Conference, Record of the Annual Meeting*, vol. 2, pp. 775-781, 2002.
4. Mohammad Mehdi Haji-Esmaeili, Ebrahim Babaei. High Step-Up Quasi-Z Source DC-DC Converter. *IEEE Trans. on Power Electronics*, vol. 33, no. 12, pp. 10563-10571, Dec. 2018.
5. Dong Cao, Fang Z. Peng, A Family of Z-source and Quasi-Z-source DC-DC Converters. *Proc. IEEE Appl. Power Electron. Conf.*, pp. 1093-1101, Feb. 2009.
6. Taro Takiguchi and Hirotaka Koizumi, Quasi-Z-source DC-DC Converter with Voltage-Lift Technique. In *IECON 2013 - 39th Annual Conference of the IEEE Industrial Electronics Society, 10-13 Nov. 2013.*
7. Dmitri Vinnikov, Ryszard Strzelecki, Step-Up DC/DC Converters with Cascaded Quasi-Z-Source Network. *IEEE Transactions on Industrial Electronics*, Vol. 59, No. 10, Oct 2012.
8. R. R. Ahrabi, H. Ardi, and M. Elmi, A Novel Step-up Multi-Input DC/DC Converter for Hybrid Electric Vehicles Application. *IEEE Trans. Power Electron*, vol. 32, no. 5, pp. 3549-3561, May 2017.

9. Y. Liu, H. Abu-Rub, and B. Ge, Front-End Isolated Quasi-Z-Source DC/DC Converter Modules in Series for High-Power Photovoltaic Systems—Part I: Configuration, Operation, and Evaluation. *IEEE Trans. Ind. Electron.*, vol. 64, no. 1, pp. 347-358, Jan. 2017.
10. J. Yang, Zh. He, H. Pang, and G. Tang, The Hybrid-Cascaded DC/DC Converters Suitable for HVDC Applications. *IEEE Trans. Power Electron.*, vol. 30, no. 10, pp. 5358-5363, Oct. 2015.
11. D. Vinnikov, and I. Roasto, Quasi-Z-source based dc/dc converters for distributed power generation. *IEEE Trans. Ind. Electron.*, vol. 58, no. 1, pp. 192-201, Jan. 2011.
12. F. A. A. Meinagh, E. Babaei, and H. Tarzamni, Modified High voltage gain-switched boost inverter. *IET Power Electron.*, vol. 10, no. 13, pp. 1655-1664, Nov. 2017.
13. Ajami, H. Ardi and A. Farakhor, A Novel High Step-up DC/DC Converter Based on Integrating Coupled Inductor and Switched-Capacitor Techniques for Renewable Energy Applications. *IEEE Transactions on Power Electronics*, vol. 30, no. 8, pp. 4255-4263, Aug. 2015, DOI: 10.1109/TPEL.2014.2360495.
14. Salil Gokhe, Umayal.C, Kanimozhi.G, Multiport DC-DC Boost Converter for Utilization in PV System with Improved Gain. *International Journal of Engineering and Advanced Technology (IJEAT)*, Vol. 9 Issue 1, Oct. 2019, pp. 2137-2142.
15. Samadian, S. H. Hosseini, M. Sabahi and M. Maalandish, A New Coupled Inductor Non-isolated High Step-Up Quasi Z-Source DC-DC Converter. *IEEE Transactions on Industrial Electronics*, vol. 67, no. 7, pp. 5389-5397, July 2020.
16. H. F. Ahmed, H. Cha, S. Kim and H. Kim, Switched-Coupled-Inductor Quasi-Z-Source Inverter. *IEEE Transactions on Power Electronics*, vol. 31, no. 2, pp. 1241-1254, Feb. 2016, DOI: 10.1109/TPEL.2015.2414971.
17. Jaisudha S., Sowmiya Srinivasan, Kanimozhi.G, Bidirectional Resonant DC-DC converter for Microgrid Applications. *International Journal of Power Electronics and Drives Systems*, Vol. 8, No. 4, Dec. 2017, pp- 1548-1561.
18. Poorali, Behzad & Torkan, Arash & Adib, Ehsan, High step-up Z-source DC-DC converter with coupled inductors and switched capacitor cell. *IET Power Electronics*. 2015. 10.1049/iet-pel.2014.0200.
19. Abirami. M, Richard Denny, Kanimozhi. G, K. Logavani, A novel hybrid DC/DC and modified Quasi Z-source converter for enhanced voltage. *Materials Today: Proceedings*, Dec. 2020 ISSN-2214-7853, https://doi.org/10.1016/j.matpr.2020.10.953.

30

Datalogger Aided Stand-Alone PV System for Rural Electrification

Aashiq A.[1]**, Haniya Ashraf**[1]**, Supraja Sivaviji**[1]**, Aadyasha Patel**[1] **and O.V. Gnana Swathika**[2*]

[1]*School of Electrical Engineering, Vellore Institute of Technology, Chennai, India*
[2]*Centre for Smart Grid Technologies, School of Electrical Engineering, Vellore Institute of Technology, Chennai, India*

Abstract
Solar energy is an infinite source of clean energy and is a growing alternative for fossil fuels nowadays. Electricity generated by solar PV panels is inexhaustible and does not pollute, thus contributing to sustainable development. A stand-alone photovoltaic (SAPV) system is one such designed setup aimed at this prospect. The need for employing a Data Logger (DL) is essential in developing countries where rural or remote areas are not connected to the power grid, since DL helps greatly in rural electrification. The proposed methodology aims to implement a cheap and energy-efficient Arduino-based DL to increase productivity. A DL is an electronic device which is used in the initial stages of solar power generation to help minimize costs through intelligent automation and energy management. DL aids in remote monitoring and controlling data, i.e., Electrical PV parameters automatically over time. This PV DL records various electrical parameters and calculates the energy yield. Such environmental conditions are then measured, documented, analyzed and validated. A user-friendly hardware prototype of a cost-efficient DL is implemented and data is recorded at specific time intervals throughout the day. A mathematical AI model is applied to the data for prediction of missing or null values in the dataset, and the results are analyzed to show the percentage error of predicted data using this model.

Keywords: Photovoltaic (PV), remote monitoring, stand-alone PV system (SAPV), Internet of Things (IoT), microcontroller

**Corresponding author*: gnanaswathika.ov@vit.ac.in

Milind Shrinivas Dangate, W.S. Sampath, O.V. Gnana Swathika and P. Sanjeevikumar (eds.) Integrated Green Energy Solutions Volume 2, (167–200) © 2023 Scrivener Publishing LLC

Abbreviations and Nomenclature

DL	Data Logger
AI	Artificial Intelligence
SAPV	Stand-alone photovoltaic
SVD	Solar Variability Datalogger
PV	Photovoltaic
PCB	Printed Circuit Board
SBC	Single Board Computer
DAS	Data Acquisition System
CBR	Case-Based Research
RTC	Real-Time Clock
MPP	Maximum Power Point
V_{mpp}	Maximum Power Point Voltage
I_{mpp}	Maximum Power Point Current
V_{oc}	Open Circuit Voltage
I_{sc}	Short Circuit Current
RES	Renewable Energy System
IEC	International Electrotechnical Commission
IoT	Internet of Things
PLC	Power Line Communication
SPI	Serial Peripheral Interface
EEPROM	Electrically Erasable Programmable Read-Only Memory
LCD	Liquid Crystal Display
LDR	Light Dependent Resistor
PIC	Programmable Interface Controller
ARM	Advanced RISC Machines
3D	Three Dimensional
Wp	Watt Peak
SQL	Structured Query Language
SD	Secure Digital
GB	Giga Byte
IC	Integrated Circuit
ADC	Analog to Digital Converter
WPAN	Wireless Personal Area Network
LABVIEW	Laboratory Virtual Instrument Engineering Workbench
MATLAB	Matrix Laboratory
ESP	Espressif Systems
JS	Java Script
DC	Direct Current

AC	Alternating Current
IEEE	Institute of Electrical and Electronics Engineers
V_{out1}	Output Voltage of Arduino
V_{out2}	Output Current of Arduino
V	Output Voltage of Solar Panel
V_d	Output Voltage of F031-06
R_1,R_2	Internal resistors of F031-06
R_s	Shunt Resistor
R_L	Load Resistor
V_o	Voltage output
VCC	Supply Voltage
DAQ	Data Acquisition
MAPE	Mean Average Percentage Error
USART	Universal Synchronous and Asynchronous Serial Receiver and Transmitter
EPANN	Evolutionary Programming Artificial Neural Network
GUI	Graphical User Interface
LAN	Local Area Network
WAPN	Wireless personal Area Network
DSP	Digital signal processing
x_{stand}	Standardised value of data
μ	Mean of the entire dataset
σ	Standardized value of data
n	number of observations
A_i	Actual value
P_i	Predicted value

30.1 Introduction

Stand-alone photovoltaic (SAPV) systems are growing popular as they are a low-cost solution for providing electricity in remote areas. This system is bringing an admirable change to the economic welfare of rural areas that are off grid connections. These systems are playing an important role in rural electrification as such places often suffer from electricity shortage and are not covered by the grid distribution system because of distance and remote location.

30.1.1 Motivation

The world is currently facing global warming and climate change at an alarmingly increasing rate. Electricity being produced by use of fossil fuels predominantly is a cause, as it releases excessive CO2 into the atmosphere. An all-time high release of carbon is being witnessed which amounts to about 8,365 million metric tons. Since the past decade a rapid increase in the global electricity demand is seen imposing excessive load on oil and coal industries to generate electricity, thus contributing to a rise in global warming and pollution. A 50% surge in energy demand globally is predicted in the next 25 years, the reason being economic development and increase in population. Hence, renewable energy is essential for a sustainable future to eradicate rural poverty and minimize dependency on fossil fuels. In our country, electricity production through SAPV systems has an enormous potential and is also an excellent alternative to commercial methods in rural electrification.

30.1.2 Objectives

The objectives of this work are:

- Designing and realizing a hardware prototype of DL using Arduino Uno which is user friendly and cost efficient for recording electrical PV parameters in SAPV systems deployed in rural areas.
- Deploying a mathematical algorithm on the data obtained from the hardware prototype to predict missing data by CBR method and verifying results.

30.2 Work Description

30.2.1 Overview of the Work

The DL aids in monitoring PV parameters like voltage, current and power in a SAPV system, which proves beneficial in especially rural locations. This helps in power prediction and fault detection which in turn aids in efficient energy management in the system. This DL consists of a solar panel to generate electricity, a microcontroller and sensors to measure data for time durations throughout the day. The null values in the recorded dataset are predicted using AI techniques.

30.2.2 Literature Review

In all non-conventional energy sources that are available for harvesting, solar energy is the most promising one—supported by renewable incentives and fueled by worries of global climate change. It is one of the fastest-growing renewable technologies, having seen considerable growth in recent years. Energy from the sun is being used to generate electricity worldwide as it is reliable, inexhaustible and the key to an up-and-coming, clean future. There is a shortage of electricity needs being met in rural areas due to the huge demand in urban areas. Moreover, many rural areas are remotely located, and therefore do not have access to the power grid. SAPV systems in rural households is a futuristic solution for rural electrification especially in developing countries, and has become more affordable nowadays owing to various government subsidies and solar power schemes in countries like India.

For monitoring the electricity generated by SPV systems for better performance, data loggers are needed. A DL is an electronic device mainly utilized for storing data continuously over a period of time, and monitoring various electrical and/or meteorological parameters in SAPV systems. A general DL is built by the usage of a microcontroller, sensors to help in recording data, and memory storage ability [1]. Most DLs interface with a local device to be a separate device, while some other data loggers can utilize software in the collection and analysis of data, having the ability to interface with a laptop or a computer [2].

Even though there was an explosion in SPV system installations recently, these systems still possess a huge margin for development in relation to various parameters like cost, efficiency and technologies adapted for its construction.

A reliable open-source embedded DL is proposed in [3]. Accuracy is achieved according to the IEC standards for SPV systems. A 32-bit ARM cortex M4 core microcontroller is used in the construction of the proposed DL in [3]. This system stores data directly into an Excel file.

The DL in [4] consists of an alarm to notify users of any failure via a Twitter account. System operation is monitored online via web server and computer. Recorded data is transferred to a web server with the help of ThingSpeak, which is a free online platform utilized for data visualization. Collection of data is achieved via radio and serial transmission. Analysis of collected data is done with the help of MATLAB software. Some issues like solar radiation block due to nearby buildings and unexpected interruptions were faced by this DL, after a period of testing for 16 days [4].

With increase in residential electricity consumption, cost reduction needs to be done for better efficiency. Each PV module should be monitored for maximum performance. This can be achieved with the help of PLC technology [5]. This technology uses existing wiring for signal transmission. High-speed data can be transferred through low voltage power lines.

Many PV systems that use PLC technology for monitoring each module, utilize communication modems that are quite costly. [5] proposes a low-cost PLC-based monitoring system for residential homes. It is built using a microcontroller unit, omitting the usual use of an expensive communication module, to cut down cost. Embedded ADC converters that are in the microcontroller unit measure temperature, current and voltage. The user-friendly system consists of an app to show system status.

30.2.3 Methodologies

All DLs have their own disadvantages and limitations. While some have unwanted complexity in terms of hardware components used, others have shortage in memory storage. Some contain extra features that are not used for simple logging and monitoring, thereby contributing to a wastage of cost and space. Hence, it is necessary to develop a data logger taking requirements into consideration, like the location of installation, type of application and time duration of monitoring. DLs can use various different technologies to achieve this. Primarily, this means the type of microcontroller used, as it is the major hardware component.

Arduino Uno is a common and popular microcontroller which is cost efficient, small in size, reliable and less energy consuming. It uses an Atmel ATmega 328 which allows for flexible design and consumes less space in terms of software code. Both analog and digital forms of data can be transmitted or received via the Arduino Uno. Command codes for DLs are written using this microcontroller.

The ATmega2560 is another popular choice of microcontroller among developers, which consumes low power and shows high performance characteristics. The ATmega2560 has a stable ADC and USART. While the Arduino Uno is more suited for beginners, and for simple applications involving lesser complexity, the ATmega2560 board offers additional features for knowledgeable enthusiasts. It has a set of extra I/O pins for applications that need them. This flexible microcontroller is also used in many DLs, DAS and PV systems. The cost is only slightly higher than the ATmega328.

ESP32S2 is yet another microcontroller that allows for more complex designs. It has an ultra-low power co-processor. Microchip PIC18F4550 Microcontroller and STM32 are some other microcontrollers that are used for designing DLs. Software used range from Arduino IDE, MATLAB, JavaScript and Python to LABVIEW.

The DC energy power logger prototype in [6] based on Arduino Uno, uses an LCD to display real-time data values. The recorded data is stored as an Excel file.

An ATMega328p microcontroller-based, low-cost SA DL design is developed in [7] which has an SD card as memory storage to store recorded data in CSV format and an RTC for time stamping data. An LCD is also used to display data. A drawback of this DL is that it can be used only for measuring low range of current and DC voltage (0 to 20 A and 0 to 30V).

The DAS designed to monitor SPV system status in [8] is based on the microcontroller AtMega328p. Micro SD cards record power outage and other electrical data.

[9] implements a DL design based on the ATMega2560 microcontroller. It can store a lot of input channel data with bigger memory storage. Raw data is converted to digital to be stored in an SD card. An RTC is used for time stamping logged data. The commercial DL Datataker DT80 is used for performance comparison. The designed DL shows an error of 0-1.4% in comparison to the average data obtained by the DT80 DL.

An AtMega2560 based complete solar emulator - DAS is constructed in [10]. Halogen light is used for uniform light dispersion. The Arduino software is used to write the command codes for data collection. MATLAB environment is used for data analysis. The Arduino MATLAB interface package is used for controlling purposes. This complete system provides an in-depth analysis and accurate working, based on different test conditions of partial shading conditions, temperature and varying light intensity.

An Arduino-based DL design with IEC standards is studied in [11]. The Arduino Uno (R3) board consisting of Atmega328 chip is used. The logged data is stored in an SD card and LCD display shows data while logging. A pyranometer, tracking and fixed PV module monitoring was used for testing (10 hours every day). The standard Wisco DL 2100 is taken for comparison with the DL in [11]. The percentage error values were seen to be from 0-25.6.

A type of technology that can be utilized for DL design is the Bluetooth technology. DLs that consume less energy and use Bluetooth can send and receive data via wireless means. Smartphones having Bluetooth enabled can be incorporated in such systems. The range of Bluetooth is over 30 feet radius. Bluetooth communication utilizes ultra-high frequency radio waves which are electromagnetic waves with frequencies about 2.4 GHz. This technology is intended to be more user friendly.

A cost-effective means of data logging can be achieved by the use of USBs in DL design. They are also portable and easy to implement, which are added advantages.

A simple, USB and Bluetooth-based DL design is proposed in [12]. It utilizes a DS3231 high accuracy RTC and an HC-05 Bluetooth module or USB port to transfer data to a computer or an SD card module to store data.

A single board DL based on an 8-bit MCU is designed and fabricated in [13]. Recorded data can be transferred via USB to be analyzed on a computer. Logging was done for three days continuously and the readings were found to be consistent with normal conditions of weather.

The design for an ultra-low powered DL based on ESP32S2 microcontroller is proposed in [14] for an SAPV system. It functions in the deep sleep mode to conserve energy, during which processing of stored data in the RTC memory is taken care of. A micro-SD card stores recorded data. The DL consumes low power up to 64 Watt-hour per year. $30 is the total cost of implementing this DL [14].

DLs are not just used for SAPV systems. Some different applications that use DLs include road-side lighting systems and monitoring systems for power plants.

A low power, small size, ATmega 8–based DL is developed in [15] for an automatic LED street light system to replace the manual method, in Jakarta, Indonesia. It uses a 64-k serial EEPROM microchip. Owing to its small size, it can be embedded inside the lamp easily. Interfacing with a computer can be done, and software is used to generate reports, view performance, check failure conditions and analyze data. The DL consumes 0.24W power [15].

A novel design has been adapted and implemented in the Czech Republic based on unique data collecting technologies in [16] and the monitoring system is named Solarmon 2.0. A DL Box collects data in regular time intervals and transmits it to a central server. The box allows only for protected communication, thereby ensuring the safety of data was installed in power plants. It not only acts as a source for scientific research, but also ensures quality management [16].

30.2.4 Optimization Techniques

Optimization and analysis techniques are required to produce an efficient output from a solar monitoring system. The following research publications examine power loss predictions as well as data forecasts with examples. Data predictions aid in determining the likelihood of future outcomes based on previously collected information.

The DL is used to analyze data in the future using a variety of methodologies. [17] Implements a real-time LED-based light analyzer that uses DL to measure the wavelength band of sunlight and differentiate atmospheric conditions such as sunlight, temperature and moisture.

The DL in [18] is combined with current and voltage sensors to record characteristics of the PV system that affect the efficiency in order to determine the maximum peak power. Computed power is then compared to the data (power), and the MPP tracking method is employed to obtain efficient peak power.

A DL with MPP tracking [20] technique is employed as a portable recorder to compare the power in various locations. The performance of crystalline silicon PV systems degrades as the temperature rises. In [19], it was discovered that simply implementing a water-film cooling system reduced non-uniform temperature distribution as well as thermal stress between the rear and front surfaces of the PV panel.

PV system monitoring at the panel level is useful for detecting repairs and pinpointing system faults. The monitoring time between the PV panels is synchronized to compare them under similar conditions. [21] proposes a PLC module at low-cost for PV panel monitoring. To detect defects, a threshold was set and compared to mean voltage. The proposed method is both user-friendly and cost-effective.

External environmental elements such as temperature, humidity, location, wind, and dust influence the effectiveness of solar panels. [22] Assessed the loss of power caused by dust collection during a 55-day period. In Jaipur, an unclean panel resulted in a 9% loss in power generation [22]. [23] contains information on the collection of meteorological and on-line electrical data and their use to gauge the working of PV arrays under various irradiation circumstances.

30.2.5 IoT and Smart Technologies

The Internet of Things (IoT) envisions a world in which common objects are connected to the internet. It is accomplished through the use of wireless networks. IoT enables things to be detected or controlled remotely using existing network infrastructure, allowing for complete physical integration with computer-based systems. It ensures increased energy efficiency, accuracy, reduced supervision time, and cost savings in solar PV monitoring systems, as well as less human intervention.

By tracking the PV system's power generation, an IoT-based remote Solar PV Monitoring System assures that the PV cells of solar panels are in excellent working order. As one receives real-time visibility into the quantity of kilowatt hours of electricity produced by the solar panels, it becomes easier to provide intelligent insights for faster and better analysis and decision-making. Access to real-time data and reporting is available all day, equipment availability is increased, and downtime and associated operation expenses are reduced.

[24] discusses a concept to construct a solar PV remote monitoring system, in which an IoT-based cost-effective Solar PV monitoring system

is built using a GPRS module and a microcontroller to provide real-time analysis of the model at regular intervals, as well as aid in problem identification if any. Data can be accessed via the internet from any location on the planet.

A smart-centric, user-friendly, remote monitoring solution presented in [25] employs an ESP 32 microcontroller-based DL system to construct a Pico solar home system in a developing country's rural location. This system saves data on a micro-SD card and presents it on the Blynk App. For analysis and verification, data can be downloaded straight from the website. Four sensors are used in the hardware prototype for humidity, voltage, temperature, and current detection. For effective monitoring and maintenance, an Android app displays all real-time data.

[26] illustrates Solar Home Systems with an efficient MCU Arduino technology. The suggested innovative, low-cost DL can accurately measure up to 14 electric and climatic parameters to the IEC61724 standard's requirements. The system's 3G and IoT connectivity enables for real-time data visualization through the web or a smartphone. Only at temperatures exceeding 40 °C does this system fail, although it can endure all other extreme temperatures. Because of its stackable hardware design, a single DL can be used as a central node to monitor multiple SAPV systems that are near enough together.

Unlike tiny household systems, the use of Arduino can be expanded to large-scale models such as solar power stations, as described in article [27]. The Arduino UNO and Arduino NANO serve as the system's brains, while the ESP8266 serves as a communication module, delivering wireless connectivity via the IEEE802.11 communication protocol. A servo motor is used to perform the angle control technique to shift the direction of the solar array to get maximum output. Sensors are utilized to detect various PV parameters. Data from the cloud can be used to improve performance. Because of its restricted bandwidth, this technology can only give limited connectivity.

As described in [28], SBC Raspberry Pi can be used for digital control and data logging for solar power plants. A double pivot sun-oriented tracker in this type successfully monitors the weather and tracks the sun, changing its position as needed to increase power yield. Pre-set times and related degrees of rotation govern the direction of the solar panels. Sensors, solar panels, two DC motors, and a motor driver IC are used to rotate the motors. It can figure out wind speed and direction, as well as humidity and temperature. The values of parameters are shown on the cloud-based ThingSpeak server, an IoT platform.

The usage of IoT integrated Control Systems or Units for tracking in [29] and [30] makes data transmission more efficient and aids in the prediction of PV system faults. The upkeep seems to be uncomplicated and inexpensive. In addition, customized daylight-based trackers with a cleaning system are used to remotely review and assess the operation of a solar PV plant in. It allows for maintenance, fault identification, and plant analysis based on historical data, as well as real-time monitoring. A DC Servo motor and a microcontroller are used to clean the device.

With the price of electricity continuing to grow and energy resources getting limited, it is critical for individuals to be aware of how to use energy most efficiently. IoT has recently experienced a considerable surge in developments, presenting enormous possibilities for changing behavior and offering effective services. Smart technologies such as IoT have made it possible for homes and companies to do so, consequently encouraging energy conservation to a fair degree.

30.2.6 Conclusion

It is clear from the literature review that DLs are essential for solar SAPV systems. As these are in remote locations, manual monitoring and maintenance are not always possible. As a result, DLs have a significant role in data logging and monitoring.

30.3 Design and Realisation of DL

30.3.1 DL Description

The DL proposed consists of the following components and software requirements, explained in detail. The list of components required to build this hardware unit are shown in Table 30.1.

30.3.2 Solar Panel

A solar PV panel is a collection of multiple series connected PV cells. Solar panels absorb energy from the sun's rays directly to generate electricity. There are primarily two kinds of solar panels – monocrystalline and polycrystalline. Monocrystalline solar panels are made from a source of a single crystal of silicon, while many silicon fragments are melted together to produce cells to make polycrystalline panels. Though the efficiency of monocrystalline panels is slightly higher than polycrystalline ones, a 10W polycrystalline panel as shown in Figure 30.1 is used in this prototype as it costs less. The PV cell as in Figure 30.2 uses a parallel connected current

Table 30.1 List of components.

Name of the component	Quantity
Solar PV Panel	1
Arduino Uno	1
F031-06 voltage sensor	1
INA169 current sensor	1
Jumper wires (M-M, M-F, F-F)	As Required
Bread Board	1
Multimeter	1
PC System (With Arduino IDE and PLX DAQ)	1

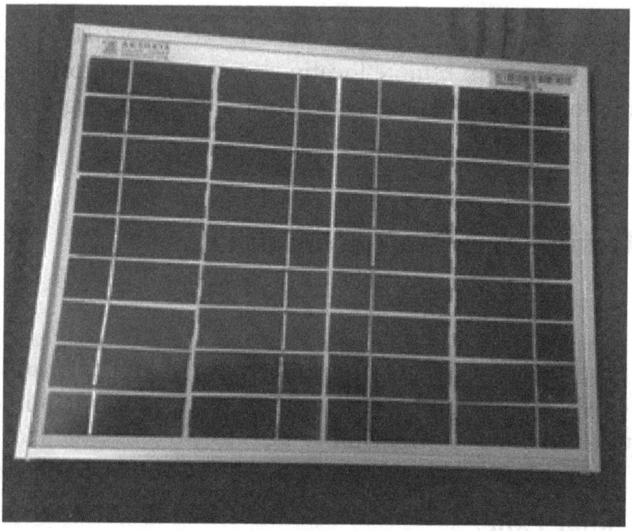

Figure 30.1 Solar panel.

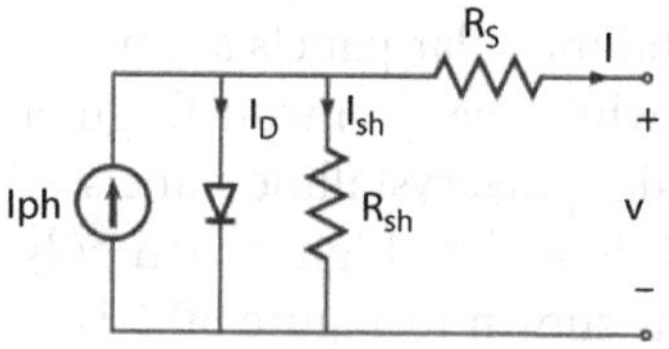

Figure 30.2 Equivalent circuit of a PV cell.

Table 30.2 Rating of the solar panel.

Specifications	**Rating**
Rated Power	10W
Maximum rated voltage	17V
Current at maximum power	0.57A
Open circuit voltage	21.00V
Short circuit current	0.69A
Tolerance	1.5%

source to a diode and a R_{sh} and in series with a series resistor R_{s1}. The specifications and rating of the solar panel employed in this paper is given in Table 30.2.

30.3.3 Arduino Uno and IDE

The Arduino Uno has a microcontroller ATmega328 integrated in it. The Arduino program allows communication between the Arduino UNO board and the PC. The board's microcontroller provides UART TTL (5V) serial communication thus enabling connection between the PC's USB and the Arduino Uno's serial ports. The PLX-DAQ window shows this as a virtual COM port.

The Arduino code uploaded to the Arduino Uno board lets the user retrieve information that was measured by the sensors from the PV panel which would then be transferred to the PLX-DAQ Spreadsheet and recorded at regular time intervals as set by the user. It is powered using a DC power source.

Arduino Uno has 14 input/output pins (0-13) and 6 analog input pins (A0-A5). This board in Figure 30.3 can be programmed using Arduino IDE software. It has an in-built ADC. The output of A0 pin (V_{out1}) and A1 (V_{out2}) of the ADC is a digital value, which is coded in 10 bits and takes the value between 0-1023.

The Arduino has an in-built ADC providing a 10-bit digital output (V_{out1}) varying between 0 and 1023. The 10-bit encoding helps in converting from analog to digital and back. The ADC of the Arduino provides a digital value (V_{out1}) which varies between 0 and 1023, because it is encoded in 10 bits so the data that is converted from analog to digital needs to be converted back. In order to do so, the following conversion factor is used

Figure 30.3 Arduino UNO.

to get the actual supply voltage and current. The analog voltage resolution and resolution of current sensor is 0.00489 (5/1023). Therefore, the minimum voltage read by the sensor F031-06 is 0.02445.

The final output voltage is calculated using (30.1).

$$V = \frac{R_1 + R_2}{R_1} * V_{out} * \frac{5}{1023} \tag{30.1}$$

The final output current is calculated in (30.2).

$$I = \frac{V_{out2} * 5}{1023} \tag{30.2}$$

30.3.4 Voltage Sensor

Voltage sensor F031-06 in Figure 30.4 measures the output voltage of the solar panel. The F031-06 is also a voltage divider which reduces the voltage to 5 times the original value. The original voltage value is reduced in order for the Arduino to read the output, as the analog limit of Arduino is 0-5V. The sensor is connected in parallel to the load and the output pin of the sensor is fixed to the ADC pin (A0) of the Arduino. The output of the sensor is calculated using the panel's output (which is divided internally) by the formula given in (30.3).

$$V_d = \frac{R_2}{R_1 + R_2} * V \tag{30.3}$$

Figure 30.4 F031-06 voltage sensor.

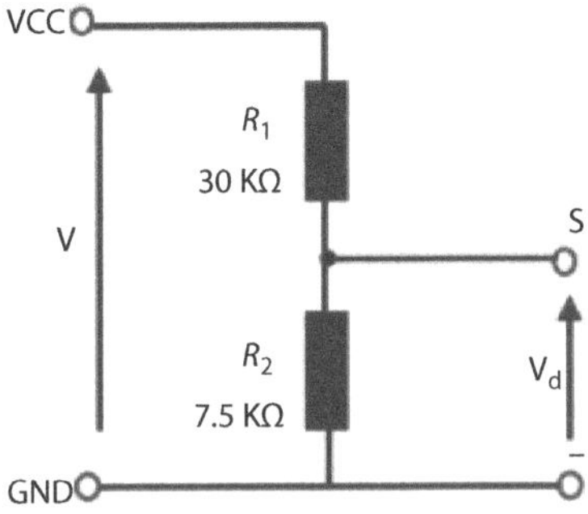

Figure 30.5 Voltage divider circuit of F031-06.

Table 30.3 Rating of voltage sensor F031-06.

Specification	Rating
Input voltage range	0V – 25V
Voltage detection range	0.02445V – 25V
Output voltage range	0V – 5V
Analog voltage resolution	0.00489V

Table 30.4 Voltage sensor F031-06 pin connection.

Pin name	Description
VCC	Connected to solar panel's positive terminal.
GND	Connected to solar panel's negative terminal and the Arduino's ground pin.
S	Connected to the Arduino's A0 pin for data transfer.
+	No connection.
-	Connected to the Arduino's ground pin.

As shown in Figure 30.5, R1, R_2 are of the value 30K ohm and 7.5K ohm. Substituting the value of R_1 and R2, the divider's ratio is calculated to be 0.2. The specifications and rating of the voltage sensor is given in Table 30.3. Pin connections, pin descriptions of F031-06 are given in Table 30.4.

30.3.5 Current Sensor

INA169 current sensor in Figure 30.6, which is a unipolar, high side component, is used to measure current. This sensor is a "current shunt monitor," that is, the sensor senses the voltage drop across a R_S as shown in Figure 30.7 which is 0.1 ohm. The output voltage across the inbuilt operational amplifier is calculated by the difference between the V_{IN+} and V_{IN-}. The internal transistor converts the amplified output voltage to current. The output current of the transistor is then converted to voltage V_0 when passing through the R_L which is 10 ohms. The measured current is given by (30.4).

$$I_S = \frac{V_0 * 1000}{R_S * R_L} \tag{30.4}$$

Figure 30.6 INA169 current sensor.

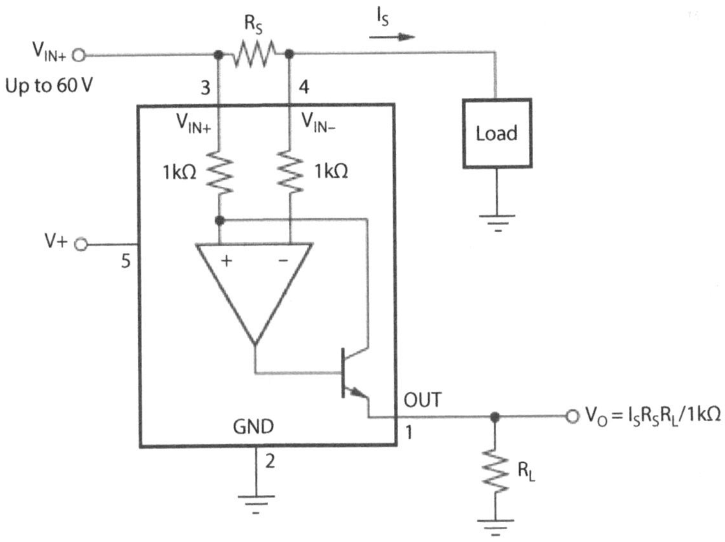

Figure 30.7 Internal circuit of the current sensor INA169 from the data sheet.

The value of V_0 depends on the VCC, which is 5V in the experiment performed. The current sensor measures up to 5 A as the maximum input difference of the amplifier is 500 mV. The Pin diagram for this sensor is shown in Figure 30.8 for connection purposes. The specifications and rating are given in Table 30.5. Pin connections, descriptions of INA169 are given in Table 30.6.

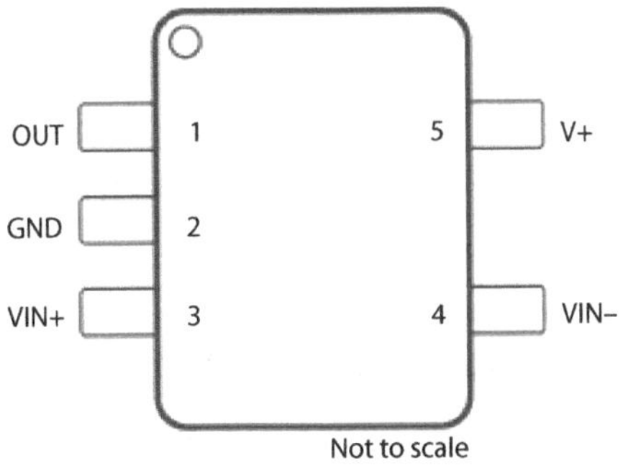

Figure 30.8 Pin diagram of the current sensor INA169 from the data sheet.

Table 30.5 Rating of current sensor INA169.

Specifications	Rating
Common mode range	2.7V – 60V, DC
Input voltage	500mV
Maximum offset input voltage	1000µV/V±
Maximum offset input drift	±1 µV/°C
Maximum nonlinearity error	± 0.1%
Maximum total output error	± 2%
Bandwidth	440 kHz, DC
Supply voltage range	2.7–60 V
Operating temperature range	– 40 to 85 °C
Quiescent current	60 µA

Table 30.6 Current sensor INA169 pin description and connection.

Pin			
Name	**Number**	**I/O**	**Description**
OUT	1	O	Connected to the analog pin of the Arduino to transfer output data.
GND	2	-	Connected to the ground terminals of the Arduino and the panel.
VIN+	3	I	Positive input voltage pin is connected to the positive of the solar panel.
VIN-	4	I	Negative input voltage is connected to the load.
VCC	5	I	Power supply voltage sources 5V supply to the current sensor INA169 from the V_{cc} pin of Arduino.

30.3.6 PLX-DAQ Data Acquisition Tool

Parallax Data Acquisition tool (PLX-DAQ) is an add-in available for Microsoft Excel. It has 26 channels for collecting information through multiple Parallax microcontrollers. The data is then dropped into columns in the excel spreadsheet as they are collected. This data becomes easy to analyze due to the simplicity in use of excel spreadsheets. The tool comes in handy while working in a lab or field for monitoring of the real-time information.

Excel can directly gather data from any of the sensors connected to the serial ports of the PC via microcontrollers. PLX-DAQ has the following features:

- ★ Real-time plotting or graphing data as it is recorded using Microsoft Excel
- ★ Record up to 26 columns of data
- ★ Mark data with time stamp
- ★ Read/Write any cell on a worksheet
- ★ Read/Set any of 4 checkboxes on control the interface
- ★ Baud rates up to 128K
- ★ Supports Com1-15

In this paper, data acquisition to an Excel Spreadsheet from the Arduino microcontroller is carried out by setting up PLX-DAQ Excel Macro. This software needs to be downloaded from a PLX-DAQ site. After installing, a folder named "PLX-DAQ" is created in the PC system with a shortcut "PLX-DAQ Spreadsheet" inside. The connection settings are then defined in the PLX-DAQ window as shown in Figure 30.9, i.e., required baud rate

Figure 30.9 PLX DAQ data logging.

and port are selected to establish communication between the Arduino microcontroller board and Excel spreadsheet.

30.3.7 Software Specifications

Parallax Data Acquisition tool (PLX-DAQ) software: version 2.11

Environment used for prediction code: Python version 3.9, 64-bit, Windows OS

Arduino IDE: version 1.8.42, Microsoft store app

30.3.8 Methodology

The work uses virtual instrumentation where captured data is logged straight into a PLX-DAQ spreadsheet from the microcontroller. The framework of setup employed in the system is shown in Figure 30.10 while the flowchart in Figure 30.11 illustrates the overall data flow process. The PV panel's current and voltage parameters are acquired through the voltage and current sensors. The sensed data is then transferred to the Arduino Uno board's microcontroller. Through UART bus communication between the excel spreadsheet and the microcontroller is permitted using PLX-DAQ Excel Macro. Real-time data is accumulated and laid out in the Excel spreadsheet throughout the data acquisition process.

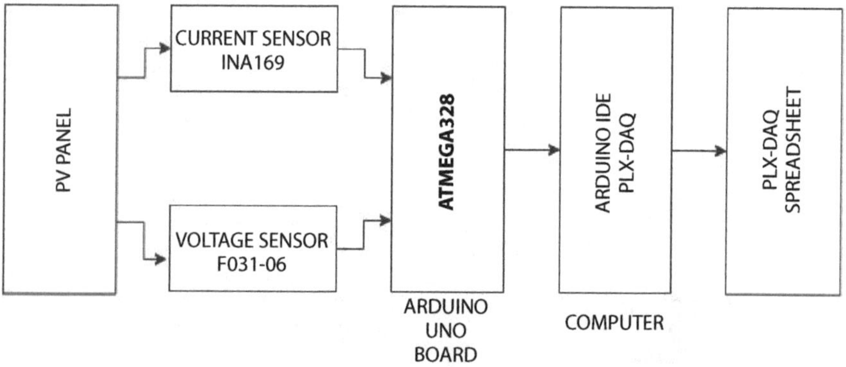

Figure 30.10 Flowchart of system design.

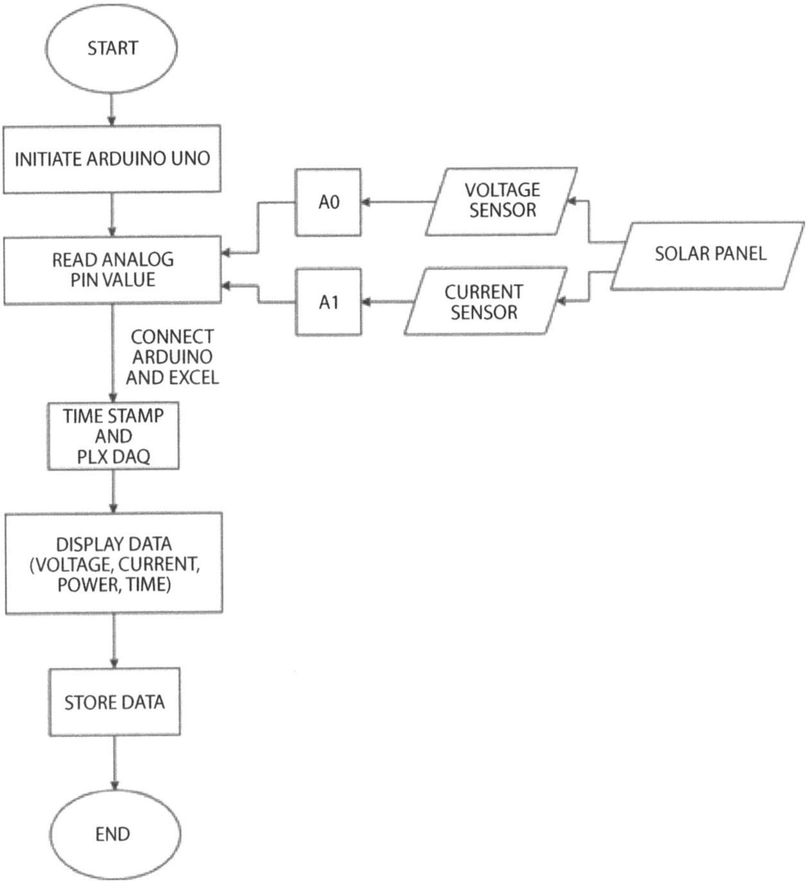

Figure 30.11 Flowchart of data logger.

30.3.8.1 Data Logging into Excel Macro Spreadsheet

Once the Arduino code is compiled and run, the current and voltage values are collected from the solar panel using serial communication. The Arduino Uno communicates via UART bus, and the data is logged onto an Excel Macro spreadsheet (format: xlsm), according to time of logging. Once connection of PLX - DAQ is established, the data is recorded continuously at an interval of 1 second for 10 to 15 minutes every hour. Another data set with a time interval of 5 seconds for 10 to 15 minutes in an hour was also logged. The file is then converted from xlsm format to csv format and saved for use later in the prediction code. The voltage and current

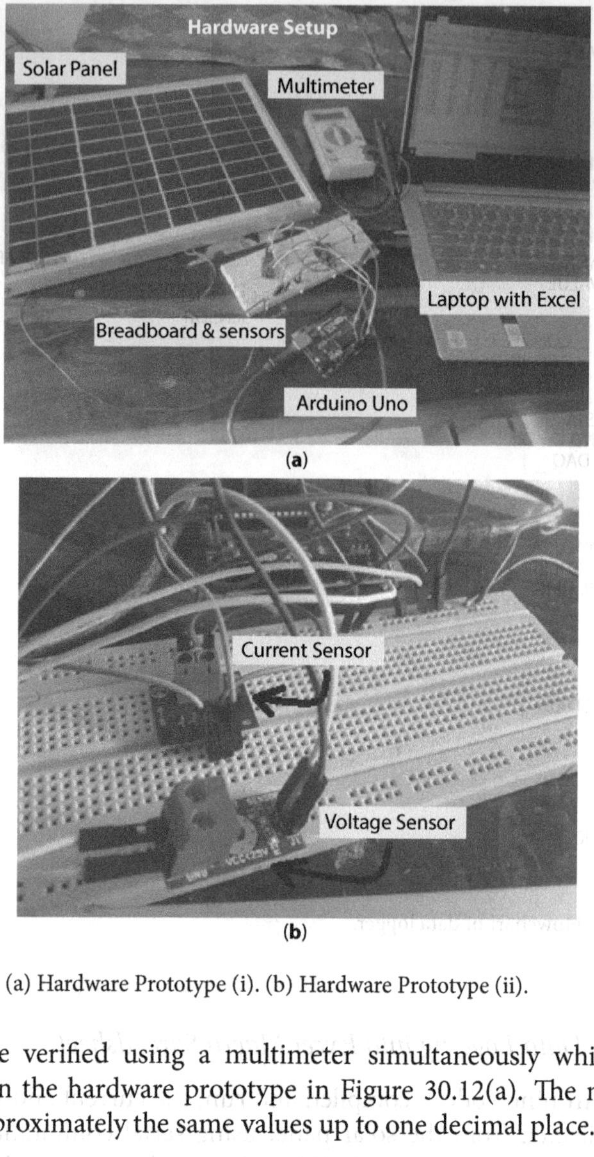

Figure 30.12 (a) Hardware Prototype (i). (b) Hardware Prototype (ii).

values were verified using a multimeter simultaneously while logging as shown in the hardware prototype in Figure 30.12(a). The multimeter showed approximately the same values up to one decimal place.

30.3.8.2 Prediction Using Mathematical Model

A fault in an electrical system is caused by abnormal conditions. Various causes of faults that can occur include equipment malfunction or instrument failures, interruption to electric flow like broken or loose connections to/from sensors in the circuit. This can lead to missing values in recorded data. The absence of values in a dataset represents a variety of problems.

Missing data leads to loss of information, reduces statistical power and causes bias in estimation of parameters, which leads to inaccurate inference being drawn to gain deeper insights about the data. Hence, prediction of null values in data helps for better decision making.

This work, in particular, focuses on predicting missing data in a recorded dataset. The method employed aids in predicting missing data in the output of PV systems. The data source is previously recorded PV electrical data, which is stored as a readily available Excel file (CSV format). The CBR method is less complex and more efficient. The direct approach to prediction can be further divided into two groups – statistical and machine learning methods. The former is based on mathematical models for prediction while the latter involves application of machine learning algorithms on the dataset. The model used is of mathematical AI type, utilizing the principle of Euclidean distance function.

The task of prediction is carried out on a dataset, where data is recorded at an interval of 5 seconds for 10 to 15 minutes every hour in a day. The dataset consists of voltage, current and power of the PV panel recorded along with the timestamp.

The algorithm chosen is CBR. The fundamental idea of CBR is the process of problem solving by adapting previously effective solutions to similar problems (cases).

The search for the best possible solution case is based on the hypothesis that similar problems have similar solutions. Two situations are possible here: Either entire features of both cases – the query and the candidate cases – may be compared, or, relevant parts of cases may be compared. Then, the associated solution is either accepted in the given form or is modified to be useful – adapted – according to the given problem.

In such systems, the quality of the results is mainly dependent on similarity measures which are used to retrieve similar cases for predicting a solution. The process of comparing two cases with each other and calculating their similarity relation degree is called matching.

Before applying the Euclidean distance principle to calculate similarity, all the case vectors should have the same range of values. Standardization, as given in (30.5), is the first step that is done to normalize the ranges of data values while applying this model. It scales each data to a range where they can easily be compared, specifically in the range of 0.0 to 1.0.

$$x_{stand} = \frac{x - \mu}{\sigma} \tag{30.5}$$

For n-dimensional vectors with case $x = (x_1, x_2, \ldots x_n)$ and $y = (y_1, y_2, \ldots y_n)$, the similarity $d(x, y)$, is then measured with the help of Euclidean distance function which is computed as in (30.6).

$$d(x,y) = \sqrt{\sum_{k=1}^{n} (x_k - y_k)^2} \tag{30.6}$$

Each data case is defined as three-dimensional vectors case $x = (x_1, x_2, x_3)$ where, x_1, x_2, x_3 are voltage, current and power respectively. The more similar the vectors are, the lesser the distance between them. Hence, farther vectors are eliminated this way.

If $d(c_i, q) \leq d(c_j, q)$
then, *case i > case j*
which means that the similarity relation of case i to case q is more than case j to case q, hence case j is preferred.

A case c_i in case base is grouped and determined as non-similar with respect to parameters k and α, if all k cases in that case base are greater than the minimum distance α from c_i.

Random values were selected and initialized to null in the dataset, while their actual values were noted down separately. This model was successfully applied for missing values of voltage, current and power in the entire dataset. The predicted values were then appended to a list and printed.

The predicted values are recorded and tabulated along with its actual values. Percentage of deviation of predicted values from actual data was calculated and MAPE was determined for each. Equation (30.7) shows the formula of MAPE, which is a commonly used metric for prediction model evaluation. Python is used as the environment for executing the prediction code.

$$MAPE = \frac{100\%}{n} \sum_{i=1}^{n} \frac{|A_i - P_i|}{|A_i|} \tag{30.7}$$

30.4 Results

30.4.1 Prediction Results

The code is run using the previously recorded dataset. The outputs for each Voltage, Current and Power are shown in Figures 30.13, 30.14 and 30.15,

```
C:\Users\Hani>python predict.py
Number of rows: 649
Number of columns: 4
373.0 => 20.45
375.0 => 20.55
376.0 => 20.55
381.0 => 20.53
382.0 => 20.45
401.0 => 19.01
403.0 => 19.67
459.0 => 20.28
463.0 => 20.5
474.0 => 19.89
493.0 => 20.28
496.0 => 20.01
532.0 => 20.19
537.0 => 20.63
588.0 => 20.11
589.0 => 20.31
35.0 => 20.21
60.0 => 19.87
68.0 => 19.75
215.0 => 19.89
363.0 => 19.65
Press ENTER to exit
```

Figure 30.13 Predicted voltage values.

```
C: \Users\Hani>python predict.py
Number of rows: 649
Number of columns: 4
393.0 => 0.7
394.0 => 0.67
398.0 => 0.65
454.0 => 0.71
504.0 => 0.68
550.0 => 0.68
551.0 => 0.66
1.0 => 0.69
5.0 => 0.7
7.0 => 0.7
82.0 => 0.69
89.0 => 0.68
128.0 => 0.65
130.0 => 0.7
188.0 => 0.67
216.0 => 0.68
217.0 => 0.66
218.0 => 0.65
326.0 => 0.66
342.0 => 0.65
364.0 => 0.64
Press ENTER to exit
```

Figure 30.14 Predicted current values.

```
C: \Users\Hani?python predict.py
Number of rows: 649
Number of columns: 4
372.0 => 13.95
376.0 => 14.02
379.0 => 13.93
389.0 => 14.05
424.0 => 13.86
464.0 => 14.08
473.0 => 13.3
488.0 => 14.0
522.0 => 13.08
555.0 => 14.12
591.0 => 14.05
41.0 => 12.84
106.0 => 13.52
184.0 => 13.76
219.0 => 13.18
273.0 => 13.5
274.0 => 13.5
300.0 => 13.45
310.0 =>  12.95
328.0 =>  13.16
329.0 =>  13.25
361.0 =>  12.46
Press ENTER to exit
```

Figure 30.15 Predicted power values.

respectively. The output shows the size of the input data set, the corresponding row and predicted value for null values.

Data set size: 649 rows x 4 columns

30.4.2 Performance Metrics

One of the commonly used metrics to evaluate prediction models is MAPE, which measures deviation of predicted data from the actual data.

30.4.2.1 MAPE

Error is defined as the difference between actual value and predicted value. MAPE is the mean of absolute percentage errors of predicted data. The absolute percentage error in MAPE helps avoid the problem between negative and positive errors cancelling out each other. The smaller the MAPE, the better the prediction. Table 30.7 shows MAPE for power calculated is approximately 0.485%. Table 30.8 shows MAPE calculated for voltage 1.04% approximate. MAPE for current is calculated as 0.45% from Table 30.9.

Table 30.7 MAPE for power.

Actual value	Predicted value	Error	Percentage error
14.07	13.95	0.12	0.852878465
14.04	14.02	0.02	0.142450142
14.03	13.93	0.1	0.712758375
14.05	14.05	0	0
13.8	13.86	0.06	0.434782609
14.1	14.08	0.02	0.141843972
13.38	13.3	0.08	0.597907324
13.94	14	0.06	0.430416069
13.06	13.08	0.02	0.153139357
14.24	14.12	0.12	0.842696629
14.17	14.05	0.12	0.846859562
12.9	12.84	0.06	0.465116279
13.4	13.52	0.12	0.895522388
13.72	13.76	0.04	0.29154519
13.25	13.18	0.07	0.528301887
13.46	13.5	0.04	0.29717682
13.44	13.5	0.06	0.446428571
13.29	13.45	0.16	1.203912716
13.04	12.95	0.09	0.690184049
13.2	13.16	0.04	0.303030303
13.21	13.25	0.04	0.302800908
12.47	12.46	0.01	0.080192462
TOTAL:			**10.65994408**
MAPE:			**0.484543**

Table 30.8 MAPE for voltage.

Actual value	Predicted value	Error	Percentage error
20.6	20.45	0.15	0.72815534
20.58	20.55	0.03	0.145772595
20.55	20.55	0	0
20.43	20.53	0.1	0.48947626
20.36	20.45	0.09	0.442043222
19.01	19.01	0	0
20.58	19.67	0.91	4.421768707
20.41	20.28	0.13	0.636942675
20.55	20.5	0.05	0.243309002
19.92	19.89	0.03	0.15060241
20.28	20.28	0	0
20.33	20.01	0.32	1.574028529
20.26	20.19	0.07	0.345508391
20.19	20.63	0.44	2.179296682
20.38	20.11	0.27	1.324828263
20.53	20.31	0.22	1.071602533
20.23	20.21	0.02	0.098863075
19.55	19.87	0.32	1.636828645
20.38	19.75	0.63	3.091265947
19.99	19.89	0.1	0.500250125
20.21	19.65	0.56	2.770905492
TOTAL:			**21.85145**
MAPE:			**1.04055**

Table 30.9 MAPE for current.

Actual value	Predicted value	Error	Percentage error
0.7	0.7	0	0
0.67	0.67	0	0
0.64	0.65	0.01	1.5625
0.71	0.71	0	0
0.68	0.68	0	0
0.68	0.68	0	0
0.65	0.66	0.01	1.538461538
0.69	0.69	0	0
0.7	0.7	0	0
0.7	0.7	0	0
0.7	0.69	0.01	1.428571429
0.68	0.68	0	0
0.62	0.65	0.03	4.838709677
0.7	0.7	0	0
0.67	0.67	0	0
0.68	0.68	0	0
0.66	0.66	0	0
0.65	0.65	0	0
0.66	0.66	0	0
0.65	0.65	0	0
0.64	0.64	0	0
TOTAL:			**9.36824**
MAPE:			**0.44611**

30.5 Conclusion

30.5.1 Cost Calculation

The list of components and their cost is given in Table 30.10.

30.5.2 Scope of Work

The DL design can be improved by incorporating various other sensors for more data collection. Weather and environmental conditions can also be recorded for more accurate analysis, depending on the location of installation of DL and PV system. Monocrystalline panels can be used as a long-term solution for remote solar PV systems, and an SD card can be included in the setup to store data for retrieval later. For the prediction part, this work has scope for further research as various other models can be applied for more varied and larger amounts of data, and the most suitable one can be used on the basis of location of installation of the DL. Power prediction can also be achieved for efficient energy management in rural households.

30.5.3 Summary

A prototype of a DL intended for stand-alone PV systems is constructed. Since DLs are necessary for automatic monitoring, it is important to make

Table 30.10 Components and cost.

S. no.	Component	Quantity	Cost
1.	Solar PV panel	1	790.48
2.	Arduino UNO	1	381.36
3.	F031-06 voltage sensor	1	60
4.	INA169 current sensor	1	500
5.	Jumper wires	Pack of 10	50
6.	Breadboard	1	90
7.	PLX-DAQ tool and Excel	1	-
8.	Python Environment	1	-
		Total Cost: 1,871.84	

a DL that is cost efficient and affordable. The paper provides a feasible solution by using Arduino UNO and other components, while the setup is cost effective. Further, data collected in MS Excel is used for source power prediction for efficient energy management. Mathematical model with the CBR algorithm is deployed to predict the missing data. The MAPE for predicted values comes around 1.04% for voltage, 0.45% for current and 0.485% for power. Overall, the prototype setup costs ₹1,871.84.

References

1. Mahzan, N., Omar, A., Mohammad Noor, S., and Mohd Rodzi, M. (2013). Design of DL with Multiple SD Cards. *2013 IEEE Conference on Clean Energy and Technology (CEAT)*, 175-180.
2. Tripathi, S. K., Ojha, P., Singh, K. A., and Baliyan, A. K. (2017). Solar DL. *IJSTE - International Journal of Science Technology & Engineering*, 3(9).
3. Engin, M., 2017. Open-Source Embedded Data Logger Design for PV System Monitoring. *2017 6th Mediterranean Conference on Embedded Computing (MECO)*, pp. 1-5.
4. Jiang, B., and Iqbal, M.T. (2019). Open-Source Data Logging and Data Visualization for an Isolated PV System. *MDPI Journals*, 8, 424.
5. Han, J., Lee, I., and Kim, S. (2015). User-friendly Monitoring System for Residential PV System Based on Low-cost Power Line Communication. *IEEE Transactions on Consumer Electronics*, 61(2), 175-180.
6. Ruzaimi, A., Shafie, S., Hassan, W., Azis, N., Ya'acob, M., and Supeni, E. (2019). Microcontroller Based DC Energy Logger for Off-Grid PV System Application. *2019 IEEE International Circuits and Systems Symposium (ICSyS)*, pp. 1-5.
7. Hadi, M. S., Afandi, A. N., Wibawa, A. P., Ahmar, A. S., and Saputra, K. H. (2018). Stand-Alone Data Logger for Solar Panel Energy System with RTC and SD Card. *Journal of Physics: Conference Series*, 1028, 012065.
8. Fanourakis, S., Wang, K., McCarthy, P., and Jiao, L. (2017). Low-cost Data Acquisition Systems for Photovoltaic System Monitoring and Usage Statistics. *IOP Conference Series: Earth and Environmental Science*, 93, p. 012048.
9. Mahzan, N. N., Omar, A. M., Rimon, L., Noor, S. Z. M., and Rosselan, M. Z. (2017). Design and Development of an Arduino Based Data Logger for Photovoltaic Monitoring System. *International Journal of Simulation: Systems, Science & Technology*, 17, 15.1-15.5.
10. Ahmed, O., Sayed, H., Jalal, K., Mahmood, D., and Habeeb, W. (2019). Design and Implementation of an Indoor Solar Emulator Based Low-Cost Autonomous DL for PV System Monitoring. *International Journal of Power Electronics and Drive Systems (IJPEDS)*, 10(3), p. 1645.

11. Rewthong, O., Boonbumroong, U., Mamee, T., Eamthanakul, B., Luewarasirikul, N., and Tabkit, N. (2019). Design of the data logger with IEC standard for PV system. *Proceedings of the 8th International Conference on Informatics, Environment, Energy and Applications*, IEEA '19, 234-237.
12. Singh, T., and Thakur, R. (2019). Design and Development of PV Solar Panel Data Logger. *International Journal of Computer Sciences and Engineering*, 7(4), 364-369.
13. Akposionu, K. N. (2012). Design and Fabrication of a Low-Cost DL for Solar Energy Parameters. *Journal of Energy Technologies and Policy*, 2.
14. Rehman, A., and Iqbal, M. (2020). Design of an Ultra-Low Powered Data-Logger for Stand-Alone PV Energy Systems. *European Journal of Electrical Engineering and Computer Science*, 4(6).
15. Purwadi, A., Haroen, Y., Ali, F. Y., Heryana, N., Nurafiat, D., and Assegaf, A. (2011). Prototype Development of a Low-Cost DL for PV based LED Street Lighting System. *Proceedings of the 2011 International Conference on Electrical Engineering and Informatics*, pp.1-5.
16. Beránek, V., Olšan, T., Libra, M., Poulek, V., Sedláček, J., Dang, M., and Tyukhov, I. (2018). New Monitoring System for Photovoltaic Power Plants' Management. *Energies*, 11(10), p. 2495.
17. Oates, M., Ruiz-Canales, A., Ferrández-Villena, M., and López, A. (2017). A Low Cost Sunlight Analyser and Data Logger Measuring Radiation. *Computers and Electronics in Agriculture*, 143, pp. 38-48.
18. Effendi, A., Dewi, A., and Ismail, F. (2018). Data Logger Development to Evaluate Potential Area of Solar Energy. *MATEC Web of Conferences*, 215, p.01014.
19. Mah, C., Lim, B., Wong, C., Tan, M., Chong, K., and Lai, A. (2019). Investigating the Performance Improvement of a Photovoltaic System in a Tropical Climate using Water Cooling Method. *Energy Procedia*, 159, pp. 78-83.
20. Zimmermann, U., and Edoff, M. (2012). A Maximum Power Point Tracker for Long-Term Logging of PV Module Performance. *IEEE Journal of Photovoltaics*, 2(1), pp. 47-55.
21. Han, J., Jeong, J., Lee, I., and Kim, S. (2017). Low-Cost Monitoring of Photovoltaic Systems at Panel Level in Residential Homes Based on Power Line Communication. *IEEE Transactions on Consumer Electronics*, 63(4), pp.435-441.
22. Gupta, V., Raj, P., and Yadav, A. (2017). Investigate the Effect of Dust Deposition on the Performance of Solar PV Module Using LABVIEW based Data Logger. *2017 IEEE International Conference on Power, Control, Signals and Instrumentation Engineering (ICPCSI)*, pp. 742-747.
23. Chowdhury, S., Day, P., Taylor, G., Chowdhury, S., Markvart, T., and Song, Y. (2008). Supervisory Data Acquisition and Performance Analysis of a PV Array Installation with Data Logger. *2008 IEEE Power and Energy Society*

General Meeting - Conversion and Delivery of Electrical Energy in the 21st Century, pp. 1-8.
24. Kekre, A., and Gawre, S. (2017). Solar Photovoltaic Remote Monitoring System Using IOT. 2017 *International Conference on Recent Innovations in Signal processing and Embedded Systems (RISE)*, pp. 619-623.
25. Tellawar, M. P., and Chamat, N. (2019). An IOT based Smart Solar Photovoltaic Remote Monitoring System. International Journal of Engineering Research & Technology (IJERT), 8(9).
26. Lopez-Vargas, A., Fuentes, M., and Vivar, M. (2019). IoT Application for Real-Time Monitoring of Solar Home Systems Based on Arduino with 3G Connectivity. *IEEE Sensors Journal*, 19(2), pp. 679-691.
27. Awais, S., Moeenuddin, S., Ibrahim, A. M., Ammara, S., and Bilal, F. (2020). IoT Based Solar Power Plant Monitoring System. *International Journal of Advanced Science and Technology*, 29(9s), 7668-7677.
28. Kadam, A., Kasar, T., Sonje, S., & Tavse, S. (2015). Digital Control and Data Logging for Solar Power Plant Using Raspberry Pi. *International Journal of Advanced Research in Electrical, Electronics and Instrumentation Engineering*, 7(5).
29. Pitchaimuthu, P., and Sridhar, K. (2019). An IoT Based Smart Solar Photovoltaic Remote Monitoring and Control System. *International Journal of Science & Engineering Development Research*, 4(5,) pp. 354-358.
30. Adhya, S., Saha, D., Das, A., Jana, J., and Saha, H. (2016). An IoT Based Smart Solar Photovoltaic Remote Monitoring and Control Unit. *2016 2nd International Conference on Control, Instrumentation, Energy & Communication (CIEC)*, pp. 432-436.

31

Working and Analysis of an Electromagnet-Based DC V-Gate Magnet Motor for Electrical Applications

G. Naveen Kumar*, **K. Indrasena Reddy and P. Ravi Teja**

Department of EEE, ALIET, Vijayawada, Andhra Pradesh, India

Abstract

Perpetual motion is generally seen as an impossible movement, although the machines based on this motion never ceased to exist. It is a motion where a body's motion continues forever. A perpetual motion machine's existence is generally not accepted as it violates the current known laws of physics. A perpetual motion machine in particular violates either or both the first and second laws of thermodynamics. It generally violates the law of conservation of energy as well. We can understand this as a machine, which can run without an energy source. Despite all these scientific facts, the machines based on this concept still draw a lot of attention. Previously, physicists demonstrated some approximately ideal perpetual machines based on different principles. Those machines were not applicable for practical use. Direct Current motors, as we know, are used for a wide range of applications since their invention in the early 1830s. Despite advances in Alternating Current motors, Direct Current motors are still a choice for user applications. In this chapter, we propose a direct current-based V-gate magnet motor, which is a perpetual motion machine. This new electric motor we propose is to outweigh the existing two such types of similar electric motor models. One model uses permanent magnets and the other model uses a cam system. The proposed V-gate magnet motor overcomes the limitations of existing perpetual motor in terms of weight and operational capabilities. Also, when in comparison to the existing conventional Direct Current motor of similar size, this V-gate magnet motor weighs less due to the absence of a permanent magnet in the stator. The permanent magnet in the stator is replaced with electromagnets in our case. The control structure

**Corresponding author*: gutha26@gmail.com

Milind Shrinivas Dangate, W.S. Sampath, O.V. Gnana Swathika and P. Sanjeevikumar (eds.) Integrated Green Energy Solutions Volume 2, (201–216) © 2023 Scrivener Publishing LLC

consists of a microcontroller-based switch. The proposed motor will be able to run single-phase electrical loads practically at a saved economy.

Keywords: Electromagnet, microcontroller, perpetual motion, V-gate magnet

31.1 Conceptual Introduction

"Perpetual Motion": many physicists, scientists and engineers have been fascinated with this term for about a hundred years. Many models were developed, designed, and patented in many ways. But still, the very idea fascinates young minds and researchers alike. Perpetual motion is a motion of bodies which continues forever in an unperturbed manner [1, 6]. A perpetual motion machine is generally viewed as a hypothetical machine. This machine is expected to work infinitely, without the assistance of any external energy source. This kind of machine in general is practically difficult to implement. This is because it would violate either or both the laws of thermodynamics [9].

Energy end users always look for a solution where their energy consumption levels meet their economic status. Customers desire efficiency with a saving of their economy. It is an accepted fact that an electric motor played an important role in the industrial revolution of late. The electric motors replaced many old methods of running a machinery. The electric motors have become more successful based on the very fact that they are compact enough and powerful for a particular application. Different types of electric motors each have different and unique characteristics themselves. These are identified from a particular application. According to the supply they take in to work, they are categorized into two types: DC motors and AC motors. Each has its own disadvantages and advantages. Many advanced and technically sophisticated motors have seen the light in modern days.

In this chapter, we are proposing a DC electric motor that is inspired from perpetual motion. The machine is a DC V-gate magnet motor. We see this as a solution to costly and heavy DC motors. A V-gate magnet motor has a unique arrangement of permanent magnets on the rotor drum, an arrangement which plays a key role in driving the motor. Perpetual motion machines refer to that particular law of thermodynamics which the machines purport to violate. When we go back into history, many perpetual machines were proposed. In the first kind of perpetual machine, we observe that the work produced is without any input of energy, thus violating the first law of thermodynamics. Another machine exists which spontaneously produces mechanical work from thermal energy. It has to be the case where thermal energy should be equivalent to the work done, which does not violate the law of conservation of energy, but practically this is not a scenario without any side effects. Another machine exists that usually, but not always, is defined as a machine which completely eliminates friction along with other dissipative forces, maintaining motion forever because of its mass inertia. It is practically impossible to have such a machine, because dissipation cannot completely be eliminated in a given mechanical system. Extracting work from these devices in general is not possible. The point we want to make is that even in this scenario, researchers continued to be fascinated by perpetual machines, and we also have been fascinated and inspired to build such a machine to act as a motor.

31.2 Existing Technologies to Review

Researchers have put forth ideas to develop a V-gate magnet motor which is closest to the perpetual motor in the past. There, the authors have utilized the idea of a V-gate and created a small-sized perpetual motor. The magnets in V-assembly are arranged on a circular drum. This had a thin pipe around which it rotates. This assembly is accelerated by a cam system to make infinite rotation of drum. In a cam system, a mechanism is designed such that it uplifts the driving magnets at specific intervals for continuous motion of drum. A machine which can give high angular velocity to do work was proposed in [8] whose possible application can be used in power plants to produce large energy, etc., to run vehicles, in robots as energy device for space discoveries. Similarly, we see perpetual motion machines in [5] where it is explained that ever since the first century A.D. there have always been relative descriptions of the said devices apart from manufactures trying to create and commercialize these machines. In [7], a model using neodymium magnets was illustrated. The generator design shown here does not run on any input. The design used disc-shaped neodymium magnets that were placed in such a way that all the south poles or the north poles are in one direction.

As illustrated from previous discussion, there exists a path to fill the gaps in motor design using perpetual motion. From this understanding, we have proposed a DC perpetual motor based on V-gate design to drive loads or in a broader sense a DC motor which can be started on-load.

31.3 Proposed Design

Designs that are currently existing have taken many modifications over the years. In the initial designs, permanent magnets were used which are arranged in a V-shape such that one side of the V-shape is the North pole and the other side the South pole. This V-shape arrangement of permanent magnets spreads all over the length of the rotor drum. One or more such V-shaped arrangement of permanent magnets can be used. A bar magnet is placed in the stator and the poles are concentrated at the ends of the bar magnet. The force is resolved into three components. The first is acting radially inward or outward. The second is acting along the length of the rotor drum and the third is acting tangential to the rotor drum. This tangential force is responsible for the rotor drum to rotate. In the design proposed as the modification to the previous one as part this chapter, the bar magnet in the stator is replaced by an electromagnet. When the electromagnet is energized, a pole is induced and the pole is placed at the center of the length of the rotor drum. With the old V-shape arrangement of permanent magnets, that is one side of the V is the North pole and the other side of the V is the South pole. With such an arrangement, the rotor will not rotate as the attraction and repulsion forces act simultaneously and equally. The V-shaped permanent magnet arrangement is made such that both the sides of the "V" arrangement are of the same pole (either North pole or South pole) for the rotor to rotate. Both the designs resulted in magnetic locking and decrease in the overall speed of the motor, which later on was rectified using the cam system in the V-gate magnet motor. By using the cam system, the bar magnet placed in the stator is pushed away at the V-gates. At this point, due to the moment of inertia the rotor would rotate little further without the presence of driving magnetic force after the V-gate crosses the stator bar magnet. The bar magnet is again placed at its position such that force is again developed to run the motor. Unfortunately, much learning material is not stored in the form of journals and books to elaborate this subject. But the material available at the disposure has been utilized to study and implement a scheme to work for electrical applications which works in the budget.

The idea we are proposing is on similar lines to the method of the cam system. If the magnetic force is absent at the V-gate due to the moment of inertia, the rotor rotates little further even with the absence of driving magnetic force. The power supply to the electromagnet is made temporarily absent. During this time, V-gate passes the electromagnet. Without the power supply there is no magnetic field in the stator electromagnet and the motor rotates without any magnetic locking.

31.4 Block Schematic

The block schematic of the proposed scheme is shown in Figure 31.1. V-gate magnet motor proposed is a DC electric motor. The current version runs on 24 volts Direct Current input supply for demonstration and testing purpose, which we want to scale up for practical implementation in the coming days. The model illustrated here essentially consists of a step-down transformer, single-phase full wave diode bridge rectifier, switching circuit consisting of MOSFET, a hall sensor and a controller unit which is driven by a microcontroller scheme. A full wave bridge diode rectifier employed here converts a complete sinusoidal waveform into rippled DC output.

The DC supply converted from an AC supply is given to the V-gate magnet motor through power MOSFET switches. These are necessary to avoid magnetic locking. Magnetic locking that was observed previously in a cam system type of model has been overcome in this model at the electromagnet set. These switches are designed to be implemented to switch off the supply whenever the V-gate passes through the stator electromagnets. This is necessary because, as the motor is running, the supply to the electromagnets should be turned on and turned off multiple times within a second. Operating a switch at that rate is not humanly possible. Hence, we use a power electronic switch to control the switching operations. Hall sensor is used to feed back the position of the V-gate that is reaching the electromagnet every time.

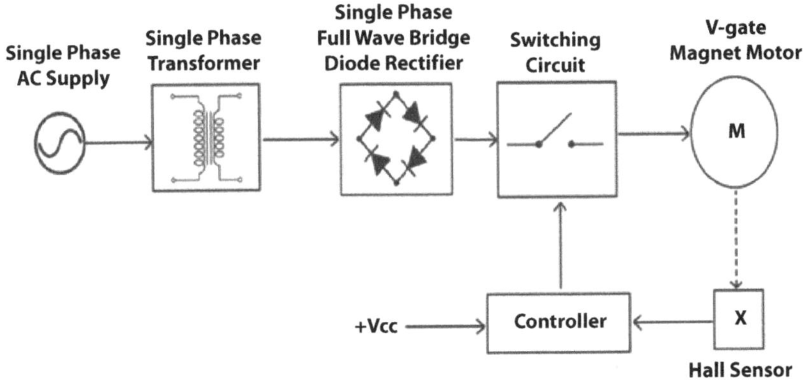

Figure 31.1 Block schematic of V-gate magnet motor.

31.5 Motor Assembly and Control Structure

The complete assembly of the motor is given in Figure 31.2. The control circuit is shown in Figure 31.3. The main component of our subject is Neodymium Magnet [2, 3]. Developed and released independently in 1984 by General Motors, and offering a cost-effective alternative to other rare-earth magnets present during that time, neodymium magnets are the strongest type of permanent magnets that have since been commercially available in the market. These were also released by Sumitomo Special Metals in the same year. A neodymium magnet also known as Neo magnet is the most widely used rare-earth magnet. This in general is a permanent magnet.

This is made from an alloy of neodymium, iron, and boron, which results in $Nd_2Fe_{14}B$ tetragonal **crystalline structure.** They are available in cylindrical shape and disc shape. Neodymium magnets have magnetic properties which far exceed all other permanent magnet materials. They are moderately priced, high in magnetic strength. They can perform well in ambient temperatures [4]. We are using disc type of neodymium magnets arranged in "V"-shaped assembly on two PVC cylinders acting as rotor drums as evident from Figure 31.2.

These rotors are aligned in line with the electromagnets powered through switching circuit by the rectifier unit. When the V-gate is at the electromagnet, the power supply to the electromagnet should be off to avoid the magnetic locking. The information of V-gate that reaches the electromagnet

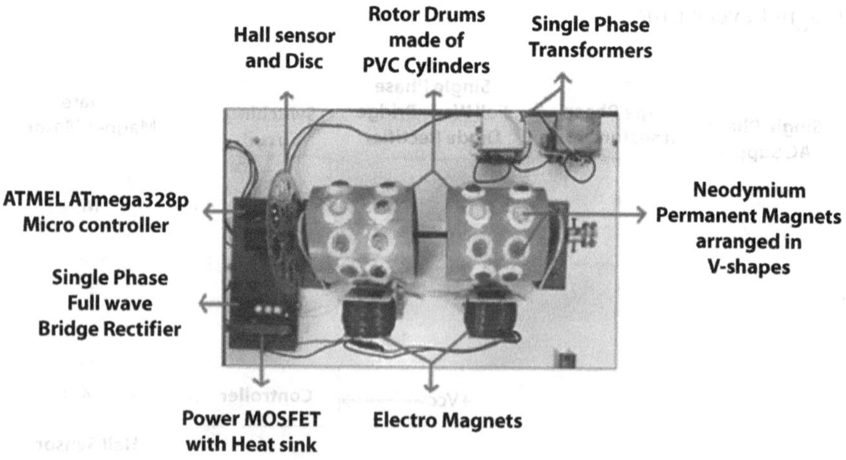

Figure 31.2 V-gate magnet motor.

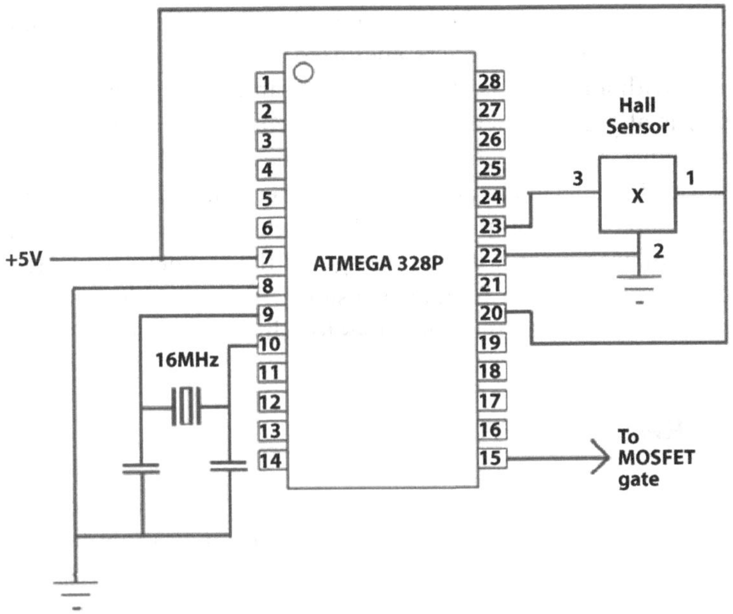

Figure 31.3 Control circuit of the V-gate magnet motor using ATMEGA 328P.

will be given to micro controller for switching off and switching on of the MOSFET. For this purpose, a hall sensor is used. The Hall sensor is aligned on the disc which is in line with the rotor drum as shown in Figure 31.2. Pin 20, Pin 22, and Pin 23 of the microcontroller are connected to the Hall Sensor for the purpose of sending signals to the microcontroller so that MOSFET is switched on and switched off multiple times to avoid magnetic locking. This can be seen in the control circuit of Figure 31.3.

31.6 Control Operation of the V-Gate Magnet Motor

The power supply to the electromagnet is given through a static semiconductor switch. An n-channel MOSFET is used as a semiconductor switch. Due to its very low threshold voltage, the n-channel MOSFET can be controlled by a very low power microcontroller. We are using ATMEGA 328P microcontroller for this control operation. Pin 15 of the microcontroller is given as input to the MOSFET switch. The microcontroller is programmed using an Arduino Uno board and Arduino software. The signal from the Hall sensor is sensed by the microcontroller and turns off the gate signal to the MOSFET. This cuts the power supply to the stator electromagnet.

When the V-gate crosses the electromagnet, the MOSFET is again triggered, and electromagnets are re-energized.

A disc with a magnet is attached to the same shaft as that of the rotor. The disc is aligned in such a way that, whenever the V-gate reaches the electromagnet, the magnet in the disc reaches the Hall sensor. When the Hall sensor is subjected to an external magnetic field, it produces an electromotive force in the output terminal. This electromotive force in the output varies in proportion with the strength of the magnetic field. Hence the output of the Hall sensor is an analog signal. This signal is used to trigger the MOSFET on and off multiple times to achieve the feat.

31.7 Results and Analysis

Electrical motors have a very wide range of applications. Each motor is specified for a particular set of applications. The applications for which an electric motor is designed can be determined by its characteristics. After having been designed and constructing a new motor, the motor needs to be tested and analyzed. The characteristic graphs are drawn from the results obtained from the testing of the motor. The analysis of the motor using the test results can help in determining the set of applications, the new motor suits. The current motor works on a DC supply of up to 24 volts. Its rated speed was identified to be varying between 800 r.p.m. to 1000 r.p.m.

Table 31.1 to Table 31.7 are the readings obtained through measurements and calculations made for determining the characteristics of this motor. Elaborating the results, we understand that this motor has characteristics similar to that of a DC series motor and DC cumulative compound motor to an extent of a 75% match. Load test was performed on the V-gate magnet motor. Figure 31.4 gives us the characteristics of the output versus efficiency. The efficiency of a motor is a very important parameter. From the graph we understand that the efficiency is better to an extent, but not up to the expected level, though the efficiency of the motor is increasing as

Table 31.1 Output-efficiency characteristics.

S. no.	Output (Watts)	Efficiency (%)
1	43.4	85.12
2	42.1	83.81
3	37.75	75.17

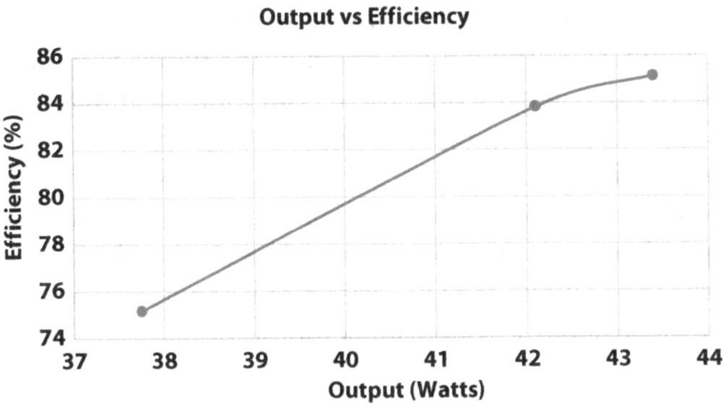

Figure 31.4 Output-efficiency characteristics.

the output. There is not a separate current division for this motor as in the case of a conventional motor. The total current input given as input to the electromagnets is the current considered to draw the current versus voltage characteristic in Figure 31.5 and torque versus current characteristic in Figure 31.9. We understand that the torque parameter could be increased or decreased with an increase or decrease of input current, respectively.

Table 31.2 Voltage-current characteristics.

S. no.	Voltage (Volts)	Current (Amperes)
1	13.35	3.82
2	13.4	3.75
3	13.5	3.72

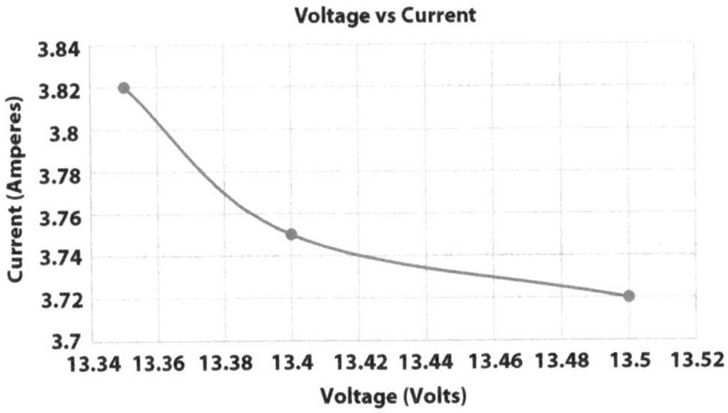

Figure 31.5 Voltage-current characteristics.

Speed versus flux and losses versus efficiency characteristics are seen in Figures 31.6 and 31.7. Though the speed of the motor is increasing with the increase of the magnetic flux, there is an indication that motor speed is affected with increasing loads though the motor can be started with load on. But we also get an inference that leakage flux losses are minimum in this motor. And talking about the speed-current, torque-current and speed-torque characteristics from Figures 31.8, 31.9 and 31.10, we understand that the applications of a DC series motor can be well pulled off using an electromagnet-based V-gate magnet motor.

Table 31.3 Speed-flux characteristics.

S. no.	Speed (r.p.m.)	Flux (Webers)
1	845	45.62
2	410	44.91
3	245	40.28

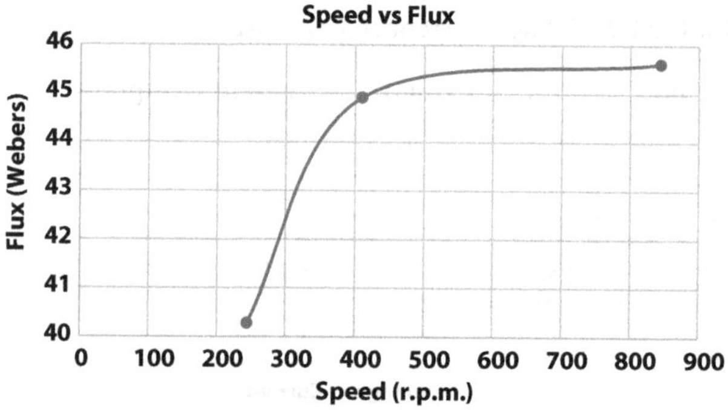

Figure 31.6 Speed-flux characteristics.

Table 31.4 Losses-efficiency characteristics.

S. no.	Losses (%)	Efficiency (%)
1	3.82	85.12
2	3.75	83.81
3	3.72	75.17

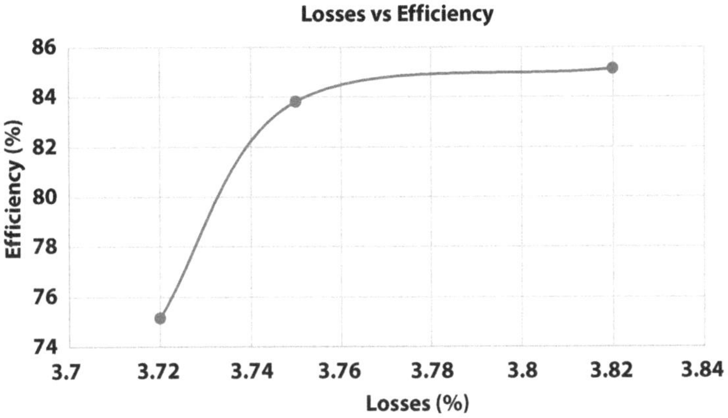

Figure 31.7 Losses-efficiency characteristics.

Table 31.5 Speed-current characteristics.

S. no.	Speed (r.p.m.)	Current (Amperes)
1	845	3.82
2	410	3.75
3	245	3.72

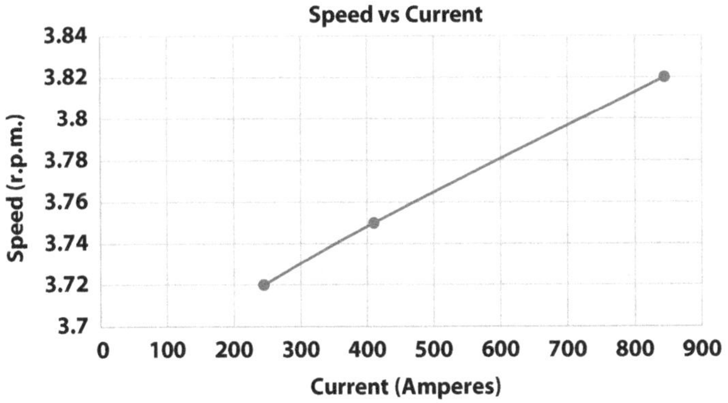

Figure 31.8 Speed-current characteristics.

Table 31.6 Torque-current characteristics.

S. no.	Torque (N-m)	Current (Amperes)
1	0.0049	3.82
2	0.0098	3.75
3	0.0147	3.72

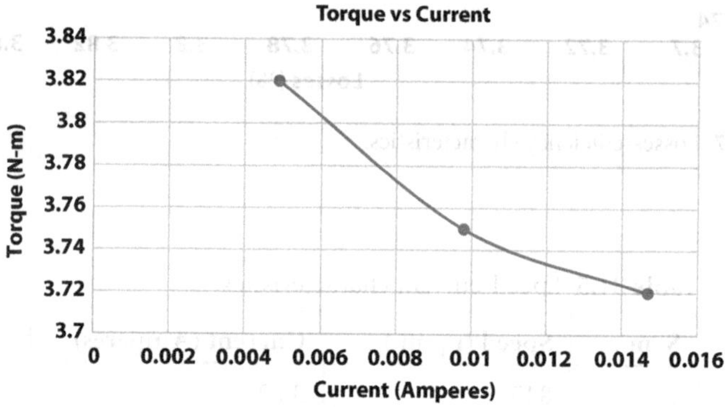

Figure 31.9 Torque-current characteristics

Table 31.7 Speed-torque characteristics.

S. no.	Speed (r.p.m.)	Torque (N-m)
1	845	0.0049
2	410	0.0098
3	245	0.0147

Before we discuss the possible applications of the proposed motor model, readers must understand that the motor was tested in laboratory conditions with loads in compromise with the current ratings of the motor. After performing the necessary tests and analyzing the test results of V-gate magnet motor about its characteristic graphs, we understand that V-gate magnet motor can be started with loads as in the case of a DC series motor, although the efficiency of the motor cannot be compared to the motor of the same rating. Suggested applications of the V-gate magnet motor can

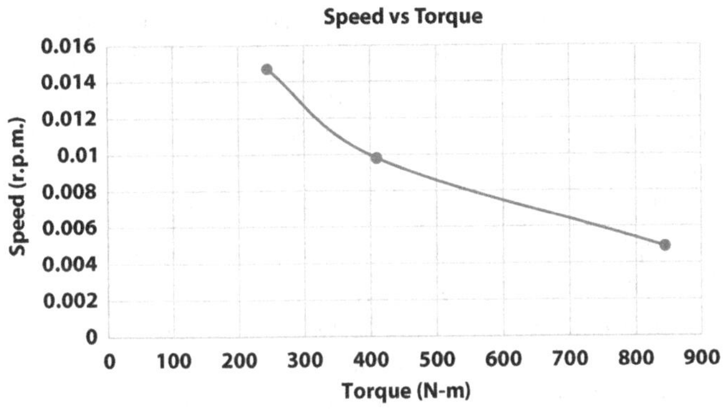

Figure 31.10 Speed-torque characteristics.

include the following, but these cannot be restricted as rigorous testing is necessary to commercialize this motor scaling up its input voltage to be suitable for domestic and commercial applications.

The possibility of applications can include:

- Fans
- Blowers
- Sample mixers in laboratory
- Portable Vacuum cleaners, Food mixers
- Conveyors
- Elevators
- Weaving machines
- Spinning machines
- Air compressors

31.8 Conclusion and Further Scope of Research

The original perpetual motion machine design was not suitable for any applications. We made this V-gate magnet motor for electrical applications. A V-gate magnet motor for electrical applications is designed and developed from scratch at benefitting economy by taking inspiration from perpetual motion machines. This model replaced the bar magnet in the stator with an electromagnet. The arrangement of poles in the rotor is replaced by V-shape neodymium magnets. The existing model had a problem of magnetic locking. The magnetic locking problem is resolved in our case using a closed loop system. The closed loop achieved by ATMEGA 328P microcontroller is utilized to control the MOSFET switch which is used to avoid magnetic locking. We have tested the motor and obtained the characteristics of the motor which shows its suitability for on-load domestic applications.

We hope that the V-gate magnet motor we designed can definitely be viewed as a replacement in the coming days for applications where conventional motors are currently being utilized. The complete setup of this small-scale motor costed less than 6000 INR. This definitely shows a big saving in the economy for the end user. We feel that this motor is efficient enough in this budget. Talking about the weight of the motor, this motor is slightly less than half the weight of the regular DC motor of similar rating. This is for the reason of using neodymium magnets which weigh less and of course the rotor drum which is coreless and made out of polyvinyl chloride. This motor is also up to 75% corrosion free.

The characteristics of this motor are different. This new unconventional motor developed is cost effective and is unique to a particular set of applications. The efficiency of V-gate magnet motor presented here, however, is not reaching the efficiency of conventional motors. Further research has to be done on this motor for improving the efficiency and control besides reducing the cost of construction. There is definitely scope for improvement in this motor and we intend to continue rigorous research to further develop a full-scale motor which is feasible both commercially and economically.

References

1. S. W. Angrist, "Perpetual Motion Machines", *Scientific American*, Vol. 218, pp. 114-122, 1968.
2. S. R. Trout and Y. Zhilichev, "Effective use of neodymium iron boron magnets, case studies," *Proceedings: Electrical Insulation Conference and Electrical Manufacturing and Coil Winding Conference*, pp. 437-440, 1999.
3. Robert C. O'Handley, *Modern Magnetic Materials: Principles and Applications*, Wiley-Interscience Publication, 2000.
4. Nicola A. Spaldin, *Magnetic Materials: Fundamentals and Device Applications*, Cambridge University Press, 2003.
5. D. Tsaousis, "Perpetual motion machine", *Journal of Engineering Science and Technology Review*, Vol. 1, Issue 1, pp. 53-57, September 2008.
6. Upadhyay Abhishek, Aman Kumar V., Arun Kumar S., Shilesh Kumar D., "Perpetual Motion Machine of first kind", *IJSRD*, Vol. 4, Issue 3, 2016.

7. Vinoth M. A., Shiva Sankar P., Linga Raj N., "Experimental Design and Optimization of Free Energy Generator by using Neodymium Magnets", *IJAERMATE*, Volume 3, 9 March, 2017.
8. Fulsundar Shubham, Kokate Vishwajit, Shirke Siddhesh, Zole Swapnil and Shinde Sachin, "Perpetual motion machine", *IJAERD*, Vol. 5, Special Issue 4, pp. 1-4, 2018.
9. Mahesh B., "Self-Flowing Generator", *International Journal of Science and Research*, Vol. 7, pp. 259-261, 2018.

32

Design and Realization of Smart and Energy-Efficient Doorbell

Shubham Pandiya[1], Saurabh Shukla[1], Saransh[1], Anantha Krishnan V.[1] and Gnana Swathika O.V.[2*]

[1]*School of Electrical Engineering, Vellore Institute of Technology, Chennai, India*
[2]*Centre for Smart Grid Technologies, School of Electrical Engineering, Vellore Institute of Technology, Chennai, India*

Abstract

COVID-19 alarms us in all ways. The used technologies failed; the entire system collapsed. It has impacted us a lot. The one thing it teaches us is to adopt modern technologies. Before COVID-19 nobody thought of having online classes at any moment of life. But now we are seeing that this way is also possible. Energy efficiency is also a trending topic these days and it is essential for a product to be energy efficient to compete in this market. So, with the forgoing situation we adapt ourselves for a better outcome. Seeing the same we decided to make a Smart and Energy-Efficient Touchless Doorbell system to tackle the present scenario.

Wherever we go, we need to touch things to do our work as some things require physical interactions. The surveys found that more than 80% of Indians enter their home without sanitizing their hands. They sanitize them in the washbasins which are mostly near the bathrooms or wash areas and are located away from entrances. So, they need to walk the whole way and during this, they touch various things. This is a much bigger issue than it seems. You may have the virus on your hands and belongings. So, we need to do proper sanitization before entering the home.

Doorbells are one of the most used electrical appliances in the world. And hence many outsiders and family members touch the doorbell countless times in their day. Thus, the doorbell becomes the device that people interact with most. Do we need to sanitize it after each touch? The answer is yes. But what if the doorbell works without being touched and just by gestures? During the COVID-19 lockdown, doorbells were the most ignored things and were not sanitized, and so

**Corresponding author*: gnanaswathika.ov@vit.ac.in

Milind Shrinivas Dangate, W.S. Sampath, O.V. Gnana Swathika and P. Sanjeevikumar (eds.) Integrated Green Energy Solutions Volume 2, (217–230) © 2023 Scrivener Publishing LLC

it is now. This disease may end, but we need to adopt some hygiene habits so that we do not face any problems in the future too. There are a lot of doorbells in the Indian market but they are not that advanced in the case of energy management, hence, we need to make a bell which uses energy efficiently.

This paper includes a doorbell which is operated touchless and a smart system which makes your hands, as well as belongings, sanitized. It will not just be a doorbell, it will be a smart machine monitoring your movement, a machine which is energy efficient as it uses the mains supply and stored solar energy in case of backup. It will also make you do the sanitization process to make sure you enter your home without any germs or viruses. As the safety of our loved ones is more important, we need to make sure we do not put their lives in danger.

Keywords: Energy management, smart doorbell, sanitization, COVID-19, hygiene

32.1 Introduction

Looking at the pandemic situation humankind is in right now, the best option for everyone is social distancing and masks which are the best ways to prevent COVID-19 [7]. According to the World Health Organization it is strongly recommended to stay at home [1]. But emergency visits cannot be avoided at homes; also, COVID warriors and other officials have to go door to door for their work. Coronavirus stays on surfaces for 24-48 hours [2]. Doorbells are one of the most used devices in the world and they become vulnerable when they are used by several people while entering the house. When some infected person presses the button, the virus sticks on it and can infect a healthy person who can act as a chain initializer [9]. We can avoid this danger by using the smart doorbell. This work includes a doorbell which is operated touchless with a smart sanitization system enabling one to get sanitized, and it provides the feature of getting their belongings sanitized as well [3]. It is not just a doorbell; it is a smart monitoring machine which tracks a person's movement in front of it and makes them do the sanitization process to make sure they enter the home without any germs or viruses. It can manage its own energy as it can switch to convectional energy whenever a power cut happens. So, it also works when there is no power supply from the convectional grid.

32.2 Methodology

The smart doorbell system will track movement in front of it and make the person do the sanitization process to make sure they enter the home without any germs or viruses on their hands and belongings. When any person comes within 6 feet of the ultrasonic sensor [4] it shows the message on its

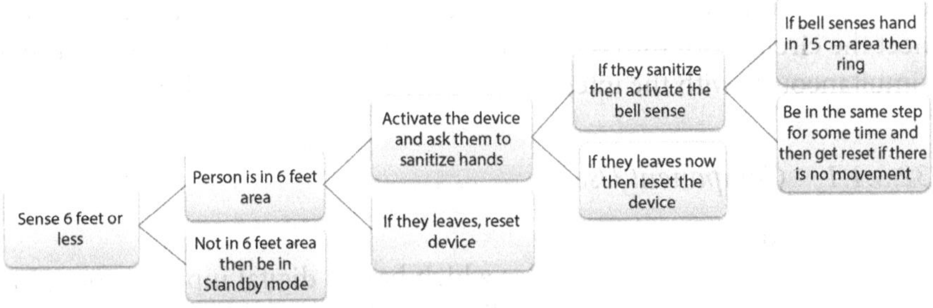

Figure 32.1 Flowchart of the working algorithm.

16x2 LCD (liquid crystal displays) screen, "Sanitize your hands" and "Keep your hands under the arrow". Unless the person sanitizes their hands, the bell will not ring. The presence of hands is determined via a PIR (Passive infrared) sensor placed under the arrow marked on the prototype. The presence of hands triggers the pump motor which pumps sanitizer and dispenses it in the hands of the visitor [5]. After the sanitization, they will be able to ring the bell. Now to ring the bell the person must move his or her hands under 15 cm of the ultrasonic sensor in front of it. It can also be used as a sanitizer machine; if anyone wants to just sanitize their hands, they can come in front of the machine, sanitize the hands, and leave without ringing the bell. The device will reset when a person leaves the radius of 6 feet around it.

Also, if a visitor wants to sanitize their belongings, they can keep their belongings in UV (Ultraviolet) box, the UV light will sanitize their belongings like mobile phone, wallet, etc. [6] which are very much in need of being sanitized as they are also those materials which are quite often used outside the home. The UV light senses belongings using a push button placed underneath the base of the box which gets pressed when any weight is applied on it. UV remains on for 10 sec and then automatically gets off. Now the visitor can take out their belongings. One can understand the entire working in a single flow diagram given in Figure 32.1.

The doorbell is also capable of switching to renewable power (solar panels, etc.) when there is a power cut in conventional grid via a relay switching mechanism.

32.3 Design and Specification

32.3.1 Software-Based Approach

To design and simulate in the best way and then implement the same onto the board to make it a real-life application and more viable product, the circuit has been designed on Tinkercad platform where one can easily connect the circuit with the controller, i.e., Arduino and do the programming simultaneously with the integration of the components and simulate it.

32.3.1.1 Component Used

Arduino Uno R3

Arduino Uno is a microcontroller which has 14 digital input, output pins (6 can be used as PWM (Pulse Width Modulation) outputs, 6 as analog inputs). It has a 16 MHz ceramic resonator, a USB (Universal Serial Bus) connection, a power jack, an ICSP header and a reset button. Connecting it to a computer can be done with a USB cable. It is the device which controls all the processes in the circuit of a Smart Doorbell. It can be coded easily and its functioning can be modified by code.

LCD 16 x 2

LCD is used as an electronic display module. They are preferred for multi-segment LEDs and seven segments. The benefits of using this module are inexpensive, easily programmable, and there are no restrictions for custom characters, special animations, etc. Here it is used to display messages so that operation and use of the Smart Doorbell becomes easier.

Ultrasonic Distance Sensor

It is a device that measures the distance of a target object by emitting ultrasonic sound waves, and converts the reflected sound into an electrical signal. They have two main components i.e., the transmitter and the receiver.

The formula for calculating distance is $\mathbf{d = \frac{1}{2}\, t \times c}$ (where d is the distance, t is the time, and v is the speed of sound ~ 343 meters/second).

For the Smart Doorbell it calculates the distance of the person from the doorbell and the distance of the hand while ringing the bell.

Voltage Multimeter

Voltage Multimeter are not auto ranging. One needs to set the range of multimeter that it can measure. For example, 2V measures voltages up to 2 volts. Here they are used to check the voltages on different segments of circuit.

PIR Sensor

It allows one to sense motion, always used to detect if someone has moved in or out of the sensors range. They are small, inexpensive, low-power, easy to use. Due to these reasons, they are commonly used. They are commonly known as PIR, "Passive Infrared", or "IR motion" sensors. For the Smart Doorbell it senses the movement of hands under the marked arrow on prototype so that sanitizer can be dispensed.

Slide switch
Whenever the slider is moved, the shell contacts the movable see-saw part of the switch and rests on its periphery where the switch is supposed to be ON or OFF. Here slide-switch is the replacement of push button which is used under the base of UV Box for triggering of UV light.

32.3.1.2 Circuit Diagram

The entire connections on the Tinkercad have been done and Figure 32.2 shows the same.

32.3.2 Hardware-Based Approach

Our hardware model is implemented on a PCB board with various stages involved like circuit designing, testing of the circuit with the necessary controller, troubleshooting (if required) and finally the integration of the components on to the board with taking proper safety measures as well.

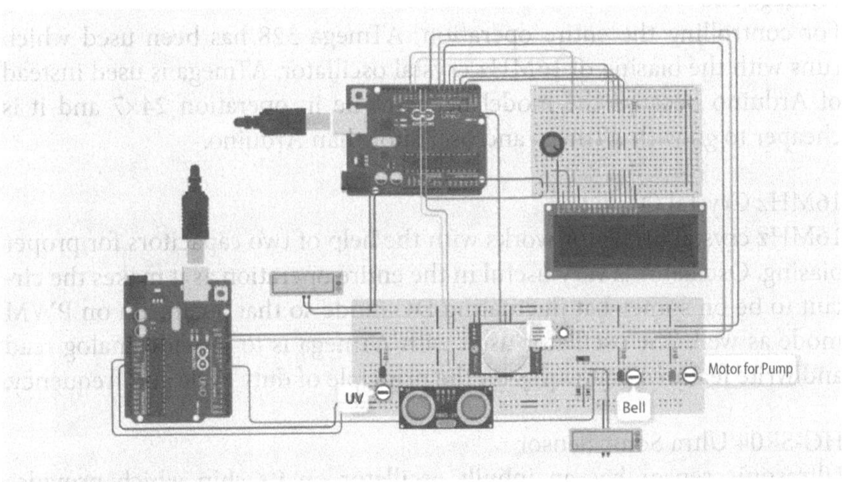

Figure 32.2 Tinkercad simulation circuit diagram of "Smart and Energy Efficient Doorbell" containing LEDs for output signal indication.

32.3.2.1 *Components Used*

5V triggering Relay Modules

Relay board and module have been used so that no back current passes through the circuit (module has a diode to prevent back current) and thereby saving from burning of the module. It has a transistor for the switching of the relay which is more efficient than using ATmega as the transistor works directly with +5V supply to provide switching with pin configuration of +5V, common ground and signal.

Phillips UV Lamp

UV lamp is used to sanitize the belongings which needs the help of choke to function properly. It is a TL-mini 4W lamp and thus the rating of choke is also kept 4W. The small 16 mm diameter of the lamp allows for a small system design and design flexibility. TUV TL Mini lamps offer almost constant UV output over their complete lifetime, for maximum security of disinfection and high system efficacy.

230V 50Hz AC Doorbell

It is a doorbell which rings after the sanitization process in this model.

5V driven DC motor

5V DC pump motor is used to pump sanitizer from the dispenser and sanitize the hands of the visitor. This motor requires 5W of power to work efficiently.

ATmega 328

For controlling the entire operation, ATmega 328 has been used which runs with the biasing of 16MHz crystal oscillator. ATmega is used instead of Arduino because the model needs to be in operation 24×7 and it is cheaper to go with ATmega and oscillator than Arduino.

16MHz Crystal Oscillator

16MHz crystal oscillator works with the help of two capacitors for proper biasing. Oscillator is very useful in the entire operation as it makes the circuit to be on somewhat fluctuating DC mode so that it can run on PWM mode as well. The oscillator used with ATmega is to provide analog read and write features as it works on the principle of duty cycle and frequency.

HC-SR04 Ultra Sonic Sensor

Ultrasonic sensor has an inbuilt oscillator on its chip which provides the sensing features for hand sanitization. It has four pins echo, trigger, +Vcc, and ground. It has good noise cancellation features too. Echo sends ultrasonic signal and trigger collects it back and thereby calculates the distance of the hand from the sensor and acts accordingly. The ultrasonic sensor is exactly accurate in operations.

Zero PCB Large
Zero PCB Large is a circuit board on which the components are integrated and it increases the flexibility of the model with the components on it.

28 Pin Base
28 Pin Base are often used to protect the controller from burning and also act as a base for controllers. The main benefit of using pin base is that we can remove the controller whenever we want and place it again.

5V-2A(10W) SMPS
To convert the supply receiving at home into usable form and range 5V-2A(10W) SMPS has been used. The idea behind using 10 W SMPS was that we need to sanitize and therefore we have to run a motor and to withstand the entire operation and 2A as it can provide enough amount of current to the other components so that they can perform easily and uninterruptedly.

IR sensor Module
IR sensor module is used to know the presence whose sensitivity can be managed with the potentiometer built on it and contains one bright and black LEDs on it. Bright LED sends the IR radiations and the black receives it and therefore both works hand in hand. The drawback with the IR sensors is that they are photosensitive and even respond to the normal light and therefore are coated with black color. It has a threshold output voltage of 2.5V and it responds on 2.5V and above.

4W Choke
Choke is an electrical inductor. Choke is used for operation of UV lamp. Rating of choke (4W) is matched with that of lamp.

32.3.2.2 Circuit Diagram

In the hardware prototype proper soldering strips for +5V and common ground potential has been given and all the components were integrated in the same way as shown in Tinkercad. LEDs are here replaced via Relays which switches and operates output components such as DC pump motor, UV lamp Choke and Bell as shown in Figure 32.3.

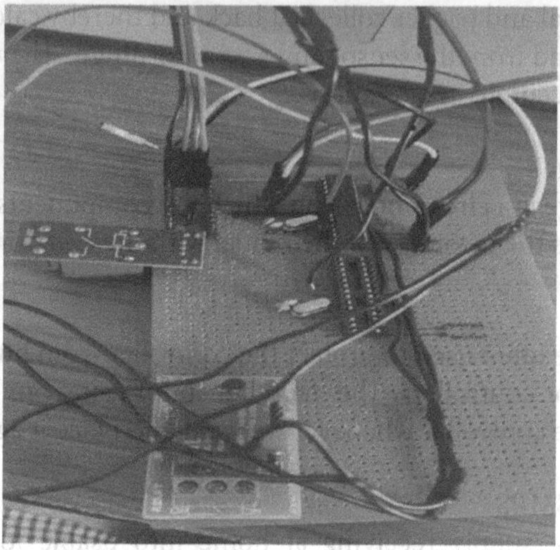

Figure 32.3 Shows the hardware implementation of the simulation circuit on 0 PCB.

32.4 Result and Discussion

Welcome message

The visitor has not arrived in the range of 6 feet from the device. The device shows a Welcome message on its display as shown in Figure 32.4. The distance is measured using US distance sensor [4].

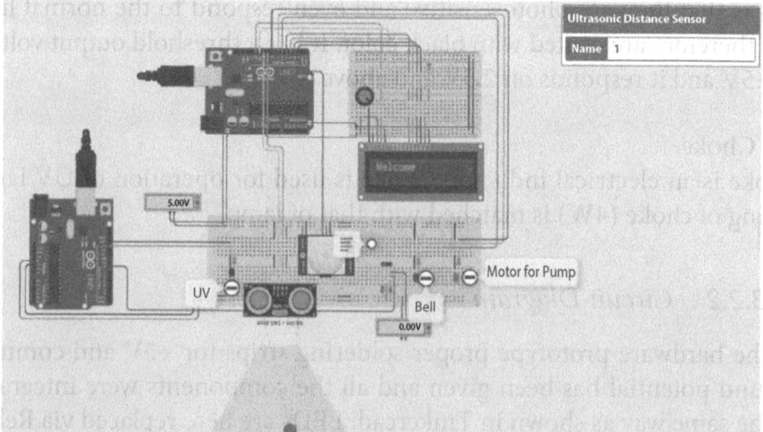

Figure 32.4 Simulation circuit showing "Welcome" message on LCD.

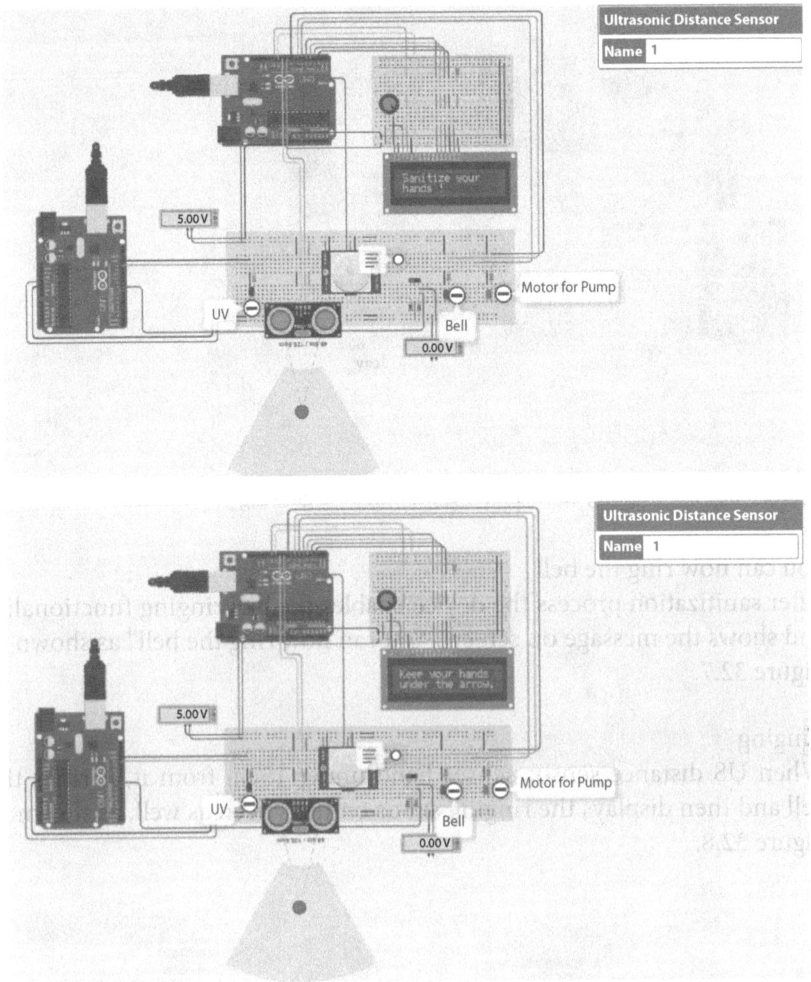

Figure 32.5 Shows the change of messages on LCD as the US distance sensor gets a reading of <=6feet & IR gets enabled.

"Sanitize your hands" and "Keep your hands under the arrow"
Presence is detected by US Distance sensor. The devices change the message from "Welcome" to "Sanitize your hands" and "Keep your hands under the arrow," to instruct the visitor as shown in Figure 32.5.

Working motor
When the visitor moves their hands under the arrow, their hands are detected by PIR sensor [5] and then the motor starts working for dispensing sanitizer [8] as shown in Figure 32.6.

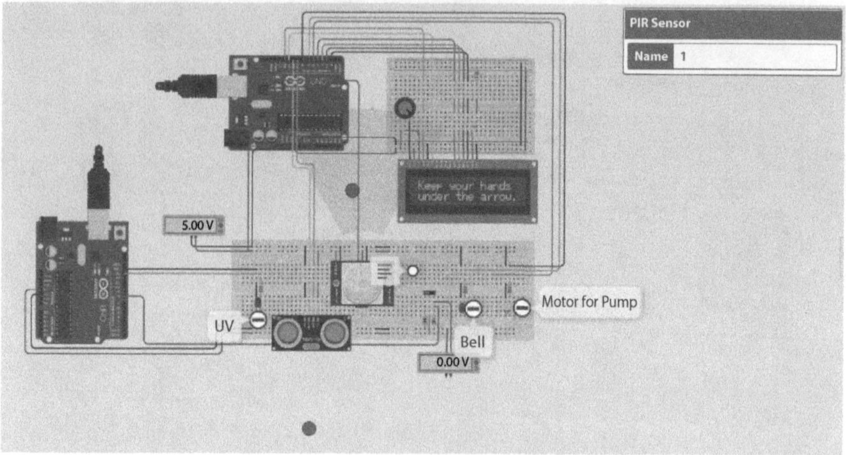

Figure 32.6 The working of the motor is indicated via glowing yellow LED.

You can now ring the bell
After sanitization process the device enables the bell ringing functionality and shows the message on screen "You can now ring the bell" as shown in Figure 32.7.

Ringing
When US distance sensor senses hand under 15cm from it, it rings the bell and then displays the ringing status on the screen as well, as shown in Figure 32.8.

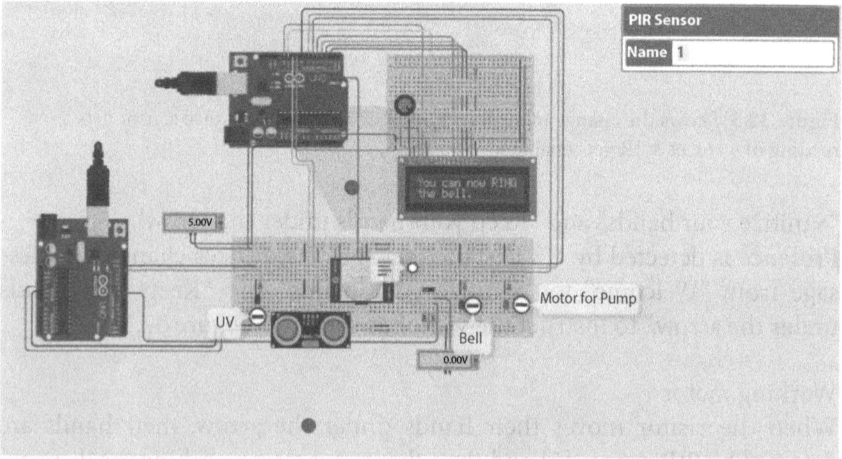

Figure 32.7 Shows change of message on LCD as the algorithm now allows to ring the bell.

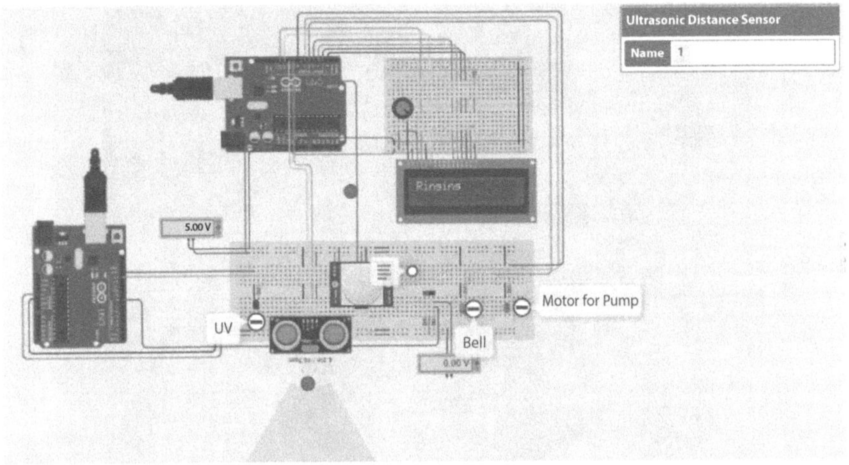

Figure 32.8 The blue LED shows that the bell is ringing and thus the LCD displays its status.

UV working

When belongings are kept in the UV box, the lamp goes on and stays in that condition for 10 seconds as shown in Figure 32.9. Here it is indicated by red LED.

Under this box as shown in Figure 32.10 contains circuitry of the Smart doorbell and its energy management system which allows it to be operated during power cuts too.

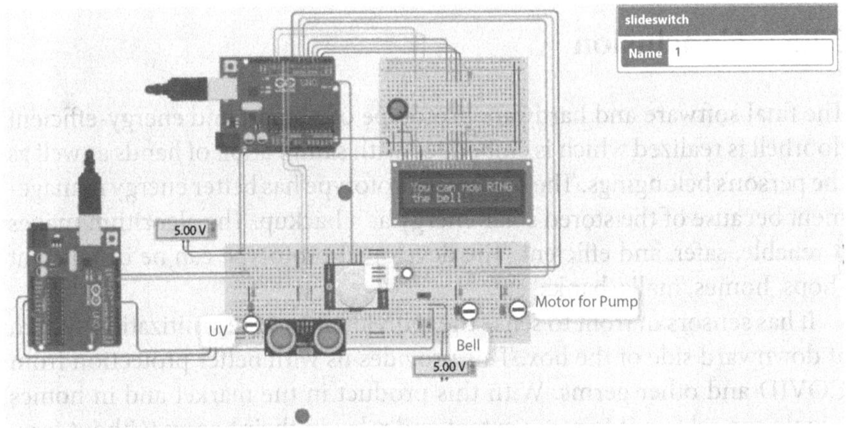

Figure 32.9 Shows the UV lamp working as some object is kept in UV box which is indicated by red LED.

Figure 32.10 This is how the doorbell looks in real when implemented on hardware. The two circular holes is the place where the Ultra Sonic distance sensor is fitted and above that is the LCD screen.

32.5 Conclusion

The final software and hardware prototype of a smart and energy-efficient doorbell is realized which is embedded with sanitization of hands as well as the person's belongings. The hardware prototype has better energy management because of the stored solar energy as a backup. The algorithm makes it reliable, safer, and efficient. The developed prototype can be installed at shops, homes, malls, banks, etc.

It has sensors in front to sense the individual and the sanitization system at downward side of the box. This provides us with better protection from COVID and other germs. With this product in the market and in homes people can rely on it to some extent and relax in their homes without fearing that they may have brought the virus into their homes with them [10].

References

1. Dr. Rohini Despande, Chethana K., Deepika N., G. Girish Reddy 2021. The Touch Less Door Bell. *International Journal of Modern Agriculture.* 10, 2 (May 2021), 3590-3596.
2. WHO Guidelines. https://www.who.int/emergencies/diseases/novel-corona virus-2019/question-and-answers-hub/q-a-detail/coronavirus-disease-covid-19-cleaning-and-disinfecting-surfaces-in-non-health-care-settings
3. Eddy Yusuf, M. N. Mohammed, Ismail I Daood, Omair Z. Ameerbakhsh, S. Al-Zubaidi, Omar Ismael Al-Sanjary, Shahad Al-Yousif, Smart Iot Technologies for Combating Covid-19 Pandemic: Automated Touchless Hand Sanitizer Dispensing System.
4. Michal Kelemen, Ivan Virgala, Tatiana Kelemenová, Ľubica Miková, Peter Frankovský, Tomáš Lipták, and Milan Lörinc, "Distance Measurement via Using of Ultrasonic Sensor." *Journal of Automation and Control*, vol. 3, no. 3 (2015): 71-74. doi: 10.12691/automation-3-3-6
5. Qusay Idrees Sarhan, "Arduino Based Smart Home Warning System", *Control Science and Systems Engineering (ICCSSE)2020 IEEE 6th International Conference on*, pp. 201-206, 2020.
6. Design and Development of a Portable Disinfectant Device by Dhananjay Kumar, Utkarsha Sonawane, Mahendra Kumar Gohil, Ravichand Pol, Anand S. Patil, Reema Mittal, Avinash K. Agarwal. Received: 10 May 2020/Revised: 9 June 2020/Accepted: 18 June 2020 / Published online: 24 June 2020.
7. News *Economic Times* (2020). https://economictimes.indiatimes.com/industry/healthcare/biotech/healthcare/no-symptomsin-80-of-covid-cases-raise-concerns/articleshow/75260387.cms?from=mdr. Accessed 21 Apr 2020
8. Qing Chun Fenga, Xiu Wang, Design of Disinfection Robot for Livestock Breeding, 3rd International Conference on Mechatronics and Intelligent Robotics (ICMIR-2019), *Procedia Computer Science* 166 (2020) 310–314.
9. G. Kampf, D. Todt, S. Pfaender, and E. Steinmann, "Persistence of coronaviruses on inanimate surfaces and their inactivation with biocidal agents," *J. Hosp. Infect.*, vol.104, no. 3, pp. 246-251, 2020, doi: 10.1016/j.jhin.2020.01.022
10. M N Mohammed, S F Desyansah, S Al-Zubaidi, E Yusuf, "An internet of things-based smart homes and healthcare monitoring and management system: Review", *J. Phys. Conf. Ser.*, 2020d, doi: 10.1088/1742-6596/1450/1/012079.

33

Optimal Solar Charging Enabled Autonomous Cleaning Robot

Aastha Malhotra[1], Anagha Darshan[1], Naman Girdhar[1], Prantika Das[1], Rohan Bhojwani[1], Anantha Krishnan V.[1] and O.V. Gnana Swathika[2*]

[1]*School of Electrical Engineering, Vellore Institute of Technology, Chennai, India*
[2]*Centre for Smart Grid Technologies, School of Electrical Engineering, Vellore Institute of Technology, Chennai, India*

Abstract
The progress of various fields of robotics improves people's quality time and their environment. Garbage is a major problem in worldwide attenuation. Developing waste collector robots is currently a field of research on efficient solutions to the present-day problem of unclean surroundings which degrade the environment. Due to the difficulties faced in keeping the surroundings clean manually, this project proposes a solution by developing an autonomous robot with the additional feature of solar charging for its batteries. Solar photovoltaic advancements using a renewable source of energy have become prominent in today's world due to the scarcity of energy resources globally. The robot will draw out energy from the solar-powered batteries needed to run its mechanical and electrical components to perform the cleaning task efficiently. The power output of the solar batteries is optimized using Maximum Power Point Tracking (MPPT) Algorithm for efficient power management for the robot.

***Keywords*:** Autonomous, RCNN, solar battery charger, ESP32, Node-RED, rack and pinion, P&O algorithm

33.1 Introduction

This project is planned so it will help society and the climate and support a perfect and better life. The name of the project has been derived from the

**Corresponding author*: gnanaswathika.ov@vit.ac.in

Milind Shrinivas Dangate, W.S. Sampath, O.V. Gnana Swathika and P. Sanjeevikumar (eds.) Integrated Green Energy Solutions Volume 2, (231–248) © 2023 Scrivener Publishing LLC

main idea of the project's concept which is to optimally collect the trash autonomously. Previous data related to the disposal of trash/litter especially plastic waste has been analyzed before designing this project.

Optimal Solar Charging Enabled Autonomous Cleaning Robot is a novel approach towards solving the issue of pollution mainly caused by litter/garbage on uncertain geographical terrain with the least possible human interaction. For the movement of the robot, a particular algorithm has been derived so that a particular area can be cleaned with the least possible movement and energy consumption. This robot detects the garbage autonomously and then collects it using a mechanical gripper. For the efficient functioning of this robot, solar power apparatus has been used to recharge the battery using optimized algorithms. With all the technicalities involved, this project is mainly focused to derive a low-cost, robust solution for litter cleaning which is feasible for day-to-day use. Therefore, for having better control over this robot, IoT applications have been used for better maneuvering and speculation. The project is designed such that it is further expandable to future advancements.

Authors in [1–5] have demonstrated use cases and application aspect of the various electronic sensors. Authors in [7] and [14] have designed a beach cleaning machine consisting of an engine that runs on fossil fuel. Different types of mechanisms such as Raking and Sifting have been proposed. For the material, a combination of mild and stainless steel is used as steel is readily available at cheaper prices. A system that consists of a robotic arm controlled by the microcontroller to pick the garbage and place it in the bin has been proposed in [1] and [8]. The practical aspect of the same has been demonstrated in [2]. Next, authors in [9] proposed a beach cleaning robot that that uses a camera and converts each frame from Red-Green-Blue (RGB) color system to The Hue-Saturation-Lightness (HSV) color space in order to detect and identify waste more accurately. The obtained images are then converted to binary data and further processing is done to identify the position of objects. Authors in [10] and [11] have employed a neural network-based approach in order to detect garbage. Further, authors in [12] designed a charging system for LiPo batteries with the help of tracked solar panels. A microcontroller along with light sensors are used to identify sunlight and track the solar panels to the right direction to get maximum power. Authors in [5] has derived a system wherein open CV is used on an embedded platform to detect multiple object. Authors in [13] and [14] have used Ramos Smart Host Microcontroller for maintaining optimal battery of robot vehicle operated using solar power.

33.2 Methodology

The vehicle designed in this project is made such that it can clean rough terrains specially beaches without any human interference. Mask R-CNN is used for Object Detection and Segmentation to identify plastic litter. The dataset used to train our model is TACO. The robot has a mechanical structure that suits the beach. DC motors connected to motor drivers and microcontrollers operate the 4-wheel drive of the robot. For the movement of the bot, the length, breadth and the time delay are initially given as input and hence all the motors have a constant speed while scanning along the length of the area. Once the robot reaches the end of its length, it turns 180 degrees. For turning left, the speed of the set of right wheels increases and vice versa. The movement algorithm has been coded in Arduino microcontroller. For picking trash, it is equipped with a claw, which will be operated by the microcontroller for when the robot detects litter. Moreover, Internet of Things applications have been used to monitor this vehicle from remote areas and optimize its usage with the least possible human interaction.

The main objective is to achieve this target autonomously with least consumable energy, which gives maximum power output. Hence renewable solar energy is the best source of energy for continuous charging of Li-ion battery used to power the bot. In this project charging of the battery is done using MPPT control and buck converter. MPPT is a process to find the point from where maximum power can be extracted, called the MPP (Max Power Point), which is the product of the peak values of voltage and current obtained from solar module and given to the load. The MPPT control circuit is given a current and a voltage input from the PV panels that collect the solar energy. The P&O algorithm used in MPPT circuit outputs a duty cycle. Using a relational operator this duty cycle is compared with a carrier signal, which is fed to the gate terminal of the IGBT switch of the buck converter.

33.2.1 Design Specification

Mechanical design

The chassis frame (Figure 33.1) has been built using lightweight Aluminium 6061 Alloy, with welding as the primary manufacturing source. Because of the weight constraints and to minimize the load on the base motors this aluminium alloy proved to be the best alternative. The CAD design of the robot has been achieved on Solidworks 2018. The vehicle has been designed in such a way that mobility becomes easier on all sorts of terrain specially on tropical surfaces and to achieve this weight of the vehicle has been optimized using lighter material.

Figure 33.1 Isometric view of vehicle displaying unmanned vehicle.

Estimated dimensions in mm (L x W x H): 600x250x500
Estimated weight: 9.8 Kg (assuming all miscellaneous weights)
Drive Mechanism: 4-wheel drive

Driving mechanism
The unmanned vehicle primarily has a 4-wheel driving mechanism wherein each wheel is individually powered and controlled to optimum efficient driving of vehicle (Figure 33.2). The motor chosen for the base wheels has optimum specifications of low RPM and high torque so that the vehicle can move easily even when the vehicle is fully loaded with trash. The reason behind using four individual motors (one motor for each wheel) is to achieve high mobility even in rough terrains like that of beach.

The wheels are chosen such that they have optimum thickness to reduce Pressure per unit area. If the wheels are too thin, Pressure per unit area will be more, therefore, mobility will be restricted in the coastal terrains. Therefore, thicker wheels are used so that pressure per unit area is decreased and vehicles can move easily even with the extra trash weight.

Motor used for base wheels: https://robu.in/product/orange-12v-og555-50rpm-dc-motor-for-diy-projects-grade-a-quality/

Base motor specification:
Rated speed: 50 RPM
Rated Torque: 346.8 N-cm
Rated Voltage: 12 V DC

Base wheel specification:
Material used: Solid rubber
Dimension (DxW) in mm: 200x45

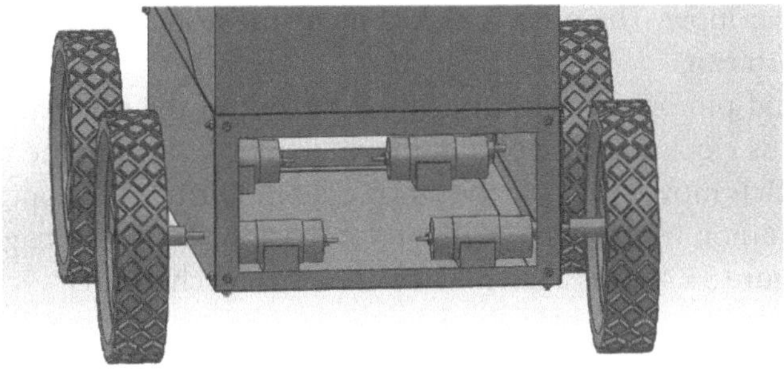

Figure 33.2 Isometric view of motor clamping on chassis of vehicle.

Figure 33.3 shows the wheels of the vehicle equidistant to each other i.e., the distance between two wheels on the same axle is the same as the distance between the two axles (front and back) of the vehicle. This is done to optimize the neutral steering for change in course path.

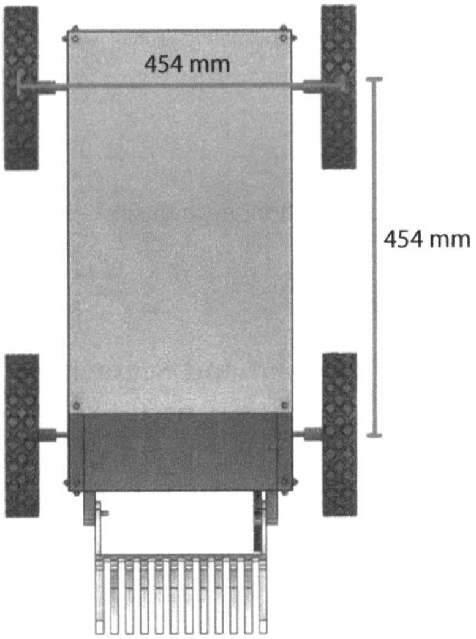

Figure 33.3 Illustration of equidistant length and width of axles.

Picking mechanism

For picking up the trash from the sand we have used a grip with 1 degree of freedom moving linearly making the maximum angle of 104°. The gripper is made such that it has slits in between so that when the trash is picked up from the beach surface, the sand gets filtered out and only the trash is left on the gripper. The trash is picked up by the gripper and placed in the inbuilt trash can.

Rack and pinion mechanism is used to control the gripper. When the motor turns clockwise, the pinion moves forward, shifting the rack backwards, which moves the gripper down. The motor used for operating the rack and pinion mechanism is a high torque, low RPM square geared DC motor. Figure 33.4 gives a lateral view into this mechanism.

Motor used for Rack and pinion mechanism:
https://robu.in/product/orange-mg555-12v-30rpm-square-gearbox-dc-motor-for-diy-project/

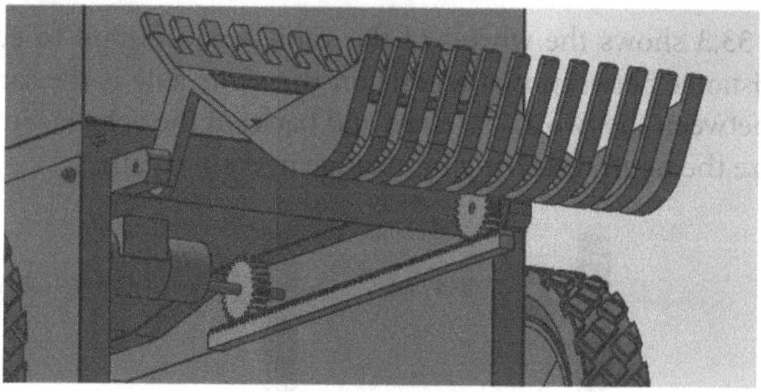

Figure 33.4 Lateral view of rack and pinion mechanism.

33.2.2 Trash Detection

Mask R-CNN for Object Detection and Segmentation has been used to implement trash detection. The dataset used to train our model is TACO which is an open image dataset containing 1500 images and 4784 annotations that can be used for garbage detection and segmentation. The dataset is available at http://tacodataset.org/

TACO Dataset
The data set contains high-resolution images of different types of litter found around us. It contains images of litter taken in common environments like woods, roads and beaches. These images were manually annotated and segmented according to a hierarchical categorization to train and evaluate object detection algorithms.

R-CNN Implementation
R-CNN is implemented with ResNet network algorithm. The ResNet algorithm is used to extract features from the images and create a feature map. The images are resized and padded to create an input layer of

size 1024×1024 px. Models were trained on TACO with a batch size of 2 images and a learning rate of 0.001. Weights are chosen based on the model trained COCO dataset [6].

Neural network models work more accurately when trained on a large dataset. To increase the size of the dataset, the data augmentation technique was used that creates modified versions of the existing images in the dataset. During training, Gaussian blur and AWG noise was added. The exposure and contrast of the images was changed and several rotations between −45° to 45° were done. Finally, the images were cropped around the annotated bounding boxes so that the litter part in the image is always in focus as shown in Figures 33.5 and 33.6 respectively.

Figure 33.5 Image cropping.

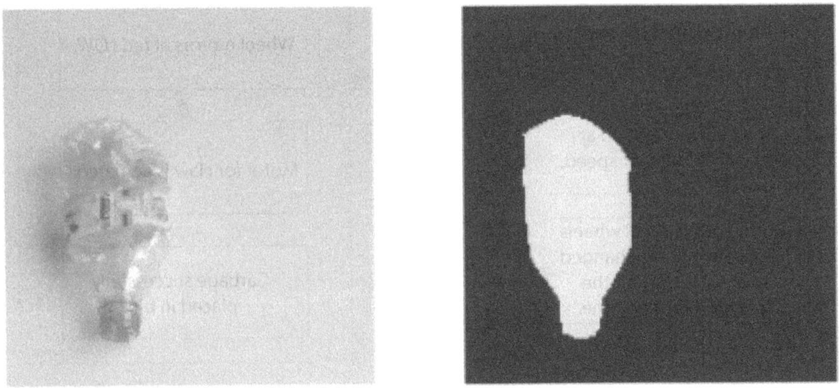

Figure 33.6 Image augmentation.

All the objects in the model are classified as one class: Litter. Our system will capture each frame of webcam feed and detect litter in real time. This is implemented using the OpenCV (VideoCapture) function.

33.2.3 Movement Algorithm

The robot needs to have movement in a specified closed area for cleaning. Figure 33.7 shows the algorithm flow for the overall movement. This algorithm can be divided into two parts, Wheel movement and Claw movement.

The movement of the robot is controlled by the four-wheel DC motors and the claw movement by one DC motor. These motors are supplied with voltage from the 12V Lithium-ion battery and controlled by the microcontroller. The robot will move over the area in a specified pattern to cover all areas uniformly and stop when it locates trash so it can be picked up by using the claw.

Wheel movement

The four DC motors attached at the wheels control the robot movement over the area. The algorithm used will make the robot scan the area lengthwise. Figure 33.8 depicts the movement algorithm pictorially.

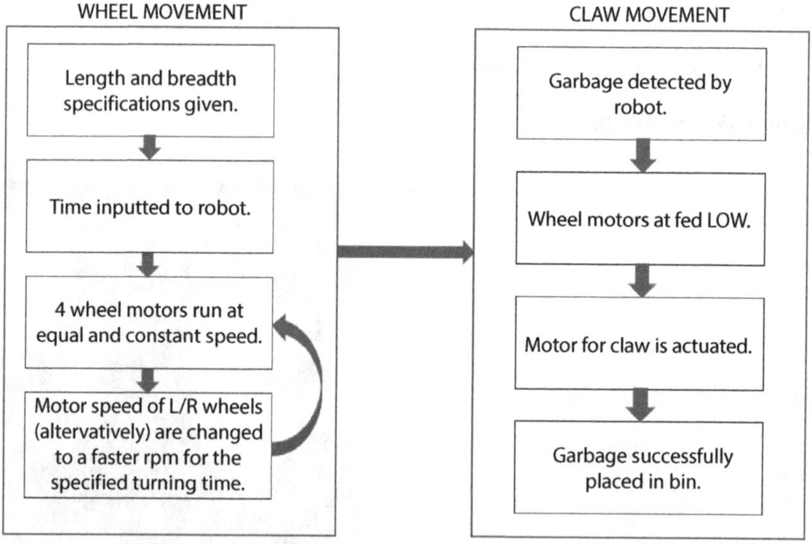

Figure 33.7 Algorithm flow for robot movement.

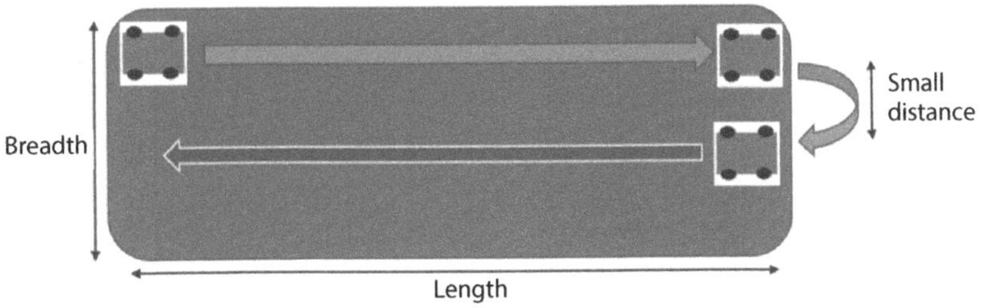

Figure 33.8 Pictorial representation of path followed by robot.

The algorithm for wheel movement is explained in the following steps:

- The length and breadth of required area is initially given as input.
- Time required to cover length is calculated as length/constant motor speed.
- Within a loop, robot is given a command to move forward parallel to the length for the time delay calculated. After moving along the length, it will turn 90 degrees twice. The first 90-degree turn will cover a small distance along the breadth, the second turn will align the robot parallel with the length (the robot will take 2 rights and then 2 lefts alternately).
- The robot will continue to move in this loop till the total breadth is covered. This can be calculated as, no. of loops required equals to breadth/small distance taken by the robot to turn around the end.

No. of loops required = breadth/small distance taken by robot to turn around the end.

A proteus simulation for the movement of the bot has been done as shown in Figure 33.9 to control the speed of the base motors. The speed of the two motors is controlled using PWM technology by Arduino micro-controller and L293D (motor driver).

Claw movement

Claw movement is executed using a DC motor controlled by the micro-controller. When the robot detects a trash item on its path, the driving motors stop bringing the robot to a halt. Now, the motor connected to the arm is activated to pick up trash, after which the arm motor stops and the robot will continue to move on its path.

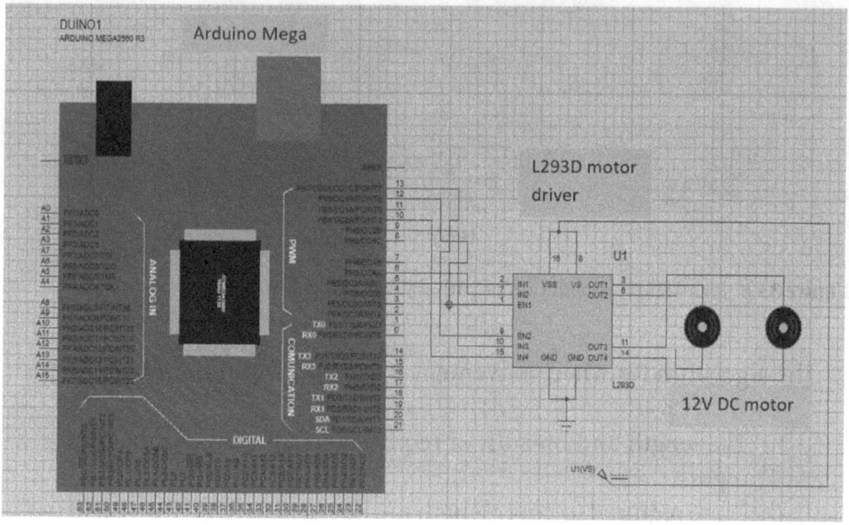

Figure 33.9 Proteus simulation of wheel movement.

A circuit simulation shown in Figure 33.10 shows this mechanism. The ultrasonic sensor is used for the demonstration purpose of trash detection and two DC motors are used to display the stopping and moving of the robot and the arm.

The Trigger and Echo pins of the ultrasonic sensor are connected to the digital pins of the Arduino Uno. L293D is used as the motor controller. The input pins of both the controllers are connected to PWM pins, enable pins

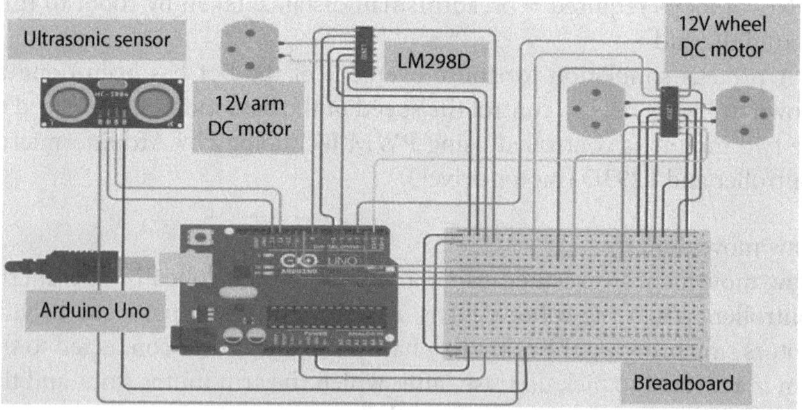

Figure 33.10 Circuit simulation for claw movement.

and power pins (5V and Ground) on the Arduino Uno. The output pins of the controllers are connected to the motors.

One arm motor and two wheel motors are used in this simulation. The two wheel motors are initially moving. When an object (trash) is detected, the wheel motors stop and the arm motor is actuated.

33.2.4 Solar Charging

In this project, solar energy is used as a renewable energy resource that will continuously replenish the battery when available. The DC that is obtained is utilized in charging of the Li-ion battery used to give power to the robot. Maximum power point tracking (MPPT) is an algorithm that will continuously adjust the impedance seen by the solar array to keep the PV system operating at, or close to, the peak power point of the PV panel under varying conditions, like changing solar irradiance, temperature, and load.

MPPT is implemented along with a DC-to-DC converter which converts the higher voltage DC output to lower voltage from solar panels. Using the buck converter equations, the inductor and capacitor values were calculated as shown below.

Calculations for designing of Buck Converter

Rated power = 213 W, Open Circuit Voltage (Voc) = 36.3V, Input Voltage (Vin) = 28-36V, Output Voltage (Vo) = 12V, Current Ripple = 10%, Voltage Ripple = 1%

Output Current = Rated Power/Vo = 213/12 = 17.750
Current Ripple = 10% of 17.75 = 1.775
Voltage Ripple=1% of 12 = 0.12V
L = (Vo (Vin-Vo))/(fsw*Iripple*Vin) = (12(28-12))/(5000*1.775*28) = 0.8653mH
C = (Iripple)/(8*fsw*Vripple) = (1.775)/(8*5000*0.12) = 369.79uF

To establish the MPPT Control, P&O control algorithm (Perturbation and Observation algorithm) is used. Perturbation and Observation method operates by periodically changing the output voltage of solar energy and evaluating the corresponding output power. P&O algorithm shown in Figure 33.11 is most widely used especially for its low-cost implementation.

A detailed simulation model for the MPPT control and P&O algorithm is created in MATLAB Simulink as shown in Figure 33.12. This procedure maximum power can be extracted under all conditions.

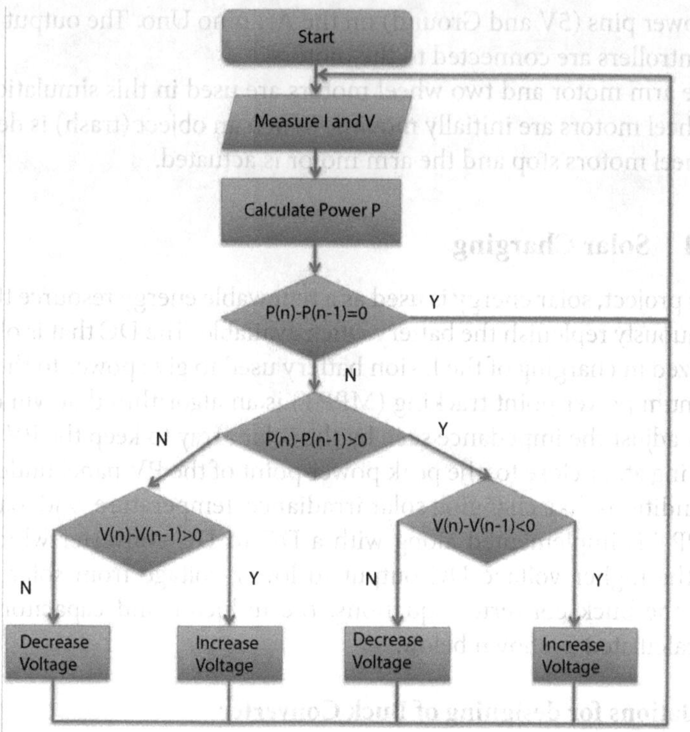

Figure 33.11 Flow chart for perturbation and observation.

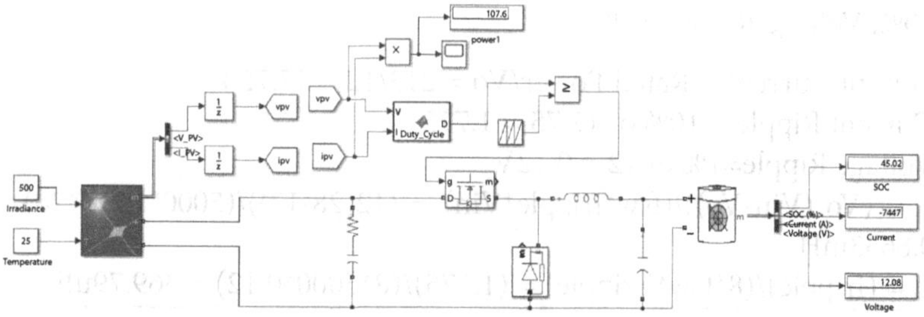

Figure 33.12 MATLAB Simulink for MPPT using P&O algorithm for solar charging.

33.2.5 Remote Monitoring

Remote monitoring feature has been implemented for the user to monitor the real time condition of the robot. We will be monitoring the real-time garbage and battery level.

Node MCU (ESP32) microcontroller, Raspberry Pi (MQTT broker), ultrasonic sensor (garbage level detection), voltage sensor (battery level

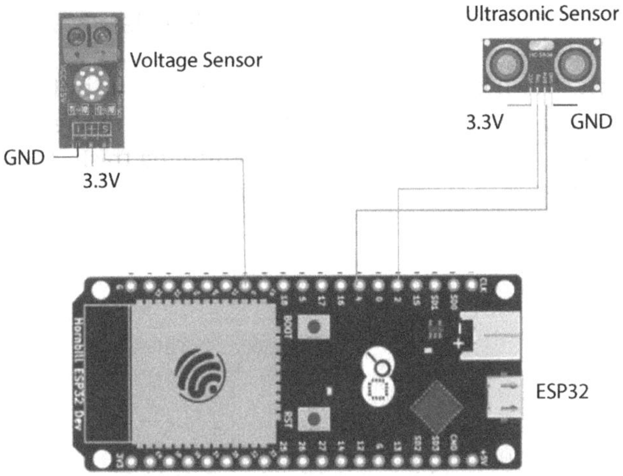

Figure 33.13 Circuit used to monitor the real-time condition of the robot.

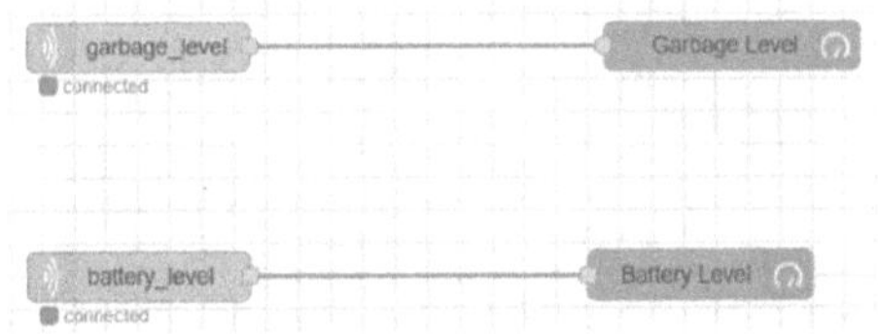

Figure 33.14 Node-RED flow for monitoring.

indicator) and a Node-RED Dashboard (output) have been used for this purpose. The connections are made as shown in Figure 33.13. The data from sensors are sent to Node-RED using ESP32 Wi-Fi module and an MQTT broker which works on a publish and subscribe model. This data can be viewed from remote places by the user.

Figure 33.14 shows the Node-RED flow which takes data from the sensors via ESP32 and MQTT.

33.3 Results

33.3.1 Trash Detection Results

The model generates bounding boxes and segmentation masks for each instance of an object in the image. The bounding boxes are computed from the masks. This makes it easier to resize, rotate, or crop images.

In order to support multiple images per batch, images are resized to one size (1024x1024).

The images shown in Figure 33.15 are collected frame by frame using the webcam feed and detect litter in real time as shown in the figure. Whenever object detection is successfully executed identifying trash, a further signal is sent to the motors to stop and pick up litter.

Figure 33.15 Examples of R-CNN prediction.

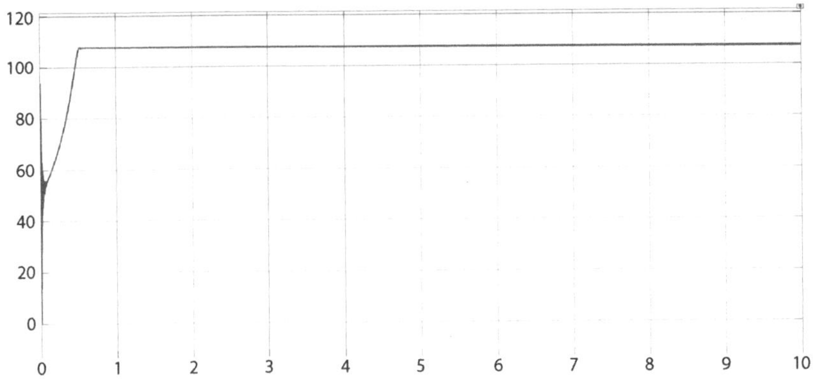

Figure 33.16 MPP (Maximum Power Point) using P&O Algorithm for sample test T=10.000.

33.3.2 Solar Charging Results

The power graph shown in Figure 33.16 gives the magnitude of power from the panels for sample test T=10.000. Value of power slowly raised to the optimal value of 107.6 Watts. The voltage across the Lithium-ion battery is 12.08 Volts and current through it is -7.477 Amps. Negative current implies that the current is flowing into the battery and is hence charging using solar power. Therefore, using the designed MPPT the maximum power available from the PV system is harvested under given irradiance of 500 $W \cdot m^{-2}$ and temperature 25 deg C conditions.

33.3.3 Remote Monitoring Dashboard

To make the bot more efficient and autonomous, remote monitoring of real-time conditions was added. the sensors sense the data and send it to the node red dashboard using ESP32 Wi-Fi module and an MQTT broker which works on a publish and subscribe model. This data can be viewed from remote places. This way, real-time conditions of the bot can be observed without any human interference.

A U/I dashboard in Figure 33.17 is made using Node-RED to show user the state of the robot remotely. The useful information of the robot (battery and garbage level) displayed will help the user to know if the garbage bin is full or if the battery is discharged completely. These situations involve the requirement of human intervention for an otherwise autonomous robot.

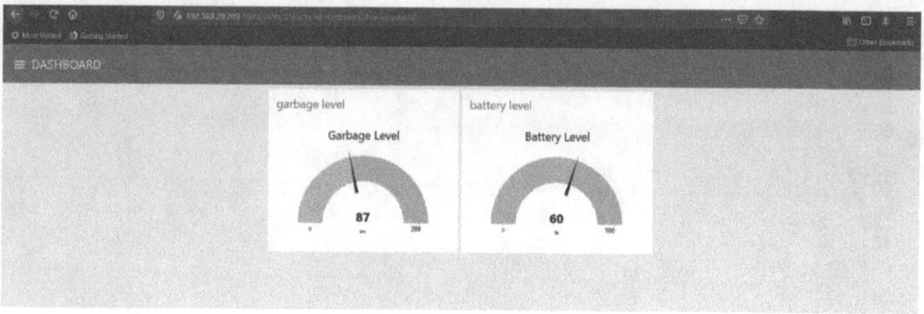

Figure 33.17 Node-RED dashboard.

33.4 Conclusion

Keeping in mind the key requirements of this robot, an Unmanned vehicle for efficient beach litter cleaning has been implemented successfully which works on solar charging and makes it more energy efficient. It will help us keep the beaches clean, reduce ocean pollution and take us one step towards sustainable renewable energy management. The system developed does not need human interference once started. The CAD model for the bot was designed pertaining to the requirements of the bot using SolidWorks. Further, image segmentation and object detection using R-CNN and OpenCV library has been implemented to identify litter. With the help of an Arduino microcontroller, we were able to successfully code for the movement algorithm for the robot, including the claw. The dashboard in Node-RED helped to monitor the battery charge and garbage bin level on the robot using IOT. The used P&O control algorithm operates by periodically changing the output voltage of solar energy and evaluating the corresponding output power. By using and designing MPPT, the bot has been made so that it can self-sustain in terms of battery and power consumption. The solar panels included in the design provide the required power backup for a complete autonomous bot which does not require human interference. It is concluded that the Optimal Solar Charging Enabled Autonomous Cleaning Robot has been able to meet the requirements of the problem statement.

References

1. Ashish Lalwani, mrunmai Bhide, S. K. Shah, A Review: Autonomous Agribot for Smart Farming, *46th IRF International Conference*, 2015.

2. Akhila Gollakota, M.B.Srinivas, Agribot, A multipurpose agricultural robot, *India Conference (INDICON), IEEE*, 2011.
3. T. Islam, S. C. Mukhopadhyay and N. K. Suryadevara, "Smart Sensors and Internet of Things: A Postgraduate Paper," *IEEE Sensors Journal*, vol. 17, no. 3, pp. 577-584, Feb. 1, 2017, doi: 10.1109/JSEN.2016.2630124
4. G. C. Meijer *et al.*, *Smart sensor systems*. Wiley Online Library, 2008.
5. S. Guennouni, A. Ahaitouf and A. Mansouri, "Multiple object detection using OpenCV on an embedded platform," *2014 Third IEEE International Colloquium in Information Science and Technology (CIST)*, 2014, pp. 374-377, doi: 10.1109/CIST.2014.7016649.
6. T.-Y. Lin, M. Maire, S. Belongie, J. Hays, P. Perona, D. Ramanan, P. Dollár, and C. L. Zitnick, "Microsoft coco: Common objects in context," in *European Conference on Computer Vision*, pp. 740-755, Springer, 2014.
7. Vivek Dhole, Omkar Doke, Ajitkumar Kakade, Shrishail Teradaleand, Rohit Patil, "Design and Fabrication of beach cleaning machine", *International Research Journal of Engineering and Technology (IRJET)*, Vol. 6, Issue 4, April 2019.
8. Bansal, Siddhant & Patel, Seema & Shah, Ishita & Patel, Prof & Makwana, Prof & Thakker, Dr. (2019). AGDC: Automatic Garbage Detection and Collection.
9. Roza, Felippe & Silva, Vinicius & Pereira, Patrick & Bertol, Douglas. (2016). Modular robot used as a beach cleaner. *Ingeniare. Revista chilena de ingeniería*. 24. 643-653. 10.4067/S0718-33052016000400009.
10. Kulkarni, Hrushikesh N., Nandini Kannamangalam and S. Raman. "Waste Object Detection and Classification." (2019).
11. Wang, Ying & Zhang, Xu. (2018). Autonomous garbage detection for intelligent urban management. *MATEC Web of Conferences*. 232. 01056. 10.1051/matecconf/201823201056.
12. Sagar Pise, Dr. Dinkar Manik Yadav, "Optimal Battery Charging in Solar Robotic Vehicle", *International Journal of Science and Research (IJSR)*, Vol. 5 Issue 11, November 2016, 618 -621.
13. Tomas de J. Mateo Sanguino and Justo E.Gonzalez, "Ramos Smart Host Microcontroller for Optimal Battery in a Solar Powered Robotic Vehicle", *IEEE/ASME Transactions on Mechatronics*, Vol. 18, No. 3, June 2013.
14. K. D. Sowjanya, R. Sindhu, M. Parijatham, K. Srikanth and P. Bhargav, "Multipurpose autonomous agricultural robot," *2017 International Conference of Electronics, Communication and Aerospace Technology (ICECA)*, 2017, pp. 696-699, doi: 10.1109/ICECA.2017.8212756.

34

Real-Time Health Monitoring System of a Distribution Transformer

Aastha Malhotra[1], Anagha Darshan[1], Naman Girdhar[1], Prantika Das[1], Rohan Bhojwani[1], Anantha Krishnan V.[1] and O.V. Gnana Swathika[2*]

[1]*School of Electrical Engineering, Vellore Institute of Technology, Chennai, India*
[2]*Centre for Smart Grid Technologies, School of Electrical Engineering, Vellore Institute of Technology, Chennai, India*

Abstract
Transformers are the most common constituents of a power system. If a transformer is damaged, it affects the whole balance of the power system. The causes of the damage are mainly overcurrent and rise in temperature, so the objective of this project is to monitor the distribution transformer's health in real time. The factors/Quantities to be monitored are current, voltage, temperature and oil level of the transformer. The data is transferred with the help of NodeMCU Wi-Fi module and can be viewed in any internet browser as a web page in which data is represented visually. This system can be scaled according to the demand and it helps in improvement of present monitoring systems.

Keywords: Incipient, IoT, HTTP, Wi-Fi module

34.1 Introduction

Distribution transformers are an integral part of the grid. They are responsible for providing power supply to the general public. Any problem occurring in the distribution transformer can disrupt the power supply for a certain region. Therefore, the transformer should be operated within rated conditions to improve its longevity [1]. But this is easier said than done. Incipient faults and external faults can create unexpected situations and

**Corresponding author*: gnanaswathika.ov@vit.ac.in

Milind Shrinivas Dangate, W.S. Sampath, O.V. Gnana Swathika and P. Sanjeevikumar (eds.) Integrated Green Energy Solutions Volume 2, (249–254) © 2023 Scrivener Publishing LLC

even damage the transformer permanently [2]. These scenarios can occur at any point in time, making it difficult to predict and to respond. That is why the working conditions need to be monitored round the clock [3].

Using IoT Technology, no human intervention is required to carry out this task. The working conditions can be monitored at all times. This can give us a head start in responding to these kinds of unexpected situations. Using different sensors, many parameters of the transformer can be monitored at the same time. All these parameters will be monitored in real time, which will help us to identify a problem within the transformer to reduce further damage.

34.2 Flow Diagram

A centre tapped transformer (12-0-12 v) has been chosen as a test transformer and ESP8266 microcontroller to receive the values from sensors. For measurement of temperature, we have used DHT11, for current, ACS712 and for oil level measurement, ultrasonic sensor. The NodeMCU has a Wi-Fi Module which enables it to connect to the internet. For providing internet, Mobile Hotspot has been used. HTTP protocol has been used to connect our NodeMCU microcontroller to our web-enabled IOT platform.

We have chosen ThingSpeak IoT Platform to display the values in a graphical manner. We can set a threshold value for each parameter and if any value is crossing the threshold, an email notification is sent using webhooks and IFTTT protocol to the respective user. This email contains information regarding different parameters, like which device has a problem, date, time and other data. The body of this email can be further configured to make it more specific and informative according to the requirement.

34.3 Operating Principle

Functionality of the transformer can reduce in case of poor cooling and overloading and thus cause improper electricity distribution. Checking the condition of the transformer and noting the values of parameters manually is a long and tedious process (Figure 34.1).

So the parameters are checked and recorded in servers using IoT. The data is captured using sensors, recorded using IoT and accessed using HTTP protocol. This way there is no need to monitor the system manually and prevent error.

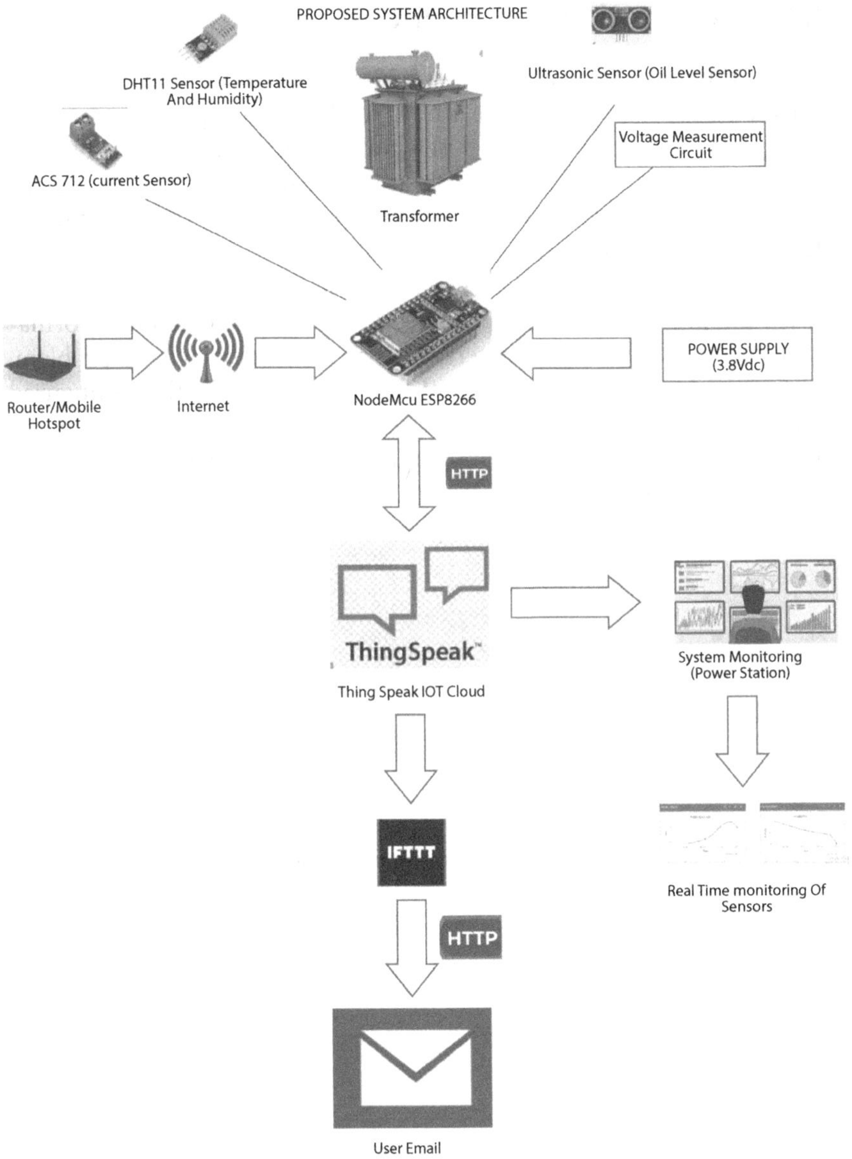

Figure 34.1 Functional block diagram.

34.4 Observation and Result

Values coming at an interval of every 30 seconds from different sensors connected to the Node MCU are sent to the IoT ThingSpeak platform via HTTP protocol and are presented in a graphical manner. This enables the user to understand the data in a much simplified manner rather than using tabular data. (This data can also be downloaded in Excel format for further data analysis and for database) (Figure 34.2).

Whenever an alert is raised, an email notification is sent using webhooks and IFTTT protocol to the respective user. This email contains information regarding different parameters, such as which device has a problem, date, time and other data. The body of this email can be further configured to make it more specific and informative according to the requirement. Figure 34.3 indicates how the temperature, humidity, oil level and current parameters are displayed in IoT ThingSpeak platform.

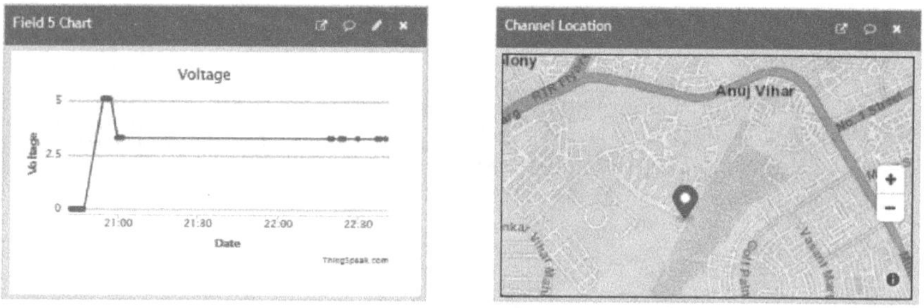

Figure 34.2 Voltage and location parameters displayed in IoT ThingSpeak platform.

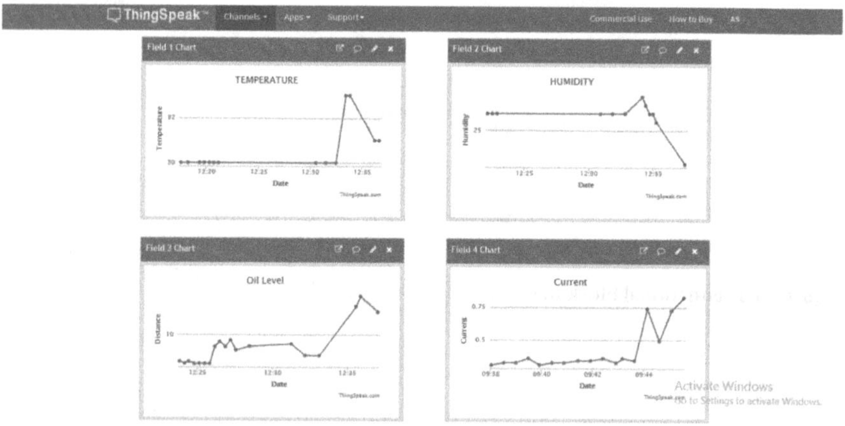

Figure 34.3 Temperature, humidity, oil level and current parameters displayed in IoT ThingSpeak platform.

34.5 IFTTT Email Notification (in case of a fault)

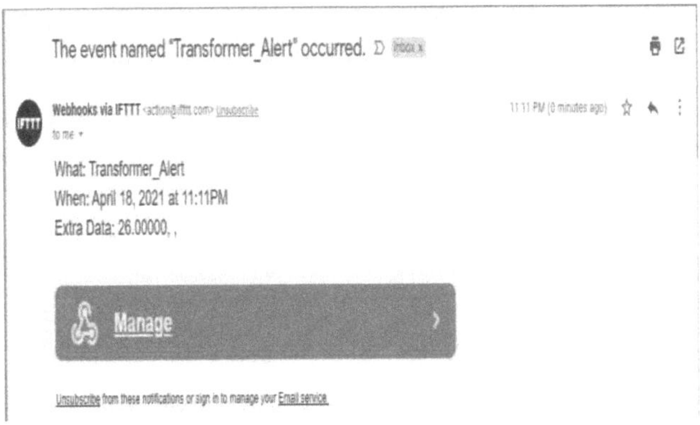

34.6 Conclusion

This project helps in day-to-day monitoring of multiple distribution transformers and is economically inexpensive. This system can be scaled according to the requirements wherein many modules of our system can be connected to one ThingSpeak platform and monitored at the same time. Synchronization and Data Redundancy are some of the challenges that have to be tackled.

References

1. A. Nduka, J. Samual, S. Elango, S. Divakaran, U. Umar and R. SenthilPrabha, "Internet of Things Based Remote Health Monitoring System Using Arduino", *2019 Third International conference on I-SMAC (IoT in Social, Mobile, Analytics and Cloud) (I-SMAC), Palladam, India*, 2019, pp. 572-576, doi: 10.1109/I-SMAC47947.2019.9032438.
2. D. Srivastava and M. M. Tripathi, "Transformer Health Monitoring System Using Internet of Things", *2018 2nd IEEE International Conference on Power Electronics, Intelligent Control and Energy Systems (ICPEICES), Delhi, India*, 2018, pp. 903-908, doi: 10.1109/ICPEICES.2018.8897325.
3. K. Sridhar, M. Assarudeen, J. Ranjith Kumar, E. Sarathbabu, "Transformer Health Monitoring and Control through Arduino", *International Journal of Advanced Engineering and Research*, vol. 5, Issue 04, 2018.

35

Analysis of Wide-Angle Polarization-Insensitive Metamaterial Absorber Using Equivalent Circuit Modeling for Energy Harvesting Application

Kanwar Preet Kaur* **and Trushit Upadhyaya**

Charotar University of Science and Technology, Changa, Anand (Gujarat), India

Abstract

In this work, analysis of a wide-angle S-band metamaterial absorber is performed with equivalent circuit modeling. The proposed absorber unit cell consists of an annular patch engraved on grounded FR4 dielectric. The thickness of the absorber unit cell corresponding to the guided wavelength is less than λ0/84 and is arranged in a periodic manner to attain an absorption of 99.64% at the frequency of 2.12 GHz. The proposed absorber is polarization-insensitive with an absorbance of above 99% for all polarization angles and absorbs more than 93% of wide (up to 45°) oblique transverse electric waves, whereas it absorbs about 93% of oblique transverse magnetic wave (up to 60°). Resistive loads are installed on the ring to study the power absorption capability of the designed absorber. The basic design is fabricated by standard PCB technology which is practically validated using the waveguide measurement method. The measurement outcome is in agreement with the numerical result.

Keywords: Metamaterials, metamaterial absorber, resonator, polarization

35.1 Introduction

Electromagnetic (EM) metamaterials (MTMs) are effective-homogeneous structures, in sub-wavelength scale, devised artificially to exhibit exotic

**Corresponding author*: kanwarpreet27@gmail.com

Milind Shrinivas Dangate, W.S. Sampath, O.V. Gnana Swathika and P. Sanjeevikumar (eds.) Integrated Green Energy Solutions Volume 2, (255–274) © 2023 Scrivener Publishing LLC

properties that are not easily available in naturally existing materials. A structure whose average cell size 'a' is much less than its guided wavelength, λ_g, is known as an effective homogeneous structure. Any metamaterial structure behaves as a natural material if the effective-homogeneity condition, $a \leq \lambda_g/4$, is satisfied [1]. The concept of MTM and its interesting properties were first proposed by Victor Veselago in 1968 [2]. MTM provides the best solution for developing novel, lightweight and thin structures for future applications. Recently, many potential applications, for example, antennas [3, 4], filters [5], electromagnetic cloaking [6], and sensors [7] are being proposed and studied widely. N. I. Landy in 2008 developed the first metamaterial absorber (MMA) with maximum numerical absorption efficiency of 96% and experimental absorption efficiency of 88% at 11.5 GHz [8]. This structure comprises numerous unit cells with each unit cell having an electric ring resonator (ERRs), dielectric substrate, and metal cut wires.

Lately, many resonances-based MMA designs with near-unity absorbance have been presented from microwave to far-infrared bands [9–13]. Subsequently, several attempts have been made in achieving polarization insensitivity [14–19], wide-angle [20, 21], and wide-band [22, 23] absorbers. These MMAs found applications in energy harvesting applications [21] and airborne radar applications [24], to name a few. Most of the MMAs are designed for C and X bands; very few MMA designs have been reported for lower S-band frequencies in literature. The reason for this is the large size of the specimen ($\sim 10\lambda_0$) which is required to practically validate the designed absorber using free-space measurement [8]. This causes the sample size to become considerably large when working on lower frequency bands.

In this work, the proposed unit cell constitutes a basic annular patch to analyze the proposed absorber using an equivalent circuit model (ECM), field, and surface distributions. The advantages of selecting a closed ring structure are discussed in Section 35.3. Herein, the suggested absorber design is numerically simulated by using TEM mode excitation via Floquet port excitation presented in the High-Frequency Structure Simulator (HFSS). HFSS is a finite element method (FEM) based full-wave simulation software. However, in [14] the numerical simulation is performed by selecting boundary conditions that replicate the practical waveguide measurement setup.

The proposed structure is simulated to exhibit nearly unity absorption with an absorption rate of 99.64% and a reflection coefficient of -24.45 dB at 2.12 GHz. Owing to the rotational-folding symmetry of the annular patch, the proposed MMA is polarization-insensitive with the absorption of about 99% for all polarization angles. The absorption rate of 93% is obtained for oblique incident angle up to 45° in case of transverse electric (TE) wave while absorption of above 98% is attained for the high oblique angle of 60° in case of transverse magnetic (TM) wave. Thus, the proposed absorber achieves a wide-angle absorption response. At the operating wavelength, the absorber structure exhibits unit cell size of less than $\lambda_0/84$ and thickness of about $\lambda_0/5$, justifying that the proposed structure is compact and ultrathin which successively satisfies the effective-homogeneity condition. Lastly, the annular patch is loaded with the different resistive loads to investigate the effect of the various values and positions of the load resistors on the absorption response. The numerical result is practically validated by the waveguide measurement technique. The simulated and measured results are agreeing with one another. This chapter is structured as follows: the theory of MMA in brief with design specifications is described in section 35.2, which is followed by analysis and discussion of the equivalent circuit model in section 35.3. Section 35.4 presents the numerically simulated results. Measurement setup and corresponding outcomes are presented in section 35.5 with the conclusion in section 35.6.

35.2 Absorber Theory and Proposed Unit Cell Design

An MTM absorber relies on matching the absorber's structure wave impedance to the impedance of the free-space for reducing reflections from the absorber surface. This matching of impedance is achieved by changing the unit cell geometries via individually regulating the electric and magnetic responses such that ε_{eff} (effective permittivity) and μ_{eff} (effective permeability) of the absorber becomes equal to each other. Though, the transmitted wave is diminished theoretically by having a bottom conductive layer whose skin depth is much less than its thickness. MMA is characterized by relation $A(\omega)=1-R(\omega)-T(\omega)$, where $A(\omega)$ represents the absorbance, $R(\omega)$ ($|S_{11}(\omega)|^2$ where $S_{11}(\omega)$ is the reflection coefficient) is the reflectance and $T(\omega)$ ($|S_{21}(\omega)|^2$ where $S_{21}(\omega)$ is the transmission coefficient) is the transmittance [8]. In this work, the effective parameter values [25] are obtained by evaluating $S_{11}(\omega)$ and $S_{21}(\omega)$ of the absorber structure using HFSS.

Figure 35.1 presents the perspective view and geometries of the proposed absorber with the direction of wave propagation and applied field directions. The absorber structure is composed of three layers of which annular patch (closed ring resonator (CRR)) forms the top layer, FR4 dielectric substrate forms the second layer and the third layer is the bottom conductive plane. The conductive plane has conductivity (σ) and thickness

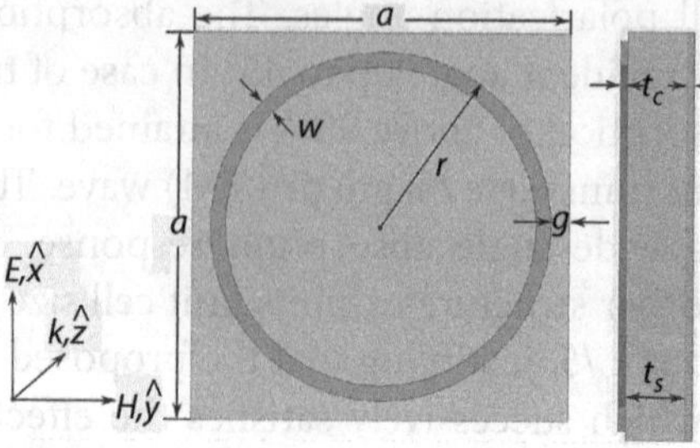

Figure 35.1 Proposed metamaterial absorber.

(t_c) of 5.8×10^7 S/m and 0.035 mm, respectively. The CRR is engraved on a 1.6 mm (t_s) FR4 dielectric substrate having tan δ (dielectric loss tangent) and ε_r (dielectric constant) of 0.02 and 4.4, respectively. The geometrical parameters of the unit cell are as follows (units in mm): a = 27.2, r = 13, w = 1.6, and g = 0.6. Eventually, the MMA prototype is fabricated on an FR4 substrate having the size of the waveguide's inner aperture. The fabricated prototype constitutes 2×4 unit cells with the periodicity of 'a'. The periodic boundary conditions are applied through Floquet port excitation of the HFSS.

35.3 Equivalent Circuit Model

The coupled line theory is employed to develop the equivalent circuit model (ECM) for the proposed CRR-based MMA. The ECM is composed of the effective capacitance (C_{eff}) and effective inductance (L_{eff}) [26] of the proposed absorber. The CRR structures are more advantageous if compared to SRR in terms of open-ended effects. The curvature effects of CRR could be neglected if the ring is narrower than 0.15 times its mean radius. Hence, CRR could be thought of as a microstrip ring that resonates when its electrical length is an integral multiple of its guided wavelength [27]. The physics involved in the structure is studied either by developing analytical formulas or by extracting the elemental values from the derived ECM. The ECMs are acquired either by using transmission line theory, lumped circuit theory, or coupled-line theory. The ECM for the proposed MMA is modeled through the even-mode and odd-mode analytical techniques [27–29]. While forming ECM the losses introduced by conductor and dielectric along with radiation losses must be considered [26].

As the EM plane wave is incident on the absorber structure, the electric field and magnetic field excitations give rise to C_{eff} and L_{eff}, respectively.

From Figures 35.2a and 35.5, it is observed that the current distribution due to the incident EM wave on the MMA causes the surface current to form a closed current loop. According to the direction of current flow, each unit cell couples to the adjacent cells, and this coupling between a pair of cells supports two different modes of propagation due to the non-homogeneous nature [28] of the MMA unit cell. For analysis, the EM wave propagating along a pair of the coupled line is expressed in terms of two modes, analogous to even-mode or odd-mode, around a plane that is superseded by a magnetic wall or an electric wall [27–29], respectively. From Figure 35.2a, the coupled pair of cells [(1, 2) & (1, 4)] are teamed through even mode while the coupled cell (1, 3) is teamed through odd mode. ECM of the proposed MMA is represented in Figure 35.2b. It is noted from Figure 35.2b that the impedance of the proposed MMA surface is expressed by the parallel combination of the top layer impedance (Z_T) and bottom layer impedance (Z_B) as shown in Eq. (35.1).

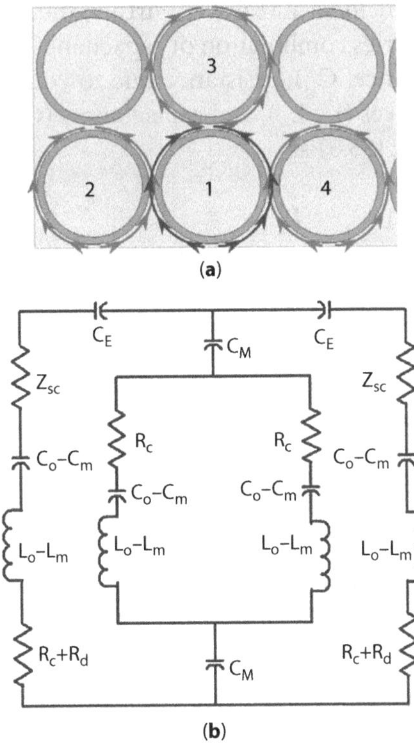

Figure 35.2 (a) Couplings between unit cells of the proposed MMA structure and (b) Equivalent circuit model of a proposed MMA unit cell.

$$Z_{SURF} = Z_T \parallel Z_B \tag{35.1}$$

In Eq. (35.1), Z_B is composed of the following series impedances viz. capacitance due to the electric wall (odd-mode capacitance, C_E), grounded short circuit transmission line impedance (Z_{SC}) [$Z_{SC} = j\sqrt{(\mu_0\mu_r/\varepsilon_0\varepsilon_r)}$ $\tan(\beta d)$], resistance due to dielectric loss (R_d) and conductor loss (R_c), self and mutual inductances (L_o, L_m) and capacitances (C_o, C_m) [27–29]. By omitting resistance due to radiation loss, the bottom layer impedance, Z_B, is defined in Eq. (35.2) as

$$Z_B = \frac{2}{j\omega C_E} + \frac{Z_{sc}}{2} + \frac{2}{j\omega(C_o - C_m)} + j\omega\frac{(L_o - L_m)}{2} + \frac{(R_c - R_d)}{2} \tag{35.2}$$

where ε_r and μ_r are the relative electric permittivity and relative magnetic permeability of the grounded dielectric substrate, respectively. ε_0 and μ_0 are the permittivity and permeability of the free space, respectively, β is the wavenumber of the incident EM wave. Further, the top layer impedance, Z_T, is regarded as a series combination of capacitance due to magnetic wall (even-mode capacitance, C_M), resistance due to conductor loss (R_c), self and mutual inductances (L_o, L_m), and capacitances (C_o, C_m) [27, 30]. Thus, Z_T is expressed in Eq. (35.3) as

$$Z_T = \frac{1}{j\omega(2C_M)} + \frac{R_c}{2} + \frac{2}{j\omega(C_o - C_m)} + j\omega\frac{(L_o - L_m)}{2} \tag{35.3}$$

The effective relative permittivity $\{\varepsilon_{re}(t)\}$ and the characteristic impedance $\{Z_c(t)\}$ [28] of the proposed MMA calculated from their physical parameters; t_s, w, and t_c, are found to be 3.17 and 68.96 Ω, respectively. The values of other elements could be calculated from [27–31] by employing the quasi-static analysis of coupled line theory.

35.4 Simulation Results

The suggested absorber presented is numerically simulated by applying periodic boundary conditions in both the x-axis and y-axis to achieve a reflection coefficient of -24.45 dB with the corresponding absorption rate of 99.64% at a frequency of 2.12 GHz, as illustrated in Figure 35.3.

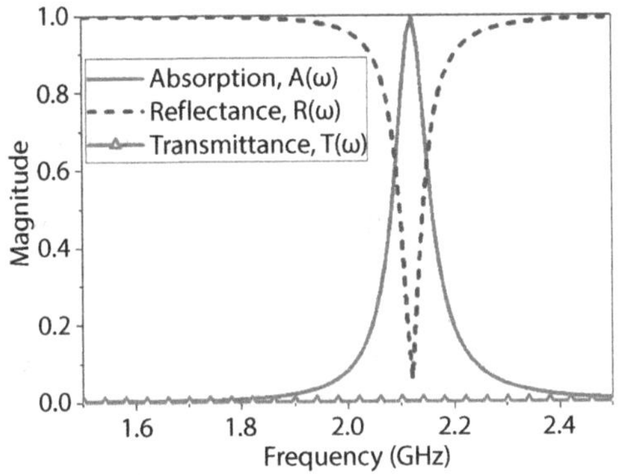

Figure 35.3 Numerical absorbance, reflectance and transmittance of the designed absorber.

35.4.1 Retrieval of the Effective MMA Parameters

As the absorber's impedance and the free-space impedance match, the real component of the absorber's wave-impedance becomes unity while the imaginary component reduces to zero at the operating wavelength. The graph of the normalized $Z_{eff}(\omega)$ (effective wave impedance) of the proposed absorber is illustrated in Figure 35.4. It is observed that the real component of the normalized impedance is nearly equal to unity, whereas the imaginary part approaches zero at 2.12 GHz. This implies that the wave impedance of the designed absorber nearly matches the impedance of the free space.

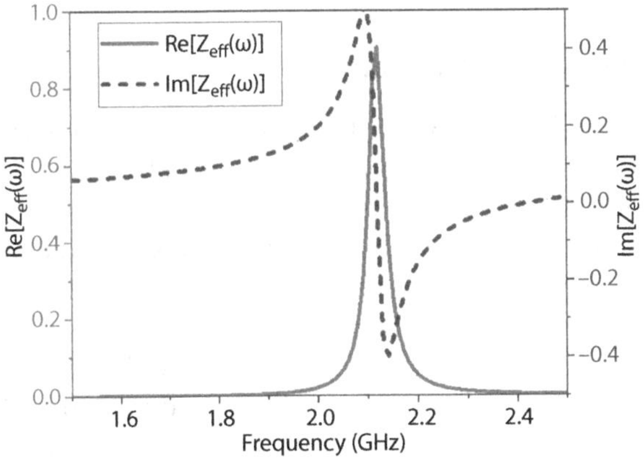

Figure 35.4 Simulated results of designed absorber of normalized impedance.

35.4.2 Absorption Mechanism

Figures 35.5a and 35.5b show the surface current distributions on the upper and lower surface of the absorber, respectively. It is observed that the surface current distributions on the upper surface are anti-parallel to the current distributions at the lower face of the absorber. These anti-parallel current forms current loops which are perpendicular to the incident magnetic field directions. Conversely, the electric excitation is induced on the upper surface of the absorber through the incident electric field component of the EM wave. This is validated by the concentrated electric field which is visible on the edges of the rings, as depicted in Figure 35.6a. The concentration of the induced magnetic field is illustrated in Figure 35.6b.

35.4.3 Polarization Angle and Oblique Angle Variations

It is noted from Figure 35.7 that in the case of normal incidence of the EM wave, the absorption rate is above 99% for all the different polarization angles (ϕ). An annular shape has maximum folding symmetry among all the possible shapes hence allowing the proposed MMA structure to have a high degree of independence against variations in ϕ angle. Practical verification of polarization independence through the waveguide measurement method is achieved by rotating the CRRs about the wave propagation axis [32]. Further, Figure 35.8 depicts the effect of oblique incidence angle (θ) variation for both TE and TM polarization cases. Under TE polarization, as depicted in Figure 35.8a, the absorption efficiency is greater than 93% up to 45° of angular variation in θ while for TM polarization, the absorption rate is observed to be higher than 98% up to 60° of angular variation in θ, as depicted in Figure 35.8b.

35.4.4 Resistive Load Variations

Figure 35.9 through Figure 35.13 depict the effect of various values of load resistors on the absorption response when installed over the ring at different locations. To investigate the designed absorber responses for the energy harvesting application, the resistive load is placed on the locations where the magnetic field and the electric field are maximal on the ring. As observed from Figure 35.6b, the magnetic field is strongly concentrated across the transverse plane of the ring, whereas the electric field is accumulated largely about the median plane as illustrated in Figure 35.6a.

As illustrated in Figures 35.9 and 35.10, initially, a single resistor is placed on the ring about the median plane (Case-I) and across the transverse

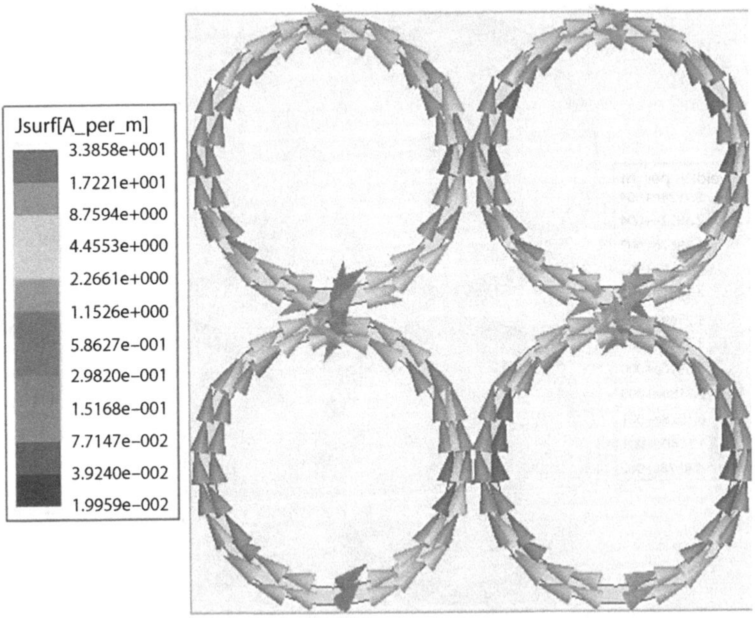

(a)

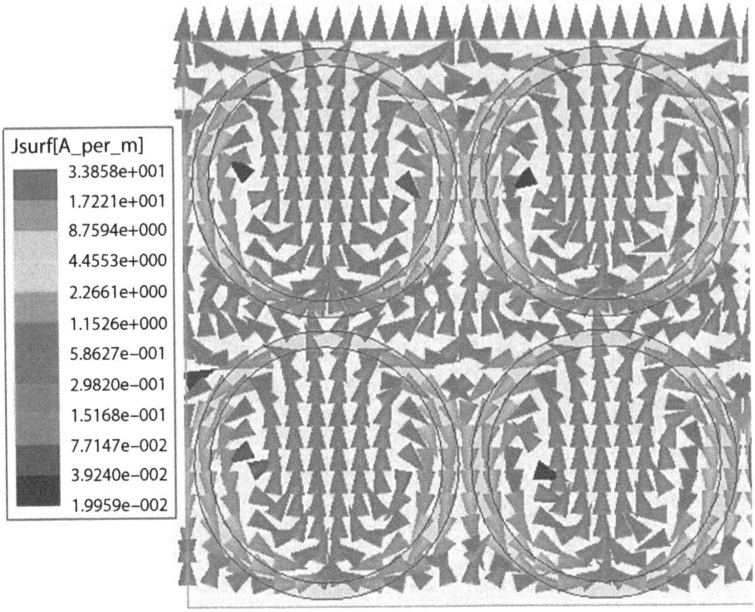

(b)

Figure 35.5 Current distributions on (a) upper and (b) lower surface of the designed absorber.

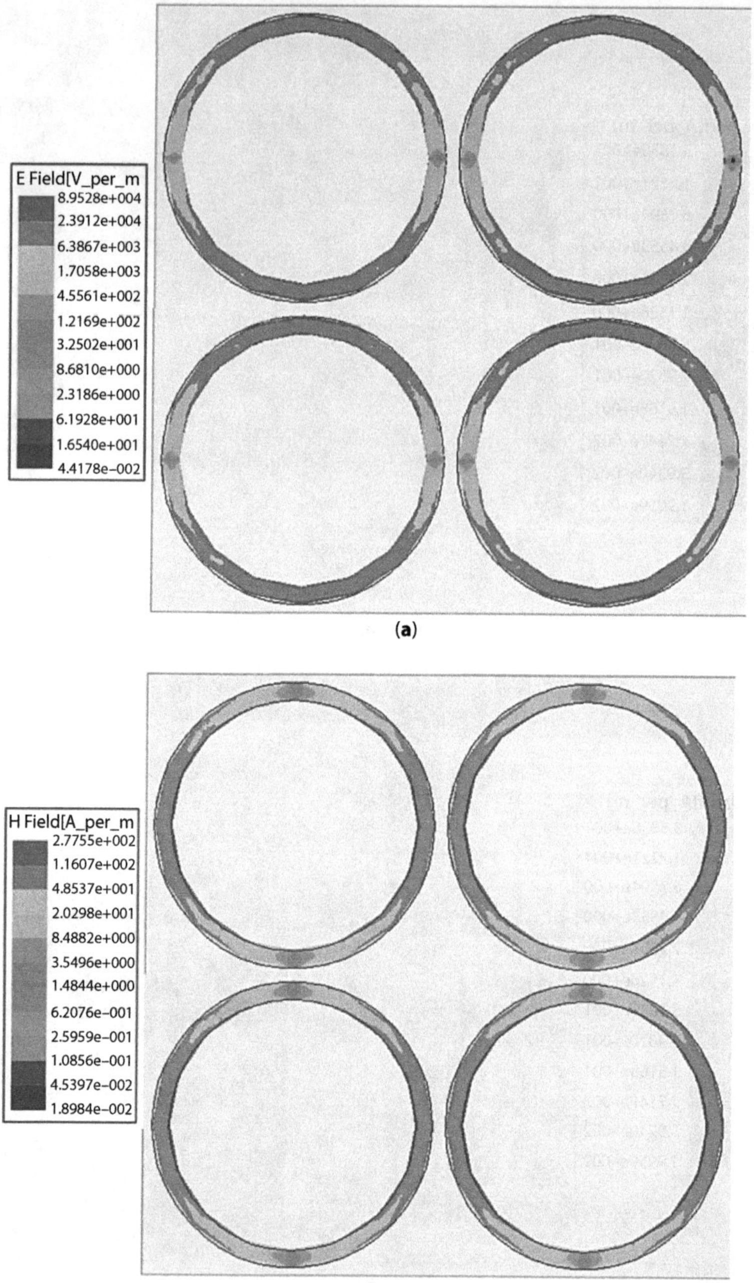

(b)

Figure 35.6 Induced field distributions (a) electric and (b) magnetic.

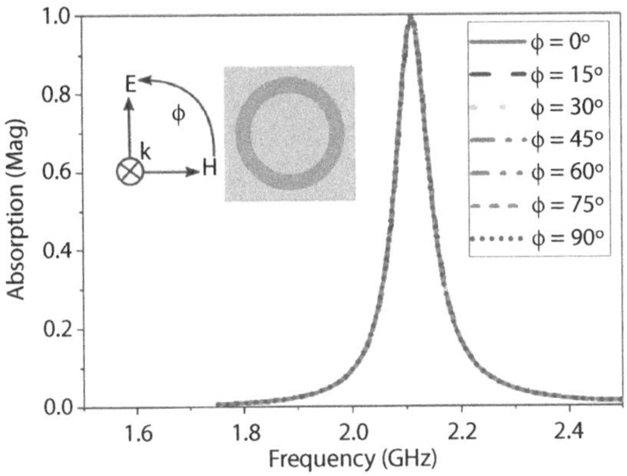

Figure 35.7 Numerical absorption results for various polarization angle (ϕ).

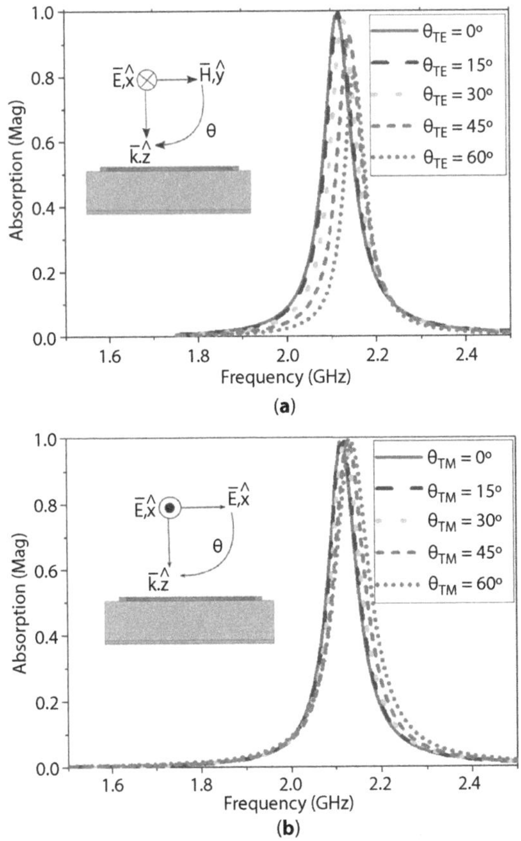

Figure 35.8 Numerical absorption response for varying θ under: (a) TE mode and (b) TM mode.

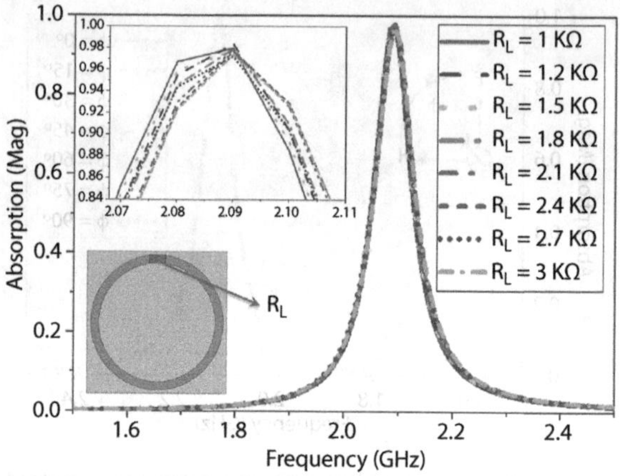

Figure 35.9 Absorption response for various single resistive loads on the median plane of the ring (Case-I).

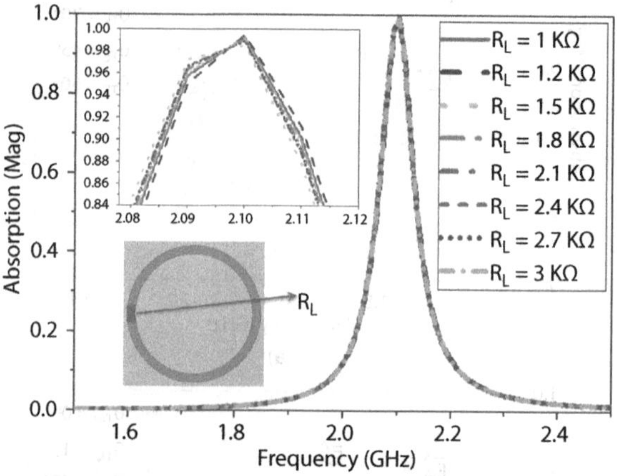

Figure 35.10 Absorption response for various single resistive loads on the transverse plane of the ring (Case-II).

plane (Case-II), respectively. It is noted that a resistive load on the median plane alters the absorption considerably in comparison to the absorption response when a resistive load is placed on the transverse plane. Maximum absorption efficiency of 98.39% is obtained at 2.09 GHz for 1.2 KΩ resistive loads in Case-I while a maximum of 99.5% absorption is attained for 1.2 KΩ resistive loads in Case-II. A single load resistor on the ring may affect the polarization of the proposed absorber. Therefore, the effect of two resistive loads on the absorption response is studied further.

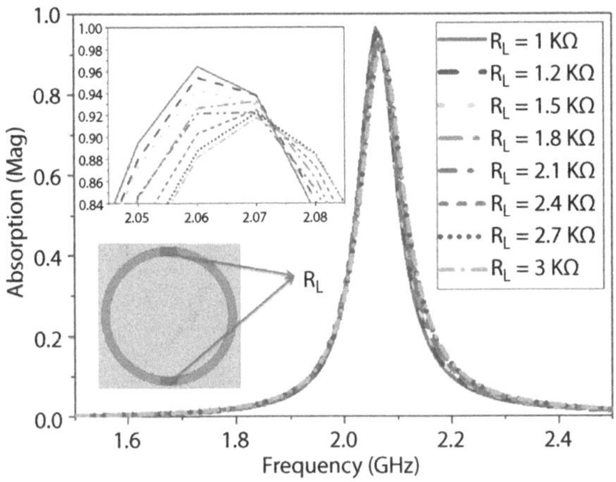

Figure 35.11 Absorption response for various dual resistive loads on the median plane of the ring (Case-III).

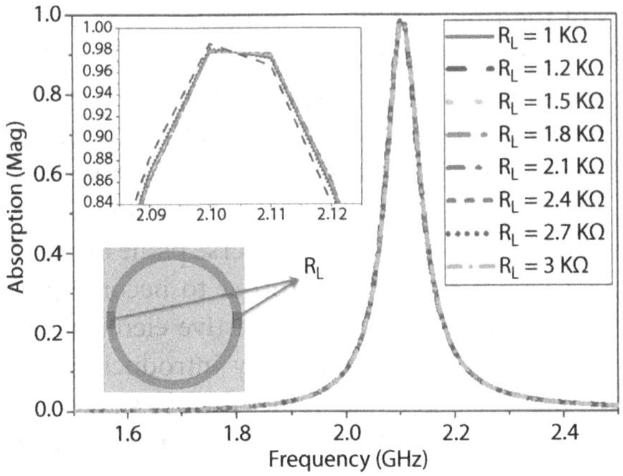

Figure 35.12 Absorption response for various dual resistive loads on the transverse plane of the ring (Case-IV).

As maximum fields are obtained on both the halves of the median plane and the transverse plane of the ring, thus, absorption response for placement of dual resistive loads on two halves of the ring is investigated. Dual resistors placed on the ring about the median plane (Case-III) and the transverse plane (Case-IV) are illustrated in Figure 35.11 and Figure 35.12, respectively. In Case-III, the maximum absorption efficiency of 96.49% is attained for 1 KΩ resistive load at 2.06 GHz, whereas in Case-IV,

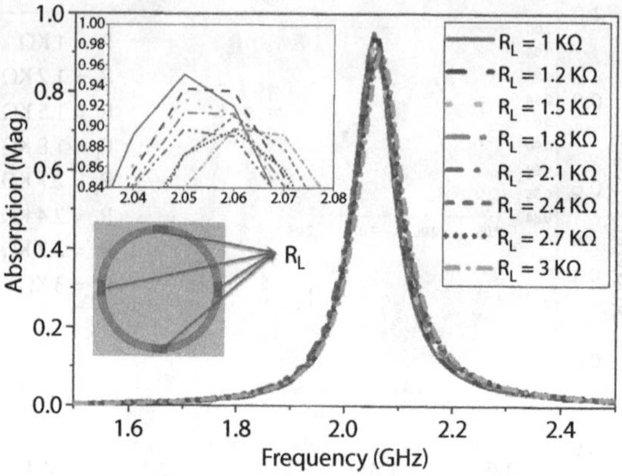

Figure 35.13 Absorption response for quad resistive loads on the median and the transverse plane of the ring (Case-V).

maximum absorption of 98.6% is achieved at 2.1 GHz for 1.2 KΩ resistive loads. Lastly, four resistive loads (Case-V) are installed on the maximum field locations of the ring as depicted in Figure 35.13. In this case, maximum absorption of 95.08% is obtained at 2.05 GHz for 1 KΩ resistive load.

Hence, to obtain maximum absorption optimum position and value of the resistive load must be determined so that maximum power transfer could take place. In the present case when the resistive loads are selected in-between 1 KΩ to 3 KΩ, the maximum absorption is obtained when a single resistive load is placed on the transverse plane of the ring. But single resistive element would cause the absorber to become polarization sensitive. Thus, it is desirable to use four resistive elements to eliminate the polarization sensitivity of the absorber by introducing symmetry in the installed load. It must be noted that in the case of a single resistor, the plane of placement (in which a resistor is placed) would negligibly change the absorption response.

35.5 Experimental Results

The proposed MMA is manufactured through a standard printed circuit board (PCB) fabrication technique and verified physically using the waveguide measurement method [33, 34]. The sample is placed inside a WR-430 standard rectangular waveguide which is covered with a metal sheet to avoid any effect of leaky waves. The waveguide along with the fabricated sample is depicted in Figure 35.14a. An Agilent N9912A vector

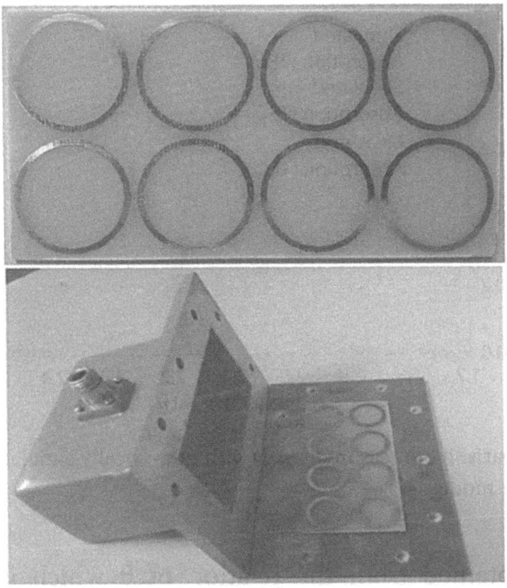

(a)

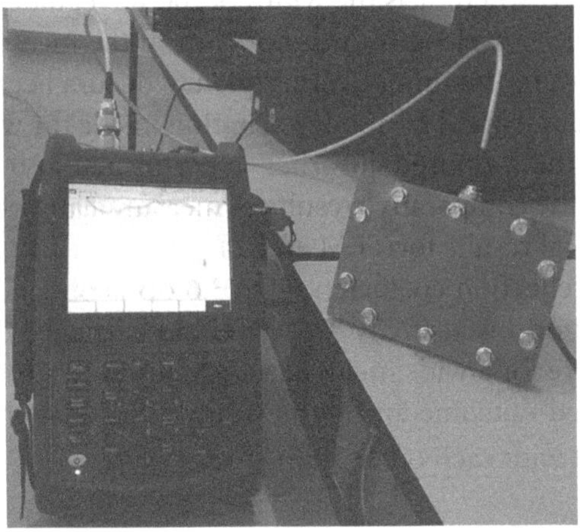

(b)

Figure 35.14 (a) fabricated proposed MMA sample and WR-430 waveguide and (b) test setup of waveguide measurement method.

network analyzer (VNA) is connected to the waveguide through Teflon coated coaxial cable (RG142) for measuring the reflection coefficient. The complete measurement setup is depicted in Figure 35.14b.

The absorber prototype, inside the waveguide, is excited by the fundamental TE_{10} mode which causes the obliqueness in the incidence angle (θ). The obliqueness of the wave inside the rectangular waveguide is 40.5°.

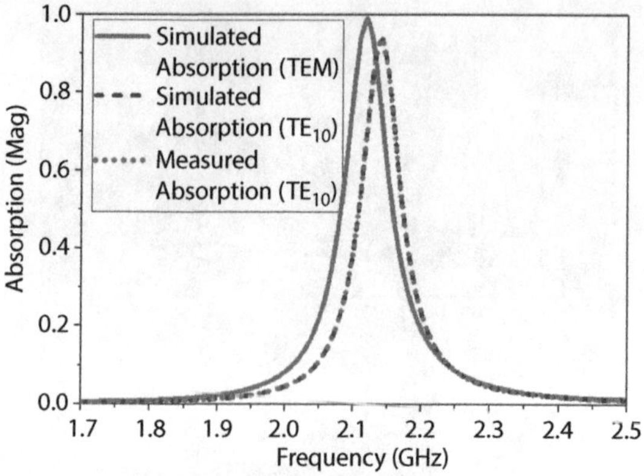

Figure 35.15 Comparison of measured result with numerically simulated results under TEM mode and TE_{10} mode.

This value is obtained from the relation of θ which is given as $\theta = \sin^{-}(\lambda_0/2a)$ [20, 34] where 'a' is the width of the waveguide's inner opening and λ_0 is the target wavelength. A minor shift in the measured absorption frequency relative to the numerical frequency occurs because of the presence of the fundamental TE_{10} mode excitation inside the waveguide. Thus, resimulation of the proposed absorber is performed at $\theta = 40.5°$. Figure 35.15 illustrates the comparative results of measured absorbance (for TE_{10}) with the numerical results for TEM and TE_{10}.

A measured reflection coefficient of −23.6 dB is obtained at 2.14 GHz. Due to polarization insensitivity and good performance of the proposed absorber structure for a wide incidence angle up to 45°, it is contemplated that the measured outcome and numerically simulated absorption result are in agreement with each other.

35.6 Conclusion

Analysis of a wide-angle and polarization-insensitive S-band metamaterial absorber (MMA) with equivalent circuit modeling is presented in this work. An annular patch shape is inscribed on low-cost metal grounded 1.6 mm thick FR4 dielectric. Periodic boundary condition under HFSS is selected to achieve near-perfect absorption of 99.64% at the frequency of 2.12 GHz. Even-mode and odd-mode analysis technique of the coupled line theory is used to obtain the values of the equivalent circuit model elements for the designed absorber. The proposed MMA is polarization-insensitive with about 99% of absorption rate and attains absorption of above 93% for TE wave and about 98% of absorption for TM wave for up to 45° and 60° of incidence oblique angle, respectively. The unit cell size and thickness of the proposed structure are less than $\lambda_0/5$ and $\lambda_0/84$, thus, achieving compact and ultrathin design. The effect of varying the number and positions of the installed resistive loads on absorption response is studied by selecting various resistive load values. The standard PCB technology is used for fabrication and the measurement is carried out via the waveguide measurement method. The measured result is in agreement with the numerically simulated outcome.

References

1. Caloz, C., Itoh, T., *Electromagnetic metamaterials: transmission line theory and microwave applications*, John Wiley & Sons, 2005.
2. Veselago, V. G., The electrodynamics of substances with simultaneously negative values of ε and μ. *Soviet Physics Uspekhi*, 10, 509, 1968.
3. Upadhyaya, T.K., Kosta, S. P., R., Jyoti, M,. Palandoken, Negative refractive index material-inspired 90-deg electrically tilted ultra wideband resonator. *Optical Engineering*, 53, 107104, 2014.
4. Upadhyaya, T.K., S.P., Kosta, R., Jyoti, M., Palandoken, Novel stacked μ-negative material-loaded antenna for satellite applications. *International Journal of Microwave and Wireless Technologies*, 8, 229, 2016.
5. Palandöken, M., Ucar, M.H., Compact metamaterial-inspired band-pass filter. *Microwave and Optical Technology Letters*, 56, 2903, 2014.
6. Schurig, D., Mock, J.J., Justice, B.J., Cummer, S.A., Pendry, J.B., Starr, A.F., Smith, D.R., Metamaterial electromagnetic cloak at microwave frequencies. *Science*, 314, 977, 2006.
7. Sabah, C., Dincer, F., Karaaslan, M., Unal, E., Akgol, O., Demirel, E., Perfect metamaterial absorber with polarization and incident angle independencies based on ring and cross-wire resonators for shielding and a sensor application. *Optics Communications*, 322, 137, 2014.
8. Land, N.I., Sajuyigbe, S., Mock, J.J., Smith, D.R., Padilla, W.J., Perfect metamaterial absorber. *Physical Review Letters*, 100, 207402, 2008.
9. Dincer, F., Akgol, O., Karaaslan, M., Unal, E., Sabah, C., Polarization angle independent perfect metamaterial absorbers for solar cell applications in the microwave, infrared, and visible regime. *Progress in Electromagnetics Research*, 144, 93, 2014.
10. Hoa, N.T.Q., Lam, P.H., Tung, P.D., Tuan, T.S., Nguyen, H., Numerical study of a wide-angle and polarization-insensitive ultrabroadband metamaterial absorber in visible and near-infrared region. *IEEE Photonics Journal*, 11, 1, 2019.

11. Luo, Z., Ji, S., Zhao, J., Dai, H., Jiang, C., Design and analysis of an ultra-thin dual-band wide-angle polarization-insensitive metamaterial absorber for C-band application. *Optik*, 243, 166785, 2021.
12. Nguyen, T.K.T., Cao, T.N., Nguyen, N.H., Le, D.T., Bui, X.K., Truong, C.L., Vu, L., Nguyen, T.Q.H., Simple Design of a Wideband and Wide-Angle Insensitive Metamaterial Absorber Using Lumped Resistors for X- and Ku-Bands. *IEEE Photonics Journal*, 13, 1, 2021.
13. Mahmud, S., Islam, S.S., Mat, K., Chowdhury, M.E., Rmili, H., Islam, M.T., Design and parametric analysis of a wide-angle polarization-insensitive metamaterial absorber with a star shape resonator for optical wavelength applications. *Results in Physics*, 18, 103259, 2020.
14. Kaur, K.P., Upadhyaya, T.K., Palandöken M., Dual-Band Polarization-Insensitive Metamaterial Inspired Microwave Absorber for LTE-Band Applications. *Progress In Electromagnetics Research C*, 77, 91, 2017.
15. Sarkhel, A., Chaudhuri, S. R., Compact Quad-Band Polarization-Insensitive Ultrathin Metamaterial Absorber with Wide Angle Stability. *IEEE Antennas and Wireless Propagation Letters*, 16, 3240, 2017.
16. Cheng, Y.Z., Cheng, Z.Z., Mao, X.S., Gong R.Z., Ultra-thin multi-band polarization-insensitive microwave metamaterial absorber based on multiple-order responses using a single resonator structure. *Materials*, 10, 1241, 2017.
17. Jiang, H., Xue, Z., Li, W., W. Ren, Multiband polarisation insensitive metamaterial absorber based on circular fractal structure. *IET Microwaves, Antennas & Propagation*, 10, 1141, 2016.
18. Bhattacharya, A., Bhattacharyya, S., Ghosh, S., Chaurasiya, D., Srivastava, K.V., An ultrathin penta-band polarization-insensitive compact metamaterial absorber for airborne radar applications. *Microwave and Optical Technology Letters*, 57, 2519, 2015.
19. Kaur, K.P., Upadhyaya, T., Dual-band Perfect Metamaterial Absorber with Polarization Independence and Wide Incidence Angle. *Indian Journal of Radio & Space Physics (IJRSP)*, 47, 42, 2020.
20. Zhai, H., Zhan, C., Li, Z., Liang, C., A triple-band ultrathin metamaterial absorber with wide-angle and polarization stability. *IEEE Antennas and Wireless Propagation Letters*, 14, 241, 2015.
21. Elsharabasy, A., Bakr, M., Deen, M.J., Wide-angle, wide-band, polarization-insensitive metamaterial absorber for thermal energy harvesting. *Scientific Reports*, 10, 1, 2020.
22. Sood, D., Tripathi, C.C., A wideband wide-angle ultrathin low profile metamaterial microwave absorber. *Microwave and Optical Technology Letters*, 58, 1131, 2016.
23. Sen, G., Islam, S.N., Banerjee, A., Das, S., Broadband Perfect Metamaterial Absorber on Thin Substrate for X-Band and Ku-Band Applications. *Progress in Electromagnetics Research C*, 73, 9, 2017.

24. Dhillon, A.S., Mittal, D., Bargota, R., Triple band ultrathin polarization insensitive metamaterial absorber for defense, explosive detection and airborne radar applications. *Microwave and Optical Technology Letters*, 61, 89, 2019.
25. Smith, D.R., Vier, D.C., Koschny, T., Soukoulis, C.M., Electromagnetic parameter retrieval from inhomogeneous metamaterials. *Physical Review E*, 71, 036617, 2005.
26. Yu, C.C., Chang, K., Transmission-line analysis of a capacitively coupled microstrip-ring resonator. *IEEE Transactions on Microwave Theory and Techniques*, 45, 2018, 1997.
27. Ghosh, S., Srivastava, K.V., An equivalent circuit model of FSS-based metamaterial absorber using coupled line theory. *IEEE Antennas and Wireless Propagation Letters*, 14, 511, 2015.
28. Garg, R., Bahl, I., Bozzi, M., *Microstrip lines and slotlines*, Artech House, 2013.
29. Bhattacharyya S., Ghosh S., Srivastava, K.V., Equivalent circuit model of an ultra-thin polarization-independent triple band metamaterial absorber. *AIP Advances*, 4, 097127, 2014.
30. Hopkins, R., Free, C., Equivalent circuit for the microstrip ring resonator suitable for broadband materials characterisation. *IET Microwaves, Antennas & Propagation*, 2, 66, 2008.
31. Garg, R., Bahl, I.J., Characteristics of coupled microstriplines. *IEEE Transactions on Microwave Theory and Techniques*, 27, 700, 1979.
32. Kaur, K.P., Upadhyaya, T.K., Wide-angle and polarisation-independent tri-band dual-layer microwave metamaterial absorber. *IET Microwaves, Antennas & Propagation*, 12, 1428, 2018.
33. Li, L., Yang Y., Liang C., A wide-angle polarization-insensitive ultra-thin metamaterial absorber with three resonant modes. *Journal of Applied Physics*, 110, 063702, 2011.
34. Zhai, H., Zhan, C., Liu, L., Zang, Y., Reconfigurable wideband metamaterial absorber with wide angle and polarisation stability. *Electronics Letters*, 51, 1624, 2015.

36

World Energy Demand

Satish R. Billewar[1]*, Gaurav Londhe[2] and Pradip Suresh Mane[3]

[1]Vivekanand Institute of Management Studies & Research, Mumbai, India
[2]Jain Deemed to be University, Bangalore, India
[3]Vasantdada Patil Pratishthan's College of Engineering and Visual Arts, Mumbai, India

Abstract

By the end of the next two decades, the developing world predicts to face the consequences of the use of fossil fuels, the production of local pollution, and the release of greenhouse gases. The USA Energy Information Administration claims that the amount of energy used by OECD countries in 2007 was equal to non-OECD countries. By 2035, the amount of energy used by OECD Countries will increase by 14%, while the amount of energy used by non-OECD countries will rise by 84%. In developing countries, China and India will expect second and third highest energy requirement in the world after the USA. Over the next 15 years, China and India will account for half of the world's total growth in energy demand. A majority of the new carbon dioxide pollution from around the world will come from China and India. According to current forecasts, India will be the third-largest oil importer in the world by 2030. China will install more new electricity generating capacity in the country between now and 2030 than the United States will have installed in its entire history.

The EU is interested in sustainability and competitiveness in the low-carbon economy. The EED guidelines reduce energy consumption in existing buildings and get as nearly zero energy buildings (NZEBs) as possible. But the developing countries have high economic inequality. Therefore, following EED guidelines for NZEB seems very difficult in developing countries. People who are relatively poor and those on the cusp of poverty will significantly impact long-term energy consumption. Therefore, modern energy sources are highly encouraged for developing countries as a replacement for traditional fuels. The developing countries are exploring new methods of integrating global concepts for rural electricity

**Corresponding author*: satish.billewar@gmail.com

Milind Shrinivas Dangate, W.S. Sampath, O.V. Gnana Swathika and P. Sanjeevikumar (eds.) Integrated Green Energy Solutions Volume 2, (275–316) © 2023 Scrivener Publishing LLC

development. There are so many international bodies working hard to analyze the trends in electrical power adoption and assess how much capital will be required to meet future demand. However, there is still a need to explore alternative solutions to problems.

Agriculture consumes a lot of energy. Numerous facets of it necessitate significant amounts of electricity and diesel, gasoline, and other fuels. Farms' energy consumption, both directly and indirectly, contributes to the total amount of energy consumed. Agriculture and food consume approximately 30% of global power. Thus, increasing agricultural energy efficiency is critical for lowering energy demand and therefore costs.

Keywords: Renewable energy, carbon emission, kinetic energy, electrification, green energy, sustainable development, green financing

36.1 Energy End Users

In this era of modernization and digitalization, we need to work on basic ideas and innovative industrial concepts of reducing efforts and developing the strategies to either construct or redevelop the plans for making humanity happy by providing several facilities around. Energy is required for delivering such facilities. To render the most feasible resources for an individual facility, we must understand the formats of energy resources.

These resources need to be highlighted and emphasized when we want to use their application in day-to-day life.

Applications of these energy patterns are discussed here as:

A. Kinetic Energy:
The energy is generated or maintained in material or medium with the help of movements of the particles/elements of the material or medium can be considered kinetic energy.

Its subcategories are widely used with the above-given figure (see Figure 36.1); this helps us elaborate the usage of the same with its application domains. The diagrammatic representation as given below in Figure 36.1.

- Electricity:
  The energy is generated with movements of electrons within the conducting material due to potential differences maintained between two different points or terminals can be termed as "Electricity".

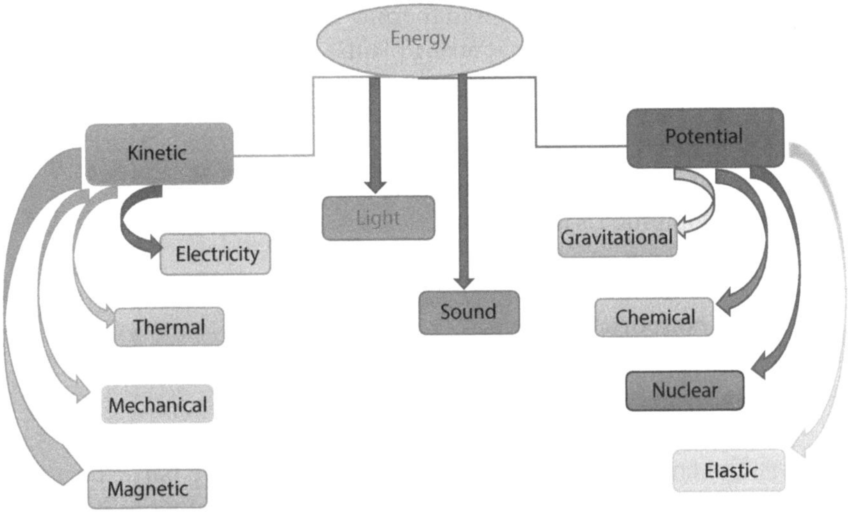

Figure 36.1 Types of energy sources.

The devices used in day-to-day life and are working with the help of such energy are termed electrical appliances.
Several users of such applications are given below for their specialized domain areas.

1. Domestic Electricity consumers
2. Industrial Consumers
3. Research Organizations
4. Agricultural requirements
5. Educational Institutions/Universities
6. Transportation
7. Digital Environment (IT applications in all sectors)

- Thermal:
  The energy generated or maintained in the material due to heat generation or dissipation is termed "Thermal Energy." This is most probably used for the initialization process as beginners for many functions, so they initiate the activity for which they are meant [3].
  This phenomenon is mostly used in several applications, as given below.

a. Electricity generation in thermal power plant
b. Mechanical manufacturing plants
c. Research Laboratories
d. Chemical plants
e. Educational Laboratories
f. Missile and rocket launcher appliances
g. Domestic Usage
h. Cooking
i. Sterilization
j. Transportation devices
k. Solar energy conversion systems

- Mechanical Energy:
  The energy which is generated due to the active participation of mechanical forces is termed mechanical energy. It is used in a plethora of applications in day-to-day life in this modern era of automatic vehicles with the help of electromechanical devices, and its usage in several instruments of different areas and its specialized applications are as given below. The diagrammatic representation as given above in Figure 36.1.

  A. Heavy Engineering devices and instruments like cranes, pokers, JCB.
  B. Hydraulic instruments and artefacts used in a plethora of applications
  C. Educational Institutions/Universities
  D. Research and Development Laboratories
  E. Mechatronics instruments and its control environments with its applications
  F. Railways
  G. Aircraft industries
  H. Construction sites of Bridges, Dams, Workshops.
  I. Steel Manufacturing
  J. Metallurgical services

- Magnetic Energy:
  Magnetic energy is the energy responsible for generating other sorts of energies only with its presence in and around;

the age of electricity within the generators is one of the examples of its kind [10].

There are many applications in different devices or instruments, which we utilize in day-to-day life for multiple platforms using magnetism as a primary energy source.

A few of the applications are mentioned here:

i. Generators for electricity generation
ii. Physics and allied Laboratories
iii. Industrial applications with electromagnetic effect
iv. Electronic devices and instruments with a magnetic effect like speakers and mic
v. Electric Fans and similar applications
vi. Research and development departments
vii. Manufacturing plants

- Potential Energy:
  The energy which possesses only with its existence is relatively higher or lower for the other elements in the vicinity. It can be considered as the existence of potential energy. The difference in their potentials between those two elements can be regarded as possible differences. An electrical current can flow between those two terminals if a conductive medium is available.
- Gravitational Energy:
  This type of energy is a form of the potential energy power of a massive object related to another enormous thing due to gravitational force. We can consider they are under the gravity of the higher vast object.
  There are several applications of gravitational energy as under:
  a. Research and development Laboratories
  b. Geothermal experimentation in geotechnical laboratories
  c. Machine design concepts
  d. All sorts of object manufacturing in all scale industries.

- Chemical Energy:
  The energy that comes out of any chemical reaction can be considered chemical energy. Again, there are a plethora of elements in the environment where we can observe such energy in several natural entities around.

Natural crops, fruits render sufficient energy, which is utilized by animals and can survive.
Some of the application areas are mentioned as under:

I. Chemical Industries
II. Pharmaceutical Industries
III. Plastic and Rubber Industries
IV. Liqueur Industries
V. Domestic usage like cooking, washing, bathing, cleaning, etc.

- Nuclear Energy:
  The energy released from nuclear reactions within any material, generally radioactive isotopes, is most helpful for such responses.
  The area of applications is shared herewith as follows:
  a. Make the hydrogen electrolytic as charge carriers to be used as future fuel as electricity.
  b. High-temperature reactors for thermochemical production
  c. Desalination process
  d. For Powerful submarine propulsion
  e. Vessels that need to be at Sea for longer duration.

- Elastic energy:
  The amount of elastic energy is coming out of any material is dependent on the elasticity of that material.

- Solar Energy:
  The energy available in photons in each quantum of energy is termed solar energy.
  These bundles of energies are emitted in specific units termed the quantum of energies and are stored in solar panels to supply energy after particular time durations.
  Application areas are:
  1. An alternative solution for fuel for missiles and satellite Vehicles
  2. An alternative solution for Automotive
  3. An alternate for electricity and mechanical energies
  4. An alternate for inverter batteries

36.2 Rural Electrification

If we cannot find power supply in rural areas in this digital era, we cannot say that we are in a digitalized environment. Until we get the power resources, the public will not be able to take advantage of the other essential resources like water at their places in enough quantity and even cannot make use of available resources food for everyone due to unavailability of electricity and hence no service of submersible pumps for water supply at farms and no utilization of machinery at their farms like crop

Figure 36.2 Rural electrification.

cutting and binding etc. The diagrammatic representation as given above in Figure 36.2.

There are several activities which require electricity power, such as

1. Temperature control for medical and domestic purposes
2. Cattle management
3. Farming
4. Cooking
5. Medical Facilities
6. Running water
7. Entertainment
8. Light
9. Dairy Management
10. Education

The several activities where we need to put a light are as given below.

The areas discussed above are shown in the Figure 36.2. Electrification is mandatory in all rural areas where people need to be educated on how to make use of the resources due to this facility and make their life with basic life requirements like food, education, house, and better clothing. The administering bodies also need to be employed for such monitoring and control of each element of it.

36.3 Residential and Non-Residential Buildings

Residential buildings and their power requirements:

In today's era, we have seen the impact of the COVID-19 pandemic. What is the condition of the market? The residents of metro cities and semi-urban areas rely entirely on electricity as breathing support or a lifeline.

By any chance, the electricity supply stops longer duration indefinitely. Most of the activities will be non-functional or inactive; people will be helpless and will be without income, due to which the entire scenario will be changed [5].

Due to a jobless environment, the market will collapse, which will lead to more joblessness, and total GDP will be down in a moment.

Medical facilities will be down. The mortality rate will be higher.

Electricity plays a vital role as a driving force to overcome these issues and keep the flow of human activities in live mode.

The need for electricity management must be considered in residential and non-residential buildings and must be noted in urban and semi-urban areas.

The electricity need of Residential buildings can be identified and can be monitored and controlled. Awareness of energy conservation also must be taught to the consumers subsequently.

36.3.1 Urban and Semi-Urban Zones Power Requirement

The areas where the electricity is most commonly used are identified as given below.

i. Light arrangements
ii. Domestic appliances
iii. Car Lifts (if any)
iv. Lift and its backup inverters
v. External amenities like Gym, Swimming pool, Sauna bath, Community Hall, Play area. etc.
vi. Security appliances like CCTV, intercom Connections, and Fire extinguisher arrangements, if any.
vii. Emergency Light arrangements, etc. The diagrammatic representation as given below in Figure 36.3.

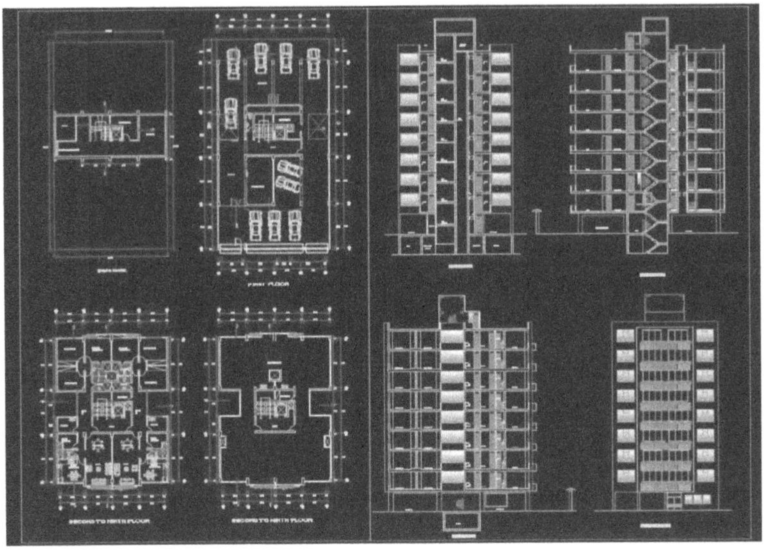

Figure 36.3 Urban residential buildings.

36.3.2 Rural Residential Requirements

According to a diagrammatic representation, the need for electrification in a rural region is determined by the residents' everyday needs, such as the needs for domestic electrical appliances used in residential zones.

The building can be used for a residential zone, as shown in Figure 36.4 below.

a) Agricultural (Submersible Pumps), water sprinklers
b) Drying and cutting machines for crops, home-based small-scale industries
c) Domestic appliances
d) Education
e) Horticulture equipment, etc.

36.3.3 Non Residential Buildings

The non-residential buildings and their power requirements are significant since their energy requirements are higher, which helps in global growth due to industrial expansion in diverse areas [7].

Figure 36.4 Rural residential building.

Here, each medium- to large-scale industry plays a vital role in revenue generation and is essential for GDP rate and similar activities. Hence the industrial appliances and their daily power requirements are higher. The diagrammatic representation as given below in Figures 36.5 and 36.6 respectively the classical and modern non-residential buildings can be seen.

Rural and Urban Non-residential buildings requirements are basically for their office use industrial purposes; hence, it will be based on their actual activities. If these are manufacturing plants like shop floors, then the power requirements are higher, or if these are corporate offices, then their needs are much lower.

Figure 36.5 Classical non-residential buildings.

Figure 36.6 Modern non-residential building.

36.4 Industry

We all tend to forget about the Industrial Revolution and the numerous benefits it brought to countries worldwide, including increased economic prosperity and improved human health. Industry's reputation as a significant source of environmental pollution has risen steadily over the last several decades. According to reports, changes in biodiversity are frequently observed due to the harmful effects of pollutants or changes in habitat quality and food webs.

The European Union had guidelines for the Water Framework Directive (WFD) in 2015. It became clear that industrial pollutants were wreaking havoc on the environment, people became motivated to enact policies and regulations to safeguard both human and environmental health from pollution's effects. Changes in legislation and the 1956 Clean Air Act have aided in alleviating respiratory illness by reducing the occurrence of illness and reducing air pollution, allowing lichens to grow in previously uninhabitable regions. Similar legislative measures have also been enacted, although the focus has been primarily on human health protection. The diagrammatic representation as given above in Figure 36.7 the Water Framework Directives can be used.

However, in recent years, industrialised countries have placed a greater emphasis on the ecological health of the human environment. Another strong point of the EU's Water Framework Directive (WFD) is that a healthy ecological state is considered in addition to the chemical environment [10].

However, in many countries worldwide, legislative implementation is limited or insufficient, leaving the receiving environment vulnerable to environmental risks. Additionally, ongoing technological advancements have resulted in the development of novel contaminants, such as those produced by nanotechnology and pharmaceuticals, and a near-complete

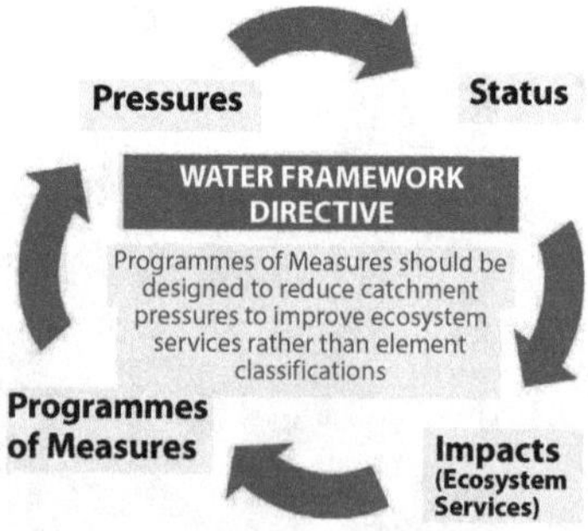

Figure 36.7 Water Framework Directive (WFD).

lack of knowledge about the long-term health effects of these contaminants on organisms. Due to the long history of industrial activity globally, harmful pollutants, including persistent pollutants, have been dispersed throughout the environment. Thus, we must recognise the contributions of these historical sources when examining the consequences of industry. Contamination from the past can still harm ecosystems today, and several additional examples include floodplain deposits, in-river sediments, and historic mine workings.

36.4.1 Industrialization, the Environment, and Pollution

All countries face interconnected environmental, economic, and social challenges. Improving human well-being, such as infant mortality reduction, food security, and life expectancy, is generally due to improved economic conditions. However, economic growth increases pressure on the environment, resulting in increased pollution, such as habitat loss and contamination of drinking water. If left unchecked, these will result in environmental degradation, reducing human well-being. One-third of the ecosystem services assessed in the Millennium Ecosystem Assessment are degraded or used unsustainable, and the poor account for two-thirds of those affected.

Along with negative mental and physical health consequences, ecosystem degradation reduces the availability and costs of raw materials, harming business and industry by altering the regulatory, economic, and social environment in which they operate. On the one hand, wealthy countries' citizens desire healthy environments, but they also require the resources generated by heavy industry to maintain their standard of living. This is the standard of living to which a large number of developing and increasingly industrialised countries aspire. This section discusses global trade in industrial products and ecosystem services.

However, it would be imprudent to prescribe precisely what flag economists should collect or what rank they should rightfully aspire to. As a result, I will abstain largely and confine myself to what I perceive to be self-evident and uncontroversial, thereby risking to breach the other pole of possible, tedious, and enforceable critical indignation. However, both ecological and economic facts are widely recognised, and what makes this even stranger is that shipments are not employed as treatments for a more sustainable and prosperous economy.

Obtaining sustainable development without also achieving sustainable industrial growth is a challenging task to do simultaneously. No country has ever attained a developed status without first investing in and expanding its industrial sector, which is valid for all countries. The development of the industrial sector is vital. Consequently, it serves as a vehicle for extensive structural change and job creation and income generation. It is also a mechanism by which ordinary people can enhance their quality of life and combat poverty.

Nearly half of all domestic consumption is sourced through industrial processes, which account for around one-fifth of worldwide revenue. Manufacturing is primarily concerned with supplying the requirements of humans in the areas of food, transportation, communication, housing, health, and recreational activities. The way people live and work in their communities has changed dramatically since the beginning of the Industrial Revolution due to waves of technological progress.

Figure 36.8 describes the Environmental heterogeneity. Environmental heterogeneity can result in biological variety across habitats, but it can also occur in contaminated environments. When naturally occurring outcrops of metal ores are present, species (particularly plants, lichens, and microbes) adapt to the presence of highly metalliferous soils. These communities are sporadic since these resources have been depleted, but they can occasionally proliferate in abandoned locations. Culinarian grasslands, for example, such as those found in the Peak District, are particularly significant. However, recent environmental regulations, industrial decline, and the possibility of altered ecological conditions cast doubt on these regions' biodiversity. In this case, it could result in the removal or remediation of metal sources from the environment, such as spoil heaps or mine runoff. Modern industrial ruins have also been shown to be excellent habitats for endangered species. A variety of contaminants has contaminated numerous sections of so-called brownfield land as the economies of

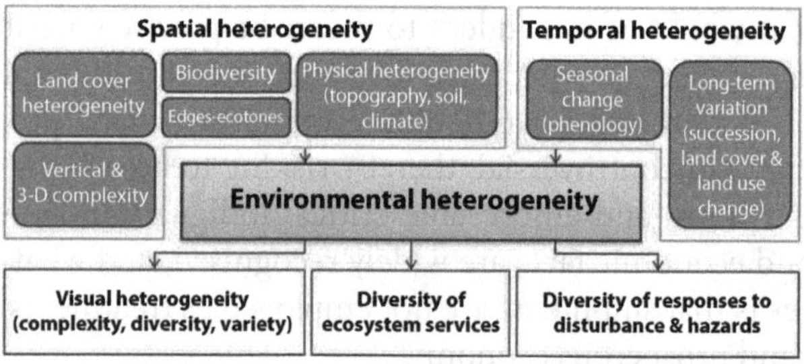

Figure 36.8 Environmental heterogeneity.

many nations throughout the industrialized world have shifted away from manufacturing and service-based economies. Due to the inaccessibility of these locations to humans, a variety of species have begun to establish themselves [8].

Researchers have examined the impact of industrial pollutants on various creatures and communities, and significant research has been conducted to ascertain the potential effects of particular pollutants on specific organisms or populations. A disadvantage of this strategy is that the resulting perspective will be limited to the primary aspect of the subject. It is becoming increasingly clear that contaminated areas contribute to biodiversity rather than merely degrade ecosystems. The organisms that thrive in such environments are highly genetically adapted and may prove invaluable in assisting with the clean-up of other contaminated areas. Another disadvantage of ecological monitoring is that it may result in measurement confusion, resulting in suboptimal remediation efforts.

Additionally, researchers' decision to separate remediation and ecotoxicology research may result in inappropriate technologies and suboptimal remediation outcomes. We chose to focus on various industries for this book to emphasize these central concepts. Industrial processes emit carbon dioxide and other greenhouse gases. However, we have chosen not to address this issue in this volume. This subject is sufficiently complex that it warrants its substantial volume. This statement is widely accepted as accurate because several of the key mechanisms underlying the impact on ecological systems and the time required for restoration are critical for understanding and forecasting climate change reactions.

Renewable energy employed around 8.1 million people worldwide in 2016. By 2020, firms, governments and households committed US$513 billion to decarbonization efforts, including the infrastructure for solar, wind, energy, carbon capture, storage and recharge. The creation of such jobs in the United States was more effective than coal or oil. Jobs in the US sector increased by 6% in 2016, whereas jobs in the oil and gas sector declined 18%. In the past, the cost of production of renewable energy has decreased significantly, with 62 per cent of the total output of renewable energy costs below the cheapest fossil fuel options by 2020. In particular, the widespread use of renewable energy has reduced the cost of generating solar energy. Levelled energy cost (LCOE) refers to the average current net electricity generation costs over the lifetime of a generating plant. In 2020, renewable energy sources will, for the first time, overtake fossil fuels as the main source of electricity in the European Union [9].

Renewable energy sources, excluding large hydroelectric plants, accounted for most new capacity additions in 2015. (There are 134 GW, accounting for 54% of the total.) The total renewable energy capacity amounted to 72 GW of wind and 56 GW of solar photovoltaics, both record-breaking and significantly higher than in 2014 (49 GW and 45 GW each). Solar accounted for 56% of total new investments, whereas 38% of winds. The diagrammatic representation as given below in Figure 36.9 helps us to understand the renewable energy sources.

Global wind energy capacity grew by 16% to 369.553 megawatts in 2014 (MW). Year after year, wind energy production rises rapidly. One of the most extensive renewable energy programmes globally is being implemented in Brazil. It includes the production of sugar cane ethanol fuel, which accounts for 18% of all automobile fuel in the country. Investments in renewable energy totalled $279.8 trillion in 2014, with China accounting for $126.6 trillion or 45% of the worldwide investment, the US $40.5 billion and Europe $40.9 trillion. The high level of liability mitigation due to greenhouse gas (GHG) emitters is responsible for the damage caused by GHG emissions.

Over the period 2010 to 2019, investment in renewable energy, which does not have a significant source of power, totalled $2.7 billion. The top five countries are China, which contributes $818 billion, the United States ($392.3 billion), Japan ($210.9 billion), Germany ($183.4 billion), and the United Kingdom ($126.5 billion). The technological advances and the advantages of mass production and market competition make renewable energy technologies more affordable. According to the IRENA 2018 report, energy costs from renewable sources are dropping rapidly and are being matched or exceeded by 2020 by non-renewable sources such as

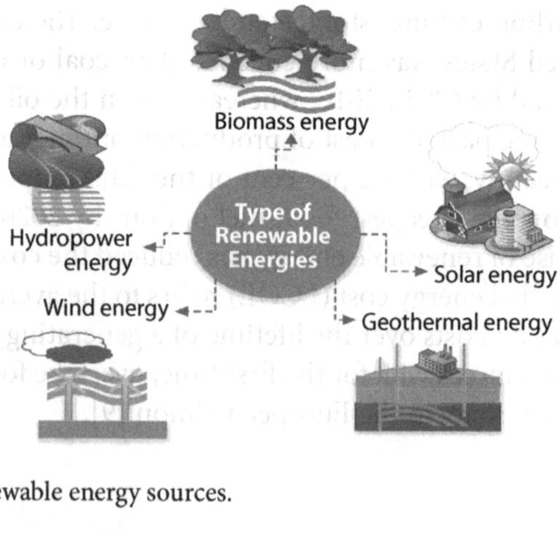

Figure 36.9 Renewable energy sources.

fossil fuels. Recent studies found that the cost of solar energy has fallen by 73% since 2010, and the cost of wind energy onshore has declined by 23%.

Current projections for future renewable energy costs are, however, inconsistent. The EIA forecasts that by 2020, renewables will account for nearly two-thirds of net energy supply additions. Hydroelectricity and geothermal energy are the most cost-effective ways to generate electricity. According to the models, recanting wind and solar energy variations on inefficiencies in fossil fuel plants resulted in additional costs of "between $0.47 and $1.28 per MWh". However, fuel savings are a maximum of $7 billion, which implies that additional costs represent no more than 2% of economic gains. The WWF debates conducted in 2014 identified significant market achievements by renewable energy companies. These principles included price competition, long-term provision of fixed prices, access to financing vehicles by third parties, and cooperation. According to statistics from the United Kingdom in September 2020, "renewable energy demand ranges from less than 4% to more than 20% for other end users".

In the early 1970s, the Asia Pacific region accounted for more than half of the area's population, with an average age of 48. Just 40% of the adult population has completed secondary education. In addition to increased resource use, pollution and waste have increased significantly, which has negatively influenced the region's environment and overall quality of life in recent years. The extensive greening of the industry is essential for the achievement of sustainable development.

It has become imperative to green the industrial sector to maintain economic competitiveness and long-term growth. Growing efficiency can be a substantial competitive advantage because resource inputs account for a significant amount of an industry's manufacturing expenses. The greening of the industry also contributes to poverty reduction by increasing safety, health, and job creation and lower costs through higher productivity and increased efficiency.

Technologies and operating practices that are old and inefficient are currently being employed in various industries throughout developing countries, and we must take action to rectify this. Furthermore, producers and consumers have adopted production and consumption patterns incompatible with the planet's existing resources, assimilative emissions capacity, and rapidly expanding population, among other factors. These concerns are essential for sustainable development, necessitating appropriate concepts and policies to address them effectively. It is a significant source of anxiety for the society that the link between natural resource consumption and economic growth is breaking down.

It has historically been challenging to make significant headway on decoupling emissions from economic growth. While most countries have witnessed a decrease in GDP-related emissions, absolute emissions have continued to climb in recent years. Long-term development of industry and expansion of industrial production in developing countries would necessitate this investment. In a nutshell, measures fostering the transition to a green economy must encourage industry's ability to "create more with less" or "produce more with less."

As the world moves toward a green economy, the economic landscape is transforming, and those who make the change swiftly are likely to enjoy a considerable competitive edge. It is impossible to achieve a green economy solely through market forces; instead, governments must create appropriate incentives, disincentives, and laws that control specific types of production to achieve this goal. There has been a resurgence of industrial strategies as a critical component of development policy debates around the world, particularly in this period of globalisation, growing environmental challenges around the world, and the onslaught of worldwide economic crises, including those involving the financial system, food security, climate change, and energy. There is no way for any of these concerns to be resolved on their own.

Sustainable green growth is considered a new industrial policy that will aid in the transition away from historical growth patterns and toward low-carbon development. All of these are directly related to the transmission of knowledge, innovation policy, competitiveness, and the facilitation of structural changes required to improve greener industrial and production processes. Industrial policy can aid in the transformation of the global green industrial structure and the transition to a more environmentally friendly economy. The organisation can assume a leading role in information and technology facilitation both nationally and internationally if it chooses to do so.

36.4.2 Green Industry Initiative

The Asia Pacific Region countries have already begun developing policies to promote industrial greening and sustainable development. Certain countries, such as Japan and Thailand, have chosen to focus on research and development and innovation, while others have committed to assisting the industry in improving its manufacturing processes. Eco-taxation, education, eco-labelling, green investment, and green procurement are all addressed in this document. G-77 governments can adjust market demand and supply balances favouring green firms by enacting economic instruments (EIs) or market-based instruments (MBIs), such as pollution levies, that encourage green company development.

Environment-friendly industries are making significant contributions to eradicating poverty and other social objectives through the development of environmentally friendly enterprises. These programmes build on the experience of the United Nations Industrial Development Organization (UNIDO) in promoting several economic and social goals, including the betterment of the lives of the urban poor. They can achieve the goal through the use of green industries in many ways.

By highlighting the beneficial role that industry plays in the green economy and sustainable development, the United Nations Industrial Development Organization (UNIDO), as the United Nations Specialized Agency for Industrial Development, contributes to the growth of green economies. The organisation founded the Green Industry Initiative in 2009 to recognise the necessity of incorporating best practices into policy reforms and raising awareness of the industry's role in the transition to a low-carbon future. Commitments and actions are being taken to reduce the adverse environmental effects of industrial processes and products through more efficient resource use; transformation of industrial energy systems to more sustainable energy sources through the expansion of renewable energy sources; phase-out of toxic substances; and improvement of industrial workplace health and safety. The growth of green industries will encourage the development and establishment of businesses that produce items and provide beneficial services to the environment.

In addition, the emergence of new firms that supply ecosystem goods and services, such as current waste management services that include collection and recycling, creates new job opportunities for the unemployed and underemployed in metropolitan areas. Greening industries results in reduced exposure to potentially harmful substances for factory workers, the surrounding communities, and consumers. Thus, healthier workplaces and communities both mitigate and promote disease and poor health conditions. The implementation of green industry initiatives requires capacity building at all organisational levels. While capacity-building focuses on the business environment and entrepreneurial skills, it also has spillover effects, including employee empowerment, increased employability, and spin-off business development.

Because of these strategies, sectors and countries will be able to estimate future energy and resource requirements and will be able to pay greater attention to the environmental effects of future structural change. Steel and iron production, nonferrous metals production, pulp and paper production, chemicals production, petrochemical production, and non-metallic mineral products are only a few examples of high-energy industrial industries. These examples show that there is significant room to reduce energy intensity.

It is also possible to use industrial politics to foresee and control worldwide trends in restructuring and diversification, not just in increased productivity but also in the direction of more environmentally friendly industries. The final phase is to establish a green industrial growth policy to boost the competitiveness of developing countries in producing environment-friendly products and services and take advantage of new prospects for trade and employment.

36.5 Transport

Transport is the most energy-intensive and CO2-emitting industry. Global energy consumption in transport rose by between 2% and 7% per year between 1971 and 2016, in line with global economic growth. Transport consumes around 19% of global energy and contributes close to a quarter of global CO2 emissions.

36.5.1 The United Nations Environment Programme (UNEP)

UNEP reports that the emission of black carbon is health hazardous as it is part of particulate material. Road crashes, air pollution, physical inactivity, family time, and the risk of increased fuel prices are all examples of social transport costs. Many of these negative consequences affect social groups that are less likely to own and operate vehicles disproportionately. Congestion costs money because the delivery of goods and services is wasted time and slowed [6].

Traditional transportation planning focuses on increasing mobility while neglecting broader consequences, particularly for vehicles. Sustainable transport was the logical extension of sustainable development, referring to transportation modes and scheduling systems compatible with broader sustainability concerns.

Sustainability includes more than the reduction of operational efficiency and emissions. An assessment of the life cycle takes into account production, use and disposal stages. A cradle-to-cradle approach is more crucial than just a factor like energy efficiency.

Energy is consumed and integrated into transportation infrastructures such as roads, bridges and railways during manufacturing and vehicle operation. Transportation systems account for 23% of global energy-related GHG emissions from road vehicles are the primary emitters of greenhouse gases. At present, petroleum accounts for 95% of transport energy [11].

36.5.2 The Initiatives of Countries

Green vehicles are designed to have less environmental impact than comparable standard vehicles, but this may not be the case if a vehicle's environmental impact is assessed over its entire lifetime. Depending on the vehicle's embodiment of energy and the source of electricity, electric vehicle technology can reduce transport emissions of CO2. In countries that rely heavily upon coal for generating electricity, there is little or no climate benefit in adapting to electric vehicles. For example, in the United Kingdom, the Nissan Leaf produced 1/3 of the greenhouse gases generated in 2019 through an average internal combustion vehicle.

Gumi, South Korea, operates a 24-kilometre roundabout. A bus is fitted to 100 kW (136 hp) of power at a maximum transmission rate of 85% while maintaining an air gap of 17 cm between the motor vehicle's underground and road surface area. Hybrid vehicles combining an internal combustion engine with an electrical engine are already widely used for higher fuel efficiency than traditional combustion engines. Natural gas remains a fossil fuel with significant emissions and is less promising [2].

During 2007, Brazil met 17% of its bioethanol transportation requirements. However, the Organization for Economic Cooperation and Development (OECD) cautioned that Brazil's success with biofuels (first generation) was due to the country's unique circumstances.

"Green transport" is a phrase that is frequently heard in the context of products that have not been proven to make a significant positive contribution to environmental sustainability. Such claims may be subject to legal challenges. The Norwegian consumer ombudsman singled out automotive companies that make claims about their vehicles being "green, clean, or ecofriendly". Manufacturers will be fined if they do not remove the words from their products.

Globally, biofuels from the first generation are not expected to be significantly higher in cost or to have a more significant impact than energy-efficiency measures on greenhouse gas emissions. However, later generations of biofuels (second to fourth generation), which do not contribute to deforestation or address food vs. fuels, have significant environmental benefits. It is produced without plants (as opposed to electricity generation) and emits minimal pollution when burned.

Researchers in developing countries, like Uganda, have looked at the factors that may influence motorcycle taxis' choice for passengers (Bodaboda). The results show that the age, gender and cycle ability of a person on average predict that they are prepared to switch to a more sustainable mode of transport. Some improvements to the transportation system that reduce perceived cycling risks were also identified as the most significant changes in cycling use.

The cities and countries that have made the most significant investments in automobile transportation systems have emerged as the most environmentally friendly fossil fuel consumption per capita. The viability of automobile engineering from a social and economic standpoint has also been called into question. In the United States, residents of large cities travel longer distances and more frequently in cars. In contrast, residents of traditional urban areas travel similar distances but rely more on walking, cycling, and public transportation.

During a meeting of the European Commission on September 30, 2009, the Sustainable Urban Mobility Action Plan was adopted. In 2007, 72% of Europeans lived in urban areas that were important for economic growth and employment. Urban areas require efficient transportation systems to maintain their economies and the general well-being of their citizens.

36.5.3 Sustainable Development Goals (SDGs)

Sustainable Development Goals (SDGs) support sustainable transport. Sustainable transport began as a grass-roots movement but became a well-known urban, national and international movement. In recent years, the trend has evolved into an environmental movement, while social equity and fairness, particularly ensuring access and services to low-income and disabled communities, including the rapidly growing population of older persons, have been emphasised. Many who are most at risk from noise, pollution and car safety are not owners or are unable to drive cars, as well as people who are severely burdened by car ownership.

Sustainable transport encompasses all social, environmental and climate-intelligent modes of transportation. Components of sustainability assessments include specific road, water and air transport vehicles, energy sources, and transport infrastructure. The evaluation also takes into account the operations of transportation and logistics. The efficiency and effectiveness of the transportation system and its environmental and climate impact all play a role in determining its long-term viability.

The purpose of transportation systems is to connect people on a social and economic level. People are quick to take advantage of mobility opportunities, with low-income households opting for low-carbon options in particular. It is estimated that transportation systems have significant environmental consequences. The direct combustion of fossil fuels was responsible for nearly all of the emissions. In comparison to any other energy sector, transportation emits greenhouse gases at a faster rate. A significant portion of local air pollution and smog is due to vehicle traffic on the roads themselves.

Although automotive travel has steadily increased throughout the 20th century, trends have become more complex since 2000. Since 2003, increases in oil prices have been due to a decline in private vehicle fuel consumption per capita in the United States, the UK, and Australia.

Many European countries are introducing financial incentives to encourage the use of more sustainable transport modes. The European Cyclists Federation has prepared a paper with an incomplete overview, focusing on daily cycling for transport. For many years, UK employers have provided employees with financial incentives. The employee rents or borrows the employer's motorcycle. You can also receive assistance. The system benefits employees by saving money and encouraging them to incorporate exercise into their day-to-day routine. Employers can anticipate tax savings, reduced sick leave and lower car park competition. Since 2010, people not driving to work in Iceland (Samgöngugreislur) have received a monthly lump sum.

It is anticipated that governments and businesses will contribute equally to this cooperative demonstration project, alongside developing country governments, multi-national development banks, and existing carbon financing channels, to cover the higher costs of carbon capture and storage in this demonstration project. This will be analogous to China's multi-stakeholder Greentech model, which involves numerous stakeholders collaborating on green technologies. A combination of government, corporate sector, and charitable organisations would contribute to establishing the various centres. Initially, they would like to assist in developing national mitigation strategies that include a varied range of stakeholders both within and outside of the geographical region under consideration. The diagrammatic representation as given below in Figure 36.10 helps us to put a light on Green financing in Asia Pacific.

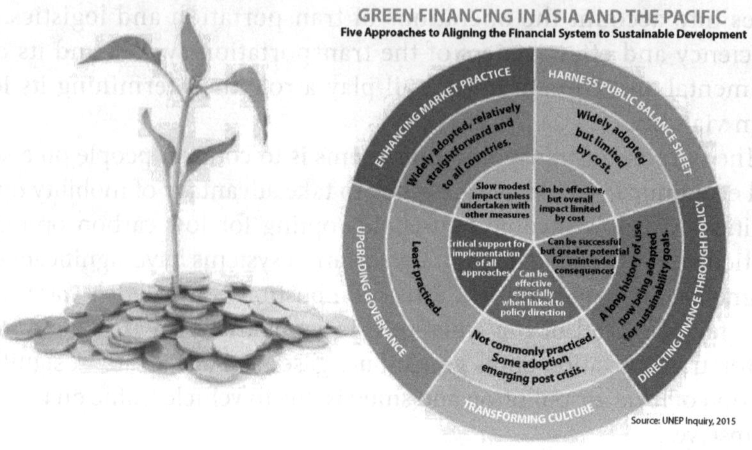

Figure 36.10 Asia Pacific Green Financing.

A strong sense of trust exists among investors in multilateral and bilateral development finance institutions, who believe they provide a secure and effective lending option for transactional purposes. Even though private investors may occasionally make significant investments in new markets and asset classes, they almost always do it with the support of a multilateral development bank (MDB).

Investment in energy systems in developing countries using a public-private strategy that combines public financing support with the regulatory capacity building has the potential to dramatically enhance investment flows in low-carbon energy systems when compared to international aid expenditures. According to a simple model, up to USD 10 billion can be generated in each region during three years using a fundamental approach (MDB Low-Carbon Challenge Funds) [6].

In the second, more ambitious approach, we anticipate that will raise capital in the range of $50 to $75 billion per region over the next three years and that the funds will be available to begin operations before the start of the second commitment period, which is scheduled to start in early 2013.

MDB-Multilateral and bilateral development finance institutions would sell proprietary rights to a number of their public finance mechanisms in exchange for a regional project package. The world's most successful (or regionally successful) management and management consulting firms would compete for the contracts, describing how they intend to use the mechanisms available to establish or enhance a new fund (or boost the flow of already existing capital). Credit support packages for these low-carbon infrastructure funds would improve their projects' risk/reward ratios.

The idea is to establish regional cornerstone funds modelled after the US Overseas Private Investment Corporation, administered by institutions established and managed in this manner. As a result, they could recruit large financial institutions and strategic partners through acquiring equity funding from a wide range of institutional investors, government and philanthropic sponsors. In addition, significant global and regional investment firms could form low-carbon energy funds, clean infrastructure funds, low-carbon construction and green technology funds. These companies are seeking the extra US $4 billion of capital from a broader universe of emerging-market investors who have made their first investment in global emerging markets to further expand their network of secondary institutional investors. Since most of the funds' project investments are infrastructure-related, they can achieve a debt-to-equity ratio of 66% for their portfolios through debt capital markets and banks. The sector now produces a considerable percentage of global CO2 emissions, and in the coming decades, the built environment is likely to rise further. According to McKinsey forecasts, building investments in North America, Western Europe and China are expected to contribute to nearly 40% of the BAU capital investment. Although construction is the most significant single investment category in the short term, throughout the 2011-2030 timeframe, growth in the power and transport sectors was nearly comparable and corresponded to the immediate requirement to upgrade a large part of the current building stock.

The green economy transition's primary goals include enabling economic growth and investment and increasing social inclusion and environmental quality. The transportation sector plays a critical role in achieving these goals. Transportation is essential for economic development because it enables people to engage in economic activities and rise from poverty.

36.5.4 Economic Sector Initiatives

The economic sector makes a direct and indirect contribution to the economy by affecting the patterns of production, trade, and investment. Privatized consumption and expenditures on public transit infrastructure and services all contribute directly to economic output. It opens up new market opportunities for consumers and producers and influences patterns of production, trade, and investment. The ability to utilize their comparative advantages and capacities, as well as to construct global supply chains, is provided by this technology. Through the use of high-quality, trustworthy, and cost-effective transportation systems, it is possible to source and manufacture raw materials from around the world and facilitate the development of responsive modern supply chains.

According to the World Bank, inefficient transportation activities and a lack of infrastructure in specific modes of transportation may have a detrimental impact on the economy's productivity. When people become more reliant on their automobiles in metropolitan settings, traffic and congestion can increase. Congestion increases the amount of time and money it takes for passengers and products to travel. Furthermore, it improves the dependability of delivery services by a significant margin. Congestion is a costly problem. In the Republic of Korea, it is estimated that congestion expenses account for 4.4% of GDP.

36.5.5 Social Sector Initiatives

Numerous empirical studies have demonstrated that transportation contributes to poverty reduction and human welfare, most notably in terms of social advantages. Rural producers profit from market expansion because it strengthens the entire rural value chain, raises the revenues of local producers, and thus contributes to rural poverty elimination. According to the Indian Management Institution's research, roughly half of all fresh produce is wasted throughout the transportation route from farm to market.

When a well-designed transportation system is combined with proper technique, it can help to reduce such waste while also providing additional cash for local producers. Prohibitively high costs frequently hamper physical access to transportation infrastructure and services. Transportation must be accessible to all people to ensure fair access.

As a result of rapid motorization, there is an increase in traffic accidents. Road collisions claim the lives of 700,000 people every year in the Asia Pacific region. Poor people are disproportionately affected by road traffic fatalities and injuries. According to the World Health Organization, more than 90% of road fatalities and injuries occur in low- and middle-income countries, with the vast majority of deaths occurring due to traffic incidents involving pedestrians and other non-motorized road users. Road traffic deaths are anticipated to exceed HIV/AIDS and tuberculosis as the world's fifth-biggest cause of death by 2030, surpassing even cancer and diabetes.

36.5.6 Environmental Sector Initiatives

The transportation industry is heavily reliant on non-renewable and finite fossil fuels for its operation. Historically, the constantly increasing cost of oil and natural gas and the contribution to climate change impacts are expected to continue in the foreseeable future. Industrial consumption of petroleum products and energy in Asia is the highest globally, with industry ranking first and third, respectively.

The industry generates various emissions due to its firm reliance on fossil fuels, including carbon dioxide, carbon monoxide, sulphur dioxide, nitrogen oxides, volatile organic compounds, and particulate matter. Air pollution originates primarily from or is preceded by transportation in the Asia Pacific region. The third-largest producer of CO2 and other greenhouse gas emissions is the transportation sector. In 2008, CO2 emissions from transport totalled 1 587,4 million tonnes, with 1 275,8 million tonnes, or 1 275,8 million tonnes, emitted in the United States. The majority of the revenue came from road transportation (more than 80%). These emissions negatively impact people's and the environment's health, which are both harmful and expensive.

36.5.7 The ASI Approach

In the Avoid-Shift-Improve (ASI) approach, three essential components are included in the overall strategy. The shift is away from less efficient and environmentally damaging modes of transportation — potential alternatives include passenger, non-motorized, and public transit. For freight transportation, rail and water transport are the most preferred forms of transportation.

The solutions are proposed that can assist in the creation of a green economy. ASI's environmental benefits are not restricted to the environment; they also have implications for sustainable transportation, economic development, and social well-being, among other things. Climate change concerns are developing in the region due to earthquakes, cyclones, and tsunamis that have occurred in the region and a larger range of temperature swings and atypical seasonal cycles that have happened in the area. Additionally, it emphasized the necessity of efficient, accessible, and integrated urban public transit and the importance of improving service, quality, and accessibility for all people in metropolitan areas. Nutrient Transfer Program (NUTP) proposed 437,7 kilometres of Bus Rapid Transit (BRT) in ten cities, with a total cost of US$1 085 million. An average daily passenger count of more than 55,000 people makes Ahmedabad the first city in the NUTP to have a fully operational BRT system launched in 2012. Major metro rail projects are being developed in major cities to accommodate the growing demand for public transportation.

Throughout India, over 24 000 kilometres of rural highways are being built or repaired, including in states such as Jharkhand, Himachal Pradesh, Meghalaya, Punjab, Rajasthan, and Uttar Pradesh, as well as in Uttarakhand, as stated in the Prime Minister's Rural Roads Programme. The project will also include a technical assistance component to help participating agencies improve capacity, enhance their business practises, and manage general maintenance more successfully. In general management, monitoring and execution, the Indian Government's Ministry of Rural Development are responsible for the National Rural Roads Development Agency.

36.6 Agriculture

Today's challenges cannot be addressed by the current paradigm of intensive agricultural production. Agriculture must develop the ability to save for thriving. Sustainable intensification is defined as agriculture that conserves and enhances natural resources, follows a nature-driven ecosystem approach, and uses the appropriate amount of external inputs in the proper time.

The Green Agricultural Revolution in the 1960s spared an estimated 1 billion people from famine throughout the vast part of the evolving world, mainly in the Asia Pacific throughout the revolution. In developing countries, introducing high-yield crop varieties, irrigation, agrochemicals, and sophisticated management practices allowed farmers to raise their food production between 1961 and 2000 from 800 million tonnes in 1961 to more than 2.2 billion tonnes.

Because of rising global energy demand, conventional energy reserves are being depleted around the world. Agriculture is an industry that benefits from renewable energy technology's numerous financial and environmental benefits. However, there are major barriers to the use of new technologies for unconventional energy harvesting. Natural Resources Canada addresses this shortfall by developing a project that helps farmers and operators to select renewable technology that is economically and environmentally sound. Farmers will be able to take advantage of developing tools and templates to help them implement and integrate renewable energy systems in their holdings.

In the transition to more sustainable patterns of energy consumption, agriculture is critical. Agriculture is a significant consumer of energy. Secondly, agriculture contributes to the supply of energy by producing biomass, particularly firewood, by-products of agriculture, animal waste, coal, other by-products and, increasingly, energy crops. The biomass part of energy consumption varies considerably globally, from 1% in Oceania to 47% in Asia. The vast majority of this consumption is accomplished through inefficient conversion systems, which negatively affect the environment and human health and present a significant opportunity to increase biomass production at a low cost while also increasing biomass produced.

Carbon storage and the use of biofuel forests can contribute to mitigating global warming by reducing net greenhouse gas emissions into the air. However, storage capacity in comparison to the massive volume of fossil fuel emissions is insignificant. Reforestation also requires land with alternative uses, such as food production, adversely affecting the global food supply. Carbon sequestration costs vary considerably globally, in developing countries ranging from a few dollars per tonne to several hundred dollars per tonne in the developed countries.

On the other hand, the decades of intensive farming have caused deterioration of rich soils, drained floods, generated pesticide epidemics, lost biodiversity, and critical ecosystem services in many countries. The World Bank has a guideline that the developing countries will have to double their agricultural output by 2050 to provide food security for almost 2.5 billion rural people, based on a world population of 9.2 billion in 2050.

Agriculture depends heavily on socio-economic and environmental factors affecting its output. It can stimulate economic growth by providing resources and biofuels, among other things, to the rest of the economy. It also can damage the ecosystem. Small farmers can contribute to food security and poverty alleviation through integration into modern value chains. Conservation agriculture, for example, can assist farmers in avoiding hidden ploughing costs such as degradation of soil and loss of nutrients as well as loss of moisture and yield. The usage of this technology has shown a 30% reduction in agricultural water consumption, energy expenditures of up to 60% and significant returns. Integrated pesticide management (IPM) helps sustain natural biological control while avoiding the use of pesticides indiscriminately. Precise irrigation enhances crop yields by fewer drops. In contrast, the exact placing and timing of fertilisers double the quantity of plant-absorbed nutrients, and accurate fertilisation boosts crop yields and reduces the amount of plant-absorbed nutrients.

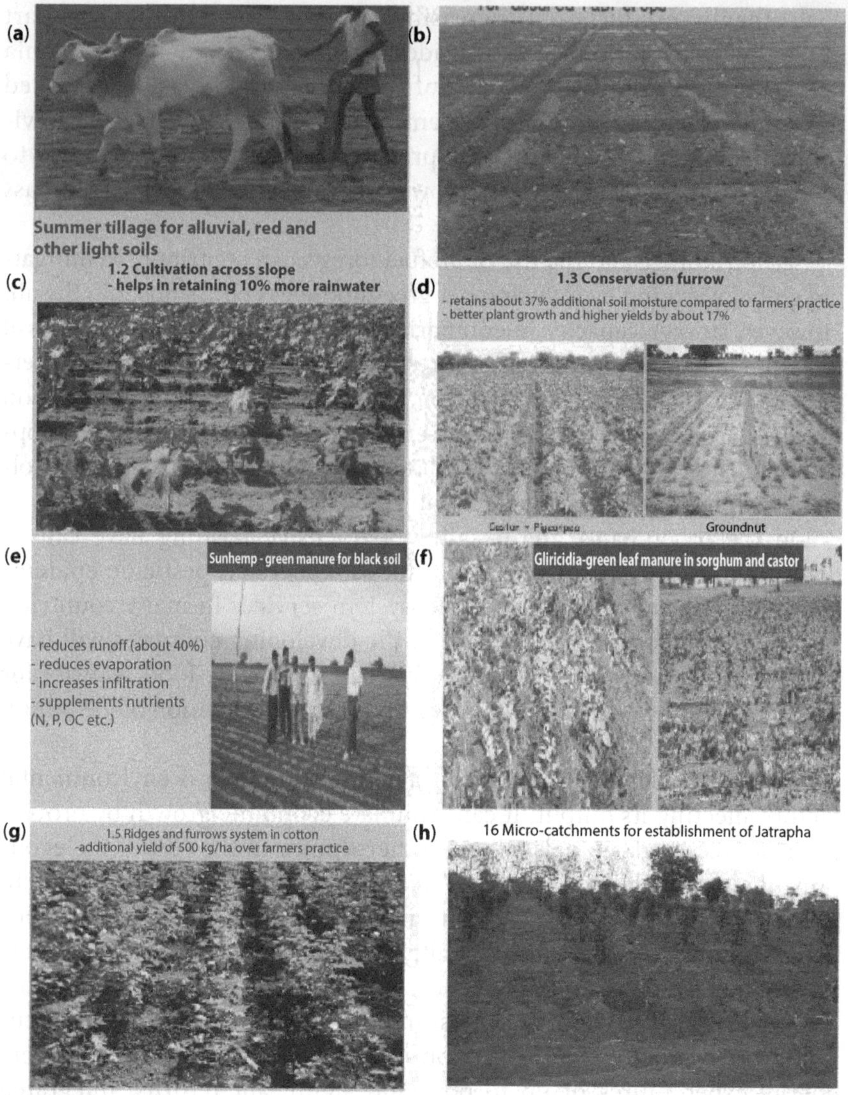

Figure 36.11 SAPI farming practises that conserve and enhance natural resource.

SAPI refers to farming practises that conserve and enhance natural resource use by relying on nature's contribution to cultivation—organic soil materials, regulation of water flow, pollination and control of natural pesticides—and by the service at the appropriate and efficient time of appropriate outside input. SAPI includes agriculture which conserves and enhances the use of natural resources by relying on nature's contribution to culture, organic soil materials, water flow regulation and pollination. It uses the technology to lead the production model to knowledge-based, often local agricultural systems that improve soil fertility, minimise soil erosion, protect soil production. The ecological economy is environmental economics. SAPI-based agriculture uses a variety of technological advancements. As per the Figure shown in 36.11 SAPI farming can be seen.

36.6.1 Soil Fertility and Irrigation

A big problem for developing countries with growing populations is reducing their fuel, fertilizer, and water usage in response to rising agricultural demands, which is a substantial source of concern. Farming practises that promote conservation agriculture attempt to achieve long-term viability and profitability while also improving farmer livelihoods. Conservation farming is founded on three principles: minimal soil disturbance, permanent soil covering, and crop rotation. It has the potential to reduce greenhouse gas emissions by increasing organic soil carbon by approximately 450 kg/ha per year while at the same time promoting fertility and moisture maintenance. Farmers have gotten more and more interested in conservation agriculture within the past 25 years.

Many techniques for reducing the environmental impacts of irrigation are being implemented, such as soil salinization and aquifer nitrate poisoning. The knowledge-based, proper irrigation system of SAPI is a vital component in the irrigation system, which allows the deployment of water and scarce irrigation and regeneration of wastewater. Incentive water conservation programmes need to be phased away over time, especially in rainfed areas where the climate threatens millions of smallholder farmers. It is only viable to adopt more drought-tolerant cultivations and water-conserving practices if the volume of rainfed items produced increases.

A holistic approach to farm planning and management will help reduce negative impacts on the environment through mapping of soil, water and soil conditions and monitoring of agricultural and drainage traffic, among other things. As a result of this decision, additional benefits can be achieved, such as improved field operations and better soil and water conservation in dire weather circumstances. Contract farming and equipment sharing are examples of agri-business schemes that incorporate communities or groups of farmers and allow for the broader overall management of agricultural land.

36.6.2 Pesticides and Biomass Pollution Control

IPM (Integrated Pest Management) is a well-known green farming strategy that helps farmers reduce health concerns and farming costs. Integrated pest management (IPM) comprises a thorough study of all possible pest control approaches and the integration of measures to prevent the spread of pest populations. In addition to lowering the hazards to human health and the environment, it limits the use of pesticides and related measures to economically sustainable levels for businesses. When it comes to integrated pest management (IPM), producing a healthy plant with the least amount of disruption to the agroecosystem as possible is paramount, and natural pest control approaches are encouraged. The Food and Agriculture Organization of the United Nations (FAO) has been advocating this as an essential component of good agricultural practices for more than two decades [4].

Incorporating livestock and crop systems is a win-win technique that results in higher crop productivity and soil fertility while decreasing costs. Agroecosystems include crop-fishing, livestock-fishing, aquaculture-fishing, poultry-fishing, and other multispecies methods. Agroecosystems perform better in terms of nutrient efficiency when compared to monocultures. Environmental concerns such as nutrient accumulation and laxation in high-intensity areas are becoming increasingly prevalent in many industrialized regions as the cattle industry and crop agriculture become increasingly specialized and unique. Soil fertility might deteriorate if there is no connection between crops and livestock. Improved fodder and animal feed are produced when excrement is returned to fields in dry locations such as northern China and the Loess Plateau in South Asia.

As a result of animal waste pollution and the need to enhance rural energy access, energy production from biogas has increased in recent years in many countries, most prominently in China and Vietnam. The use of biogas technology and its benefits in pollution prevention and energy generation is also advantageous since it is a rich source of nitrogen, phosphate, and potassium, making it a particularly effective organic fertilizer. As a result of greater commercial feasibility in comparison to the rising expense of chemical fertilizers, composting on both a big and a small scale and the manufacturing of agro-processed and urban green waste biofertilizers have all expanded in recent years.

As per World Food Security Report 2017, biodiversity in crops and livestock improves food safety by increasing and stabilizing yields and providing nutrient-dense, balance, and diverse diets; it also increases environmental stress resilience and can mitigate climate change consequences. The preservation and adaptation of genetic pools to changing ecological conditions serve as the foundation for future breeding approaches.

36.6.3 Agroforestry

Shrubs and trees closely resemble natural ecosystems are introduced into agroforestry systems to increase animal or crop production. It is possible to find agricultural forestry systems in a wide range of farming intensities. For example, they can be found in: (i) intensive cash crop systems in countries such as China and the Philippines; (ii) tea or coffee crops grown in countries such as India, Indonesia, and Papua New Guinea; and (iii) as counter raiding strips in maize/soybean systems in countries such as Sri Lanka, Australia, and the Philippines. Agricultural systems that produce non-food products such as wood and biofuel are frequently used in agroforestry systems.

Many Asian nations have adopted stock farming because it is the only technique capable of transforming energy into a form of energy directly available to humans in grassland flora, which is essential for human survival. Sustainable pasture and pasture management require selecting the most appropriate grazing season or time of year for the most appropriate number of animals on a given piece of land. Access control and rotational grazing, both necessary for maintaining a good pasture balance, are key methods for achieving sustainable grazing [1].

Increased demand for safe, high-quality food and biological dangers such as the development of infectious illnesses and invasive species will be driving forces, with negative consequences for public health, economic expenses, and the environment. Farmers' access to higher-value markets was enhanced by the growth of Asian countries' capacity to meet health and phytosanitary regulations. At the same time, protection against pest and disease outbreaks, such as highly pathogenic avian flu, was improved. Organic farming practises and recognized or certified agricultural food safety and organic agriculture practises enable farmers, cooperatives, and individual farmers to become approved suppliers on both domestic and international markets. At the same time, society benefits from improved health environmental management systems.

In the Asia Pacific region, hunger eradication has become more difficult and complex, as food prices and fuel costs, climate change, and the more significant usage of biofuel food have increased instability. The considerable growth in fertilizer consumption has now reached the lowlands while cereals are growing due to improved farming practices. Pesticide consumption is anticipated to follow a similar pattern for fertilizer consumption while insufficient regional data.

The Food and Agriculture Organization of the United Nations (FAO) has begun to promote the gathering and compilation of indicators for greening agriculture. The Asia Pacific Region has achieved considerable gains in the area of pesticide control improvement. The phase-out of hazardous pesticides was a significant milestone in the development of several countries. In response to a brown planthopper outbreak in 1986, the Republic of Indonesia banned 25 pesticides and quickly implemented integrated pest management on rice fields. Later, the Chinese government stopped the practice, which had a knock-on effect on Cambodia, the Lao People's Democratic Republic, and other countries that imported pesticides from China. Several hazardous pesticides have been prohibited in China as of recently, and other countries have followed suit. As a result of these regulatory efforts, many items have been brought into line with World Trade Organization (WTO) laws, which substantially impact agricultural exports to China and other regional countries.

The Asia Pacific region developing countries are partnering with the FAO to increase pesticide control. Take, for instance, Laos, where the government recently amended its pesticide registration to restrict the toxic but extensively used herbicide paraquat, which was previously allowed under the country's previous registration. To bring Vietnam and Cambodia into line with World Trade Organization regulations, the FAO aided both countries in upgrading their plant protection and agricultural materials legislations.

Many Asian countries are modifying their dietary practices due to rising wealth and an increasingly urban population. In recent decades, there has been an increase in demand for livestock products; for example, meat consumption in China and India has increased by 5% each year, while dairy consumption has increased by 3.5–4% per year. Increased livestock production has resulted in increased competition for land for food or animal feed production, which has resulted in land prices rising. Because of the pollution caused by intensive livestock production systems, traditionally utilized as fertilizer or biofuel, countries bordering the South China Sea are taking steps to address the problem through win-win treatment and recycling technology. For green economies to thrive, innovation (including the revival of prior innovations) is required to maximize the use of natural and economic resources while minimizing polluting the environment and benefiting society at the same time.

36.6.4 Biotechnologies

Farmers and local government officials administer the Thai government's more than 500 rural laboratories, which employ more than 10,000 people. For further development to take place, additional private sector investment is required. The government must play three roles to regulate (quality, safety, and efficacy), facilitate the environment (via patents, for example), and encourage marketing. The ASEAN Secretariat has launched a programme to promote a regulatory framework conducive to biocontrol to combat invasive species.

At the heart of the new paradigm of "saving and growing" are action research, participatory assessments, the conservation of farmer's varieties, gene management, and the awareness that a shortage of adequate farms frequently puts more significant yield constraints than crop and animal genetics. Saving and growing is a word that refers to the act of preserving and developing. According to the organization, a collaboration between the Agricultural Research Institutes Asia-Pacific Association (APAARI) and other partners has enhanced wheat productivity in the Asia-Pacific region. Capacity building has begun in the Democratic People's Republic of Korea and Myanmar to facilitate the rapid release of novel crop varieties and participatory seed selection and production, among other things. SEARICE is collaborating with a regional network of non-governmental organizations (NGOs) to conduct a project to conserve and develop indigenous rice varieties through selection and action research. The International Rice Research Institute (IRRI) uses a standardized participatory review process to evaluate novel disclosures to promote rice conservation practices.

Rice intensification is more widespread in South Asia, Southeast Asia, and East Asia rice production systems. As a result, fertilizer efficiency increases, greenhouse gas emissions are reduced, and earnings are increased. A sheet colour chart produced by the International Rice Research Institute (IRRI) and extensively used in Bangladesh and Asia helps rice growers determine the best time to apply nitrogen fertilizer to their crops.

At the FAO's 31st Regional Conference for Asia and the Pacific, which took place in 2012, various green projects were recognized, including the concept of conserving while increasing. A multi-donor trust fund has been established to mitigate the effects of climate change. It has been found in Vietnam that a pilot project will test and promote communities of practice by incorporating mitigation-enhancing approaches into agricultural practises on a modest scale as part of the country's climate change strategy. It looks at methods to improve existing integrated food energy systems and educates farmers on linking food, biogas, and biofertilizer crop production with livestock, forestry, and fishery activities. It also looks at ways to improve existing integrated food energy systems.

36.7 Performance Mapping in Conjunction with Technological Evolution

Reconfiguring the world's energy grid is critical. Energy systems based on fossil fuels must be phased out and replaced with renewable energy systems. This section discusses a strategy for decarbonizing the economy based on the efficiency of renewable energy and illustrates how the transition is taking place and how renewable energy assists in ensuring an energy supply. The contribution of solar energy in total energy consumption increased around 0.25%, to roughly 19%.

On the other hand, it must accelerate the growth of renewable energy significantly. Renewable energy must increase from 19% today to 2/3rd by 2050 to reach our future energy needs, according to the TFEC. To assist with this, global GDP must reduce its energy intensity by 2.8% per cent per year from current levels, up from 1.8% in recent years. Even with population growth and economic activity, the world's energy demand in 2050 will be roughly the same today. Carbon dioxide emissions increased in 2017 as a result of insufficient energy-efficiency improvements. Despite this, the International Energy Agency (IEA) conducted an analysis, which indicates that the global energy system's state has improved and can use numerous methods to increase efficiency.

This study discusses various strategies for accelerating the energy transition. It demonstrates the application of modern supply- and demand-side technologies, as well as the investment required. Additionally, this study examines the benefits and drawbacks of switching to renewable energy. This section emphasizes how the costs of transitioning to a more sustainable energy supply far outweigh the benefits. The potential benefits to overall welfare could be significantly increased by using a broader-based welfare measure.

In comparison, the Reference Case such as Nationally Determined Contributions (NDCs) is based on existing and future policies. In 2050, this alternative scenario predicts a 1% increase in global GDP (NDCs). Millions of new jobs would be created globally, stimulating demand and the economy as a whole.

The global energy system has to change. Carbon reduction is only one of the energy revolution's benefits. It can achieve a variety of outcomes, including increased national energy security, cost reductions associated with universal energy access, diversification of energy supplies, and improved human health. IRENA›s new Commission is researching the geopolitics of energy transition. The energy sector cannot address all of the issues concurrently. When it comes to energy transformation, we must take a holistic approach that balances the needs of society and the economy.

Renewable energy is assisting the energy transition by significantly lowering its cost over the last decade. When it comes to new renewable energy capacity additions, fossil fuel generation expansion does not even cut. According to the IRENA, global renewable energy installations totalled 167 GW in 2017, an increase of 8.3% in the previous year and an annual growth rate of 8-9% for the preceding five years. Renewable energy sources generated more net additional power capacity than conventional sources for the sixth consecutive year. Solar capacity increased to 94 GW in 2017, while wind capacity increased by 47 GW.

Simultaneously, the costs of wind and solar photovoltaic energy are increasing. When electricity costs fall, renewable energy becomes more affordable for everyone, but it also signals a trend toward using clean, renewable energy for a variety of purposes. However, significant cost savings are also found in some emerging technologies. For the first time, projects based on the offshore wind was made available at market rates with no need for government subsidies in 2017.

In 2017, renewable energy grid integration also set new records. Solar and wind energy combined to generate nearly half of the electricity generated in eastern Germany. The 50Hertz power supply demonstrated that increasing the proportion of variable renewables in power systems in this area is both economically and technically feasible (50Hertz, n.d.). For several days now, numerous locations on the planet have consumed more renewable energy than they did previously. The good news is that renewable energy systems have been demonstrated to be both reliable and necessary for long-term economic viability. These recent developments show unequivocally that renewable energy is accelerating its growth.

Regrettably, advancement has lagged far behind the power industry. 20% of electricity is used for transportation, heating, and other energy services. The remaining 20% is made up of other resources, primarily renewable thermal fuels or energy and fossil fuels. In end-use sectors, renewable energy sources such as solar thermal, geothermal, bioenergy, and energy efficiency are critical. Additionally, as renewable energy supply grows, the share of renewable energy in end-use sectors will grow.

Together with advancements in vehicle propulsion, electric vehicles are making the promise of carbon-free road transport a reality. According to estimates, global sales of electric vehicles (approximately 1.5% of all vehicle sales) reached an all-time high in 2017. America has surpassed Japan to become the world's largest market. Electric vehicle sales have grown rapidly over the last five years, with an average annual growth rate of 52%.

Construction consumes more electricity per unit of production than other sectors. For cooking and heating, mainly fossil fuels are used. For most of the population, electrification of cooking burners and contemporary cookware represents a traditional biomass alternative. In 2017, a new record was set for heat pump installation for heating. The rules are written to benefit buildings that are on the verge of becoming self-sufficient, if not entirely autonomous. Nonetheless, progress in the sector is agonizingly slow because the energy-efficiency sector is moving at a snail's pace of about 1% per year.

The industry is the toughest sector to work in, due to the high needs of some energy-intensive industries, the high carbon content of certain products, and the high emissions of some processes. As a result, novel ideas and lifecycle thinking are required. Even though new technologies for biological and renewable hydrogen feedstocks are being developed, significant advances were made in heavy industry.

CO2 EMISSIONS RELATED TO ENERGY: BRIDGING THE GAP

Several countries have increased their efforts in recent years to reduce their country's carbon emissions. The Reference Case illustrates the cumulative CO2 emissions associated with energy, new policies and programs, including NCDs. Energy-related CO2 emissions are expected to decrease by 11% in the reference case, from 1.380 Gt in 2015 to 1.230 Gt in 2050. Even though CO2 emissions increased in 2017, this progress has yet to be reflected in current emissions, which increased by 1.4%, according to the International Energy Agency (2018a).

It is unknown whether current policies will be sufficient to achieve the required carbon reductions. According to the Reference Case, the world's energy-related emission budget will be depleted under current and anticipated policies in less than 20 years. The cumulative emissions should be decreased to keep global temperatures below 2°C, by another 470 gigatonnes by 2050. (with a 66% probability). According to Figure 36.12, where current and proposed policies are compared, the present and proposed policies are significantly different [3].

According to the Reference case, the carbon emission required to keep the temperature below 2°C will be met by total energy-related emissions in less than 20 years. By 2050, approximately 470 Gt of

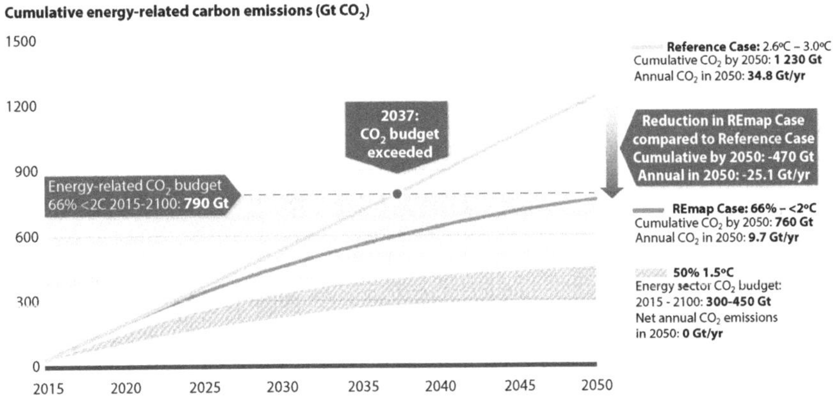

Figure 36.12 Cumulative energy-related carbon emission.

carbon dioxide equivalent emissions must be reduced to keep global warming below 2°C.

In less than 20 years, the global energy-related CO2 emissions budget required to keep warming below 2°C would be exhausted. CO2 emissions from energy and the emissions gap, 2015-2050 (Gt CO2). The Reference case assumes that CO2 emissions from energy will increase each year slightly until 2040, at which point they will stabilise near their current level by 2050 (see Figure 36.12). These figures demonstrate how the increasing affordability and performance of renewable energy influences long-term energy planning and forecasting. It is disappointing, but additional spending cuts are necessary. Annual CO2 emissions must be reduced from 35 Gt to 9.7 Gt by 2050 (a reduction of more than 70%) to meet the global goal of limiting warming to 2°C.

Nuclear energy production, we estimate, would remain as it is at 2016 levels. A significant effort will be necessary to minimise CO2 emissions from industrial operations and land use to near-zero levels by 2050. Unless and until these changes occur, the climate change goal will remain unachieved. The diagrammatic representation as given above in Figure 36.12 renders the clear picture of it.

The Paris Agreement to achieve the aspirational aim of reducing global temperature rise set forth into 1.5°C, additional emission reductions will be required, as well as a sharper curve-shaping effect. Global CO2 emissions must be zero by 2040, or at the very least zero by 2050 if negative emission technologies are deployed in the second half of the century [10].

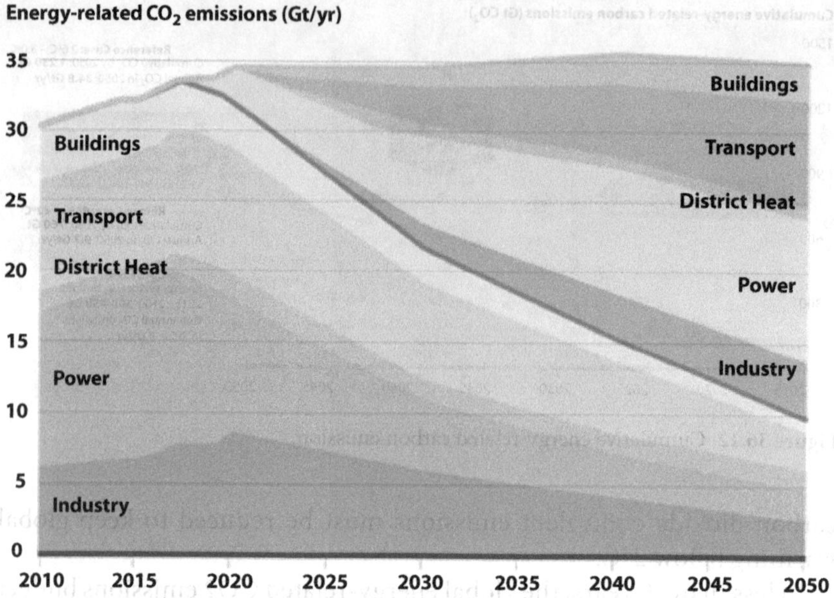

Figure 36.13 Energy-related CO2 emission.

Renewable energy and energy efficiency can account for more than 90% of the decrease in CO2 emissions associated with energy. Annual energy-related emissions are expected to remain unchanged in the Reference Case. Still, to keep temperature rise below the 2°C target, emission reductions must be at least 70% greater than in the Reference Case. The vast majority of the consumption reductions required are accomplished through various energy efficiency and renewable energy measures. The diagrammatic representation as given above in Figure 36.13 helps is to understand it.

A strategy for transforming the world's energy system
By around 2050, renewable energy will account for approximately 66% of the total primary energy supply (TPES) (see Figure 36.14 helps us to understand it). This percentage will only increase to 27% under current and projected policies. Renewable energy use would nearly double if accepted in the REmap Case. Solar and wind energy would supplant biofuels. The accounting system accounts for more than half of all renewable energy. By 2050, bioenergy will account for one-third of renewable energy.

The global percentage of renewable energy would have to climb to two-thirds by 2050, while TPES would have to stay steady. In the Reference and REmap scenarios, the TPES and renewable and nonrenewable energy proportions are shown.

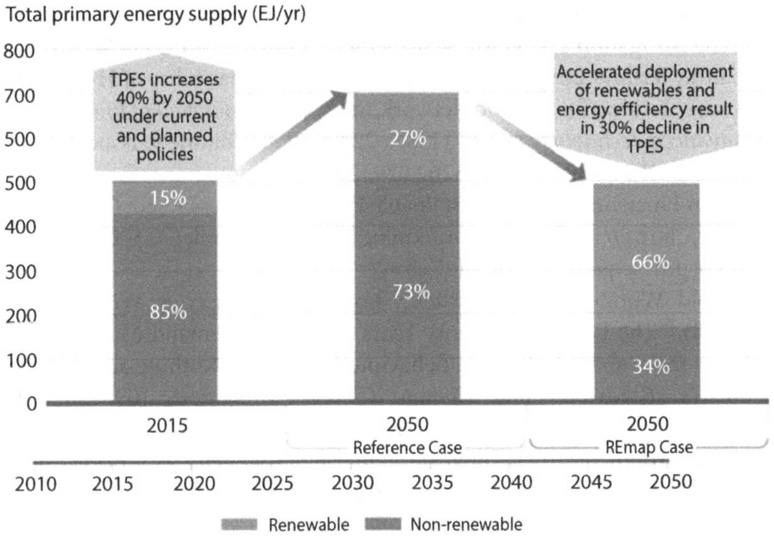

Figure 36.14 Total primary energy supply.

TPES is projected to increase by approximately 40% according to current and planned policies. A roadmap for an energy transition would include a reduction in energy efficiency and renewable energy to supply two-thirds of energy.

By 2050, the estimated increase in TPES under current and proposed regulations is approximately 40%. In the instance above, the total primary energy consumption must be reduced below 2015 levels for an energy transition (i.e., REmap).

References

1. Alfons Oude Lansink, Ekko C. van Ierland and Gustavo Best (2002), "Sustanable Energy in Agriculture – Issues and Scope", *Economics of Sustainable Energy in Agriculture*, 1–7, Kluwer Academic Publishers. Printed in the Netherlands.
2. Amela Ajanovic, Carol Dahl & Lee Schipper (2012), "Modelling transport (energy) demand and policies — An introduction", *Energy Policy* 41 (2012) iii–xiv, doi:10.1016/j.enpol.2011.12.033
3. Catherine Wolfram, Orie Shelef, and Paul Gertler How Will Energy Demand Develop in the Developing World? *Journal of Economic Perspectives,* Volume 26, Number 1, Winter 2012, pp. 119–138.

4. Claude Faucher and Julie Bastien (2006), “Renewable Energy and Agriculture: GHG Mitigation and Waste Management Strategy”, *IEEE* 1-4244-0218-2/06.
5. Gaurav Vishnu Londhe, Dilendra Hiren (2020) Feasibility of Hybrid Energy Aware Routing for Ad-hoc Network,International *Journal of Engineering and Advanced Technology (IJEAT)* ISSN: 2249–8958, Volume 9, Issue 3, February 2020. DOI:10.35940/ijeat.C6262.029320
6. Green Financing (2019), “Sustainable Finance Progress Report”, UN Environment, https://wedocs.unep.org/xmlui/bitstream/handle/20.500.11822/34534/SFPR.pdf?sequence=1&isAllowed=y
7. Harold Wilhite, Elizabeth Shove, Loren Lutzenhiser & Willett Kempton (2000), “The Legacy of Twenty Years of Energy Demand Management: we know more about Individual Behaviour but next to Nothing about Demand”, *Society, Behaviour, and Climate Change Mitigation*, 109–126, Kluwer Academic Publishers. Printed in the Netherlands.
8. Iryna Dronova (2017), “Environmental heterogeneity as a bridge between ecosystem service and visual quality objectives in management, planning and design”, *Landscape and Urban Planning*, DOI: 10.1016/j.landurbplan.2017.03.005
9. Jillian Anable, Christian Brand, Martino Tran & Nick Eyre (2010), “Modelling transport energy demand: A socio-technical approach”, *Energy Policy* 41 (2012) 125–138.
10. Londhe G.V., Hiren D. (2020) Hybrid Energy-Aware Routing (HEAR) Approach (Using Network Simulator). In: Luhach A., Kosa J., Poonia R., Gao XZ., Singh D. (eds.) *First International Conference on Sustainable Technologies for Computational Intelligence. Advances in Intelligent Systems and Computing*, vol. 1045. Springer, Singapore. https://doi.org/10.1007/978-981-15-0029-9_21
11. Ming Zhang, Hailin Mu, Gang Li & Yadong Ning (2009), “Forecasting the transport energy demand based on PLSR method in China”, *Energy*, doi:10.1016/j.energy.2009.06.032.
12. T. Giakoumis & N. Voulvoulis (2019), “Water Framework Directive programmes of measures: Lessons from the 1st planning cycle of a catchment in England”, *Science of the Total Environment*, Volume 668, 10 June 2019, pp. 903-916.

37

Education in Energy Conversion and Management

Satish R. Billewar[1]*, Karuna Jadhav[2] and Gaurav Londhe[3]

[1]*Vivekanand Institute of Management Studies & Research, Mumbai, India*
[2]*Neville Wadia Institute of Management Studies & Research, Pune, India*
[3]*Mukesh Patel School of Tech. Mgmt. & Engg., NMIMS, Mumbai, India*

Abstract

In recent years, concerns about these problems have grown rapidly, and the underlying causes are now addressed at a global level. International treaties that impose compulsory limits on industrial nations' use of fossil fuels can achieve these goals by replacing fuel, managing energy, increasing energy efficiency, using renewable energy systems, and restricting land clearance. Global warming and rapid growth have required multi-disciplinary energy research programs run by the universities to integrate energy technology research with energy policy, energy conversion awareness, climate change policy, sociology, and green economies. There is a need for universities to take the initiative to incorporate these programs in education for the students to learn about new career opportunities in renewable energy.

Rapid technological advancements put the efficacy of traditional technical education in jeopardy. New revenue streams are emerging, and technical departments' reliance on them is diminishing. We are looking for technical experts with a high level of qualification. Short-term training for current professionals and recent graduates appears to be a viable option (Bin Zhu, 2013). One significant sector that exemplifies this is renewable energy, which is distinct in its characteristics. A battle is started to advance renewable energy conversion systems. However, because it is a relatively new field, it requires highly skilled personnel. Numerous researchers have examined professional development, attempting to quantify and evaluate various approaches.

**Corresponding author*: satish.billewar@gmail.com

Milind Shrinivas Dangate, W.S. Sampath, O.V. Gnana Swathika and P. Sanjeevikumar (eds.) Integrated Green Energy Solutions Volume 2, (317–352) © 2023 Scrivener Publishing LLC

The case for a green economy is defined as a set of enabling conditions, which include public policies, fiscal spending and investment, tax regulations, training, and education. The green economy grows faster than the brown economy over time by conserving and restoring natural capital. A green economy maximizes human well-being while minimizing environmental risks and scarcities. A green economy is one that maximizes human well-being while reducing environmental hazards and scarcities. One more minor, prominent aspect of our journey is how we will arrive.

The chapter examines several references in the literature. It makes conceptual and primary criticisms of several methods and evaluations for determining the effectiveness of education in energy conversion and the management of successful implementations.

***Keywords*:** Climate change, industrial ecology, sustainable development, green economy, ecosystem

37.1 Role of University

Nation development can only be achieved by generating a developed generation, which must handle the nation and its development in all respective domains.

The entire load of the above-given role of developing the nation is solely on the shoulders of educational institutions and universities, for creating an environment where the next generation can be nurtured and nourished, through which they will be able to handle the responsibility of the country and its next generations [7].

Indeed, the university must play an essential role in the overall development of a human being in all respects. The student's story must be carried out in all domain areas equally. Only then can he or she be a well-cultured nation citizen.

Here, when we discuss the university perspective, we look at energy conversion and its management as one of its aspects [5].

We must consider that one must educate the people around us or make them aware that unless they think about energy conservation and conversion of energy from one form to another and management of the same, such development will not occur, because if we cannot help generate the energy, we must think about its conversion and management.

The diagrammatic representation as given above in Figure 37.1 can elaborate in detail about the courses that can be implemented to accomplish the university's role.

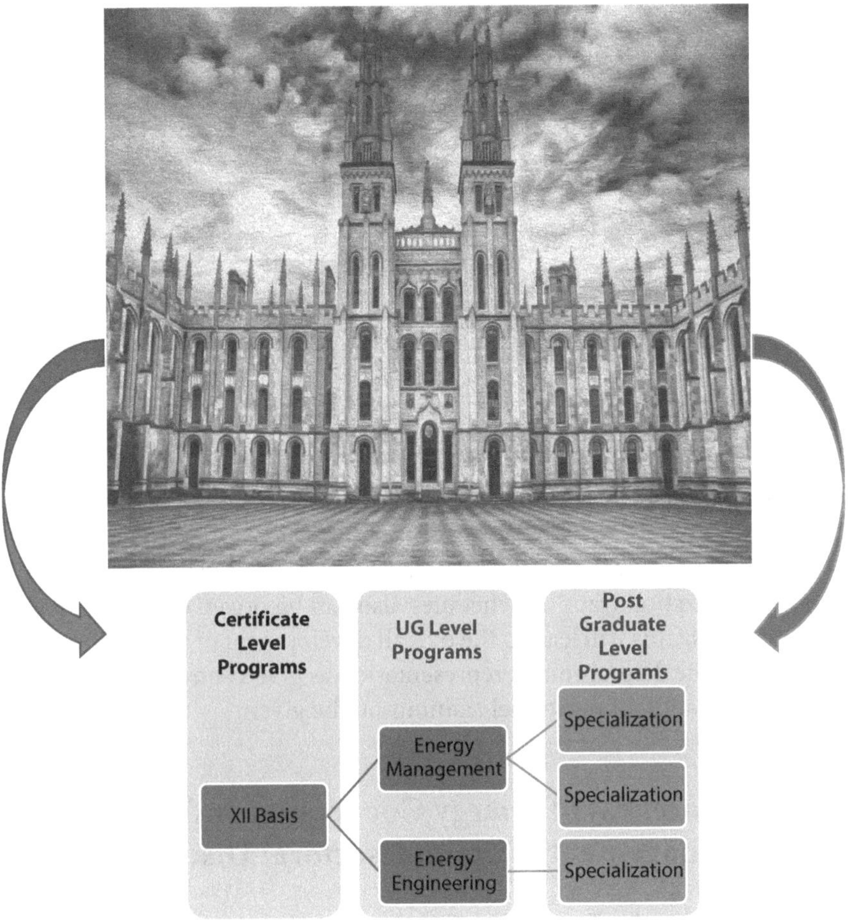

Figure 37.1 Courses can be implemented in universities.

37.2 Personnel Training

This is an important aspect that needs to be addressed before we design the strategy for development. Many activities can be planned for human resource training for public awareness and related objectives.

The above-mentioned structure with respect to the level of education shows some programs that can be implemented for training and education. Accordingly, the job prospectus can also be made available for the same [7].

Figure 37.2 Personnel training.

Category-wise training schedules can be organized for the team members. Specialization training schedules also can be organized into individual categories. This will lead to the overall development of the organization as a whole. The diagrammatic representation as given above in Figure 37.2 the representation of personnel training can be given.

37.3 Awareness of Energy Conversion and Management as an Intersectoral Discipline

As we have already discussed under the role of the university, we must make citizens aware of the fact that electricity is a very powerful, useful and only cheapest resource than the fuel we use for generation of electricity in generators in case of emergencies.

We must educate everyone about its use and conversion techniques to easily conserve it at some percentage, which will support the energy management perspective.

There are various ways in which people can be well versed in these concepts. Local government agencies and social media play an essential role in these types of activities in each country, be it a developed, underdeveloped or developing country [3].

There are intersectoral disciplines that must be involved in achieving this objective.

1. Government Agencies
2. Health care units
3. Hoardings
4. Social media services
5. News agencies
6. Posters at public places
7. Public notices
8. Involvement of social workers and NGOs

When we discuss overall national and global growth, it can never be possible automatically. Each of us needs to contribute to creating such an environment through critical thinking and policy framing.

There must be is an awareness of the need for the adoption of newly developed strategies and ways to live. Government policies must be appropriately structured to facilitate the growth of these strategies in a way that produces national GDP growth.

Government must be able to promote the policies which will be most beneficial and adopting this change smoothly, either by involving people in such activities or by facilitating them in their day-to-day life activities. This must be conveyed to people, as adopting the new policies regarding energy conservation and conversion is the only way for them to stay safe and satisfy their everyday needs.

37.4 Climate Change

Climate change is caused by human activity, including carbon dioxide and other greenhouse gases, into the atmosphere. Carbon dioxide has never been higher in the atmosphere than it is today. Carbon dioxide levels have risen by 40% over the last two centuries [4].

According to the National Geographic Society's dictionary, the long-term change in temperature and typical weather patterns of a location are commonly referred to as climate change.

These shifts may be brought on by things like the cycle of the sun. But human activities have been the main driver of climate change since the 1800s, primarily due to the release of greenhouse gases like carbon dioxide and methane due to the use of fossil fuels like coal, oil, and natural gas. There are lots of ways in which we release greenhouse gases.

People's continued use of fossil fuels: These include coal, oil, and natural gas, which generate electricity, run vehicles, and power industry and manufacturing. Burning fossil fuels creates pollution. The carbon dioxide that has been "trapped" in the ground for thousands of years can be found in fossil fuels like oil, gas, and coal. Removing these items from the land and setting them on fire allows us to release the carbon dioxide previously sequestered in these items.

Deforestation: Trees take up carbon dioxide and store it in their tissues. Forests are significant sources of carbon dioxide removal and storage. The elimination of the trees results in more carbon dioxide accumulating because there are no longer any trees to absorb it. Besides that, when we burn trees, carbon dioxide is released.

The increasing scale of intensive agriculture: This type of agriculture emits methane and nitrous oxide in the form of greenhouse gases. Rearing animals and cultivating crops emits various greenhouse gases into the atmosphere. Methane, which is thirty times more powerful as a greenhouse gas than carbon dioxide, is emitted by multiple animal species. Carbon dioxide is used in fertilizers, but the gas used in fertilizers is ten times worse and is nearly 300 times more potent. Another portion of our total carbon dioxide emissions comes from the production of cement, which contributes 2% to the problem.

Carbon dioxide and methane are greenhouse gas emissions that are causing climate change. These are made by using gasoline for vehicle propulsion or coal for building heating. Carbon dioxide is also released when land and forests are cleared. One of the significant sources of methane emissions is in landfills for garbage. Among the significant emitters of carbon dioxide are energy, industry, transport, buildings, agriculture, and land use.

Climate change is more alarming than ever. While we can take steps to make sure that our future is as prosperous as possible, we must not make the mistake of thinking we have arrived at our destination. These activities fall into one of two categories: stabilizing the climate and mitigating the effects of climate change. It is difficult to predict the exact effects of changes to the natural greenhouse, but some results are likely:

1. Throughout history, Earth has been getting warmer. Warm temperatures are something some regions will embrace, but others may resist. Since conditions are warmer, evaporation and precipitation will probably increase overall, but various locations will have different effects. Some areas will become wetter, and others drier.

2. Greenhouse warming will raise ocean temperature and melt glaciers and ice sheets, which will lead to higher sea levels. Increasing the ocean's temperature will also cause the water to expand, raising sea levels even further.
3. Higher levels of atmospheric carbon dioxide have both positive and negative effects on crop yields, depending on the greenhouse where they are found. One preliminary study found that elevated carbon dioxide levels appear to stimulate plant growth. Even if there were a gain in yield, other factors, such as climate change, ozone, and nutrient constraints, may more than counteract it. Any gains in yield previously possible will be reduced or reversed if the temperature range for some crops is exceeded.
4. Droughts, floods, and extreme temperatures can harm farmers' crops and threaten the food security of local communities. In warmer climates, wetter climates, and higher carbon dioxide levels, weeds and pests may thrive, and climate change will likely increase the number of weeds and pests.

A substantial amount of research indicates that rising carbon dioxide (CO2) reduces the nutritional value of most food crops, but, on the other hand, this causes plant growth. Changes in the climate may also contribute to the emergence of new pests and diseases, negatively affecting plants, animals, and humans. This has the potential to pose unknown risks for food security, food safety, and human health. As per the comparison of skeptic arguments stated in the Table 37.1given below represents the climate conditions.

The effects of climate change on our health, food security, housing, and overall well-being are significant. Many of us, including small island nations and other developing countries, are already more vulnerable to climate impacts. Today, problems like rising sea levels and encroaching seawater are getting worse to the point where entire communities must relocate, and long-term droughts endanger people's very survival. The number of "climate refugees" will increase in the future [6].

The climate is already visibly changing. Warming of the Earth, increasing rainfall patterns, and rising sea levels are all currently happening. By making these changes, heatwaves, floods, droughts, and wildfires could be more likely.

Table 37.1 Comparative table of skeptics' arguments and what the science says.

Skeptic argument vs	**What the science says**
Climate has changed before.	Climate reacts to whatever forces it to change at the time; humans are now the dominant force.
It's the sun.	In the last 35 years of gobal warming, sun and climate have been going in opposite directions.
It's not bad.	Negative impacts of gobal warming on agriculture, health, and environment far outweigh any positives.
There is no consensus.	Ninety-seven percent of climate experts agree humans are causing global warming.
It's cooling.	The last decade, 2000-2009, was the hottest on record.
Models are unreliable.	Models have successfully reproduced temperatures since 1900 globally, by land, in the air, and in the ocean.
Temperature record is unreliable.	The warming trend is the same in rural and urban areas, measured by thermometers and satellites.
Animals and plants can adapt.	Global warming will cause mass extinctions of species that cannot adapt on short-time scales.
It hasn't warmed since 1998.	
Antarctica is gaining ice.	Satellites measure Antarctica losing land ice at an accelerating rate.
An Ice Age was predicted in the 70's.	The vast majority of climate papers in the 1970s predicted warming.
Carbon dioxide lags temperature.	Carbon dioxide didn't initiate warming from past ice ages, but it did amplify the warming.

Many people, especially those in rural areas, could find themselves having to leave their homes due to the effects of climate change on agricultural production and human health. Certain species are at a greater risk of extinction because of their presence. Climate changes are already occurring.

Over the past century, global temperatures have risen by about 1°C. The years 2015, 2016, 2017, and 2018 were all among the hottest in recorded history. Climate change is already starting to have an impact on our lives. Rising sea levels impact people all over the world. 39% of people are at risk of flooding because sea levels are rising. Of these people, 600 million live in a "low-level coastal zone," while an additional 200 million people live in a coastal flood plain.

Even if we manage to cut our emissions, the rise in sea levels will last until 2100. However, if we reduce our carbon emissions to a low enough level, we could potentially slow the rate of growth. Some people will have to leave their homes, but the numbers will vary depending on how we act: we can reduce global emissions and improve flood defences, influencing how many people will have to leave their homes [14].

Floods can occur due to the overflowing of drainage systems or the bursting of river banks during heavy rainfall. Also, in urban areas where the ground is heavily concreted, the effects are far more severe because they cannot absorb the water directly into the soil. Severe damage to buildings and transportation is caused by flooding, which is costly and challenging to recover from.

This is one of the first studies of its kind to be conducted in South Asia, where about one-third of the land is vulnerable to droughts and floods due to climate change. The report notes that climate change can have the following severe impacts in the regions of Andhra Pradesh and Maharashtra and the state of Orissa:

As a result of these policies, dryland farmers in Andhra Pradesh may see their income decline by 20%.

1. Agricultural yields in Maharashtra could be slashed by about 25-30%.
2. Flooding is projected to rise dramatically in Orissa, causing a significant drop in rice yields in some districts.
3. Increased flooding is another possible danger. Rains that come with regularity, such as those in Bangladesh, are expected to bring even more frequent floods, putting additional populations at risk.

Other Indian climate hotspots that are vulnerable to climate change include the fragile Himalayas, the biodiverse Western Ghats, the extensive coastal areas, and the vital agricultural regions of the Gangetic Plain.

Growing enough food in some areas may be challenging as our climate warms and rainfall patterns change. Climate will change, and crops will grow differently depending on where a person lives. Even in areas that can produce new crops, crop production will be reduced in many places due to increased temperatures.

A longer growing season and higher carbon dioxide concentrations mean higher crop yields in colder countries. On the other hand, if warming persists in the long term, these side effects may not last. Wetter events could cause access to food to be disrupted, potentially interrupting transport from farms to shops and, as a result, hurting vulnerable people.

People all over the world are impacted by climate change in various ways. The impact is dependent on the local climate and national wealth. Climate change exacerbates problems, making problems that already exist much worse.

Heatwaves are an example. We believe most regions will see even more frequent heatwaves in the future. The human heat stress limits will be exceeded more often in countries already hot, which is dangerous.

To what extent will the severity of climate change be determined by how quickly we reduce dangerous greenhouse gas emissions? Even if we halted all emissions today, some irreversible changes would still occur. However, the sooner we cut emissions, the less dramatic the results of global warming will be.

37.5 Economic Policy Options

Mitigation of climate change is not easy, but there is a clear path to a prosperous future for everyone. CO2, in particular, has remained in the atmosphere for thousands of years. When it is all added up, it comes down to the fact that the increase in the temperature of the Earth's atmosphere as a result of all those emissions will have terrible consequences, and we have to drastically cut back on all the emissions, which total 51 billion tonnes per year. In order to prevent the planet from warming to greater temperature extremes, climate change mitigation includes minimizing the emission of heat-trapping greenhouse gases into the atmosphere.

Human-caused climate change mitigation consists of strategies that help keep greenhouse gas emissions low or to decrease them. Some examples are mitigation, using new technologies, increasing the energy efficiency of existing equipment, or changing business or consumer behaviour. A plan for a new city is needed for complex tasks, while simple tasks like improving a cookstove design can be accomplished with less effort. Ranging from high-tech subway systems to bicycling paths and walkways, efforts are underway all over the world.

Mitigation comprises strategies aimed at preventing global warming and its associated consequences. In addition to human emissions of greenhouse gases (GHGs), this strategy also focuses on various practices that aim to reduce the amount of these gases in the atmosphere. One of the ways to combat climate change is by implementing adaptation measures [2].

In addition to mitigating climate change, agricultural changes, reforestation, forest preservation efforts, and improved waste management practices may also be used. Meat consumption, which has a relatively large short-term impact, can be targeted by reductions in cattle.

Five climate change mitigation ideas:

1. Reuse whatever you can, but for environmental purposes.
2. Use environmentally friendly protein.
3. Utilize energy more efficiently.
4. When you get new products, think outside of the box.
5. Make a difference with your initiatives.

Structural changes in the energy system, such as shifting to low- or no-carbon energy sources such as nuclear power and carbon sequestration technologies, energy-efficiency improvements, and changes in land use, are all means of mitigating or reducing greenhouse gas emissions.

It is easy to convey the concept of mitigation but challenging to implement. We must move away from fossil fuel dependence and embrace clean and renewable energy. To keep reaching net-zero carbon emissions, we must stop deforestation and habitat loss. Once greenhouse gases are released into the atmosphere, they must be offset by capturing and storing those gases in places like tree roots. Like saving for retirement, it is in our best interest to mitigate the impact of climate change now rather than wait for its effects to take place. In the beginning, progress has been slow, but momentum is picking up.

37.6 Policy in Practice

We should stop using fossil fuels and start using renewable energy sources like solar or wind to stop climate change. At the 1992 United Nations Conference on Environment and Development (UNCED), the city of Rio de Janeiro hosted delegates from all over the world. At COP-3 in 1997, the Kyoto Protocol, the first legally binding greenhouse gas reduction targets, was adopted.

Environmental crises were addressed globally with the emergence of an activist and formal effort in the 1970s. In contrast to national climate change policy, which emphasises cooperation and establishing international guidelines to combat global warming, international climate change policy has emphasised collaboration and cooperation between nations. UNFCCC is an accepted international agreement that has continually adapted to meet new problems [1].

In domestic climate change policy, national governments have focused on adopting internal measures for reducing greenhouse gas emissions while also incorporating international agreements into domestic legislation. The

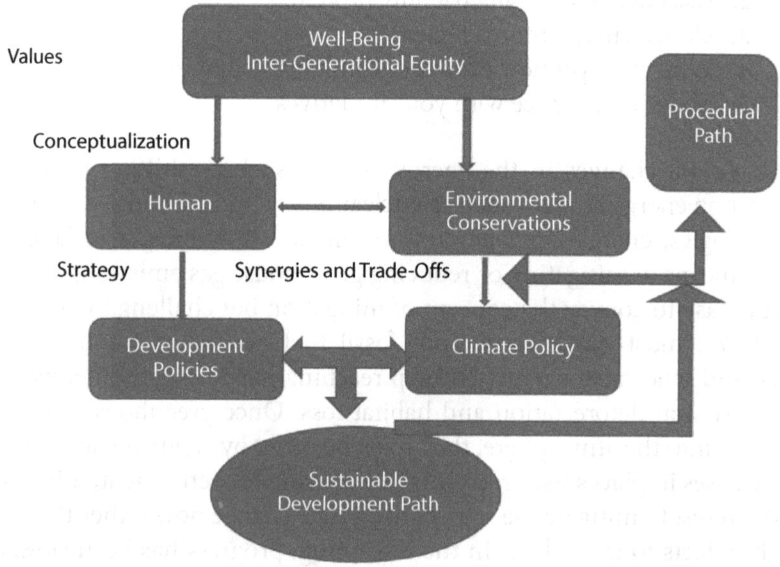

Figure 37.3 Sustainable development and climate policy.

diagrammatic representation as given above in Figure 37.3 the sustainable development can be seen in climate policy.

Abandoning efforts to lessen the impact of global warming and planning for unavoidable change to the human environment became the norm in the latter half of the 20th and early 21st centuries. The vulnerability-based policy has also arisen to help those who are most affected by environmental anomalies. Concerns have emerged over the treatment of developing nations and the lack of gender-specific action in climate policy.

India's global climate commitments should be increased by 55% for a 50% chance of averting a global temperature rise of 2 degrees or more. If it increases the commitments by 147%, it is a 95% certainty.

Table 37.2 Important events in international climate change negotiations.

Year	Location	Outcome
1992	Rio de Janeiro	UN Framework Convention on Climate Change (UNFCCC). Countries agree to reduce emissions with "common but differentiated responsibilities."
1995	Berlin	The first annual Conference of the Parties to the framework, known as COP. U.S. agrees to exempt developing countries from binding obligations.
1997	Kyoto	At the third Conference of the Parties (COP-3) the Kyoto Protocol is approved, mandating developed countries to cut greenhouse gas emissions relative to baseline emissions by 2008-2012 period.
2001	Bonn	(COP-6) reaches agreement on terms for compliance and financing. Bush administration rejects the Kyoto Protocol; U.S. is only an observer at the talks.
2009	Copenhagen	COP-15 fails to produce a binding post-Kyoto agreement, but declares the importance of limiting warming to under 2°C. Developed countries pledge $100 billion in climate aid to developing countries.
2015	Paris	At COP-21, 195 nations sign the Paris Agreement, providing for worldwide voluntary actions (NDCs) by individual countries.

A World Wide Fund for Nature (WWF) report issued in 2007 claims that the Indus River will most likely dry up because of water scarcity. Climate change is causing heatwaves to become more frequent and powerful in India. Landslides and floods due to heavy rain are predicted to occur more frequently in the states of Assam [13].

The Intergovernmental Panel on Climate Change has predicted that global warming could reduce India's GDP by up to 9%. Major crops, such as rice, will likely decrease production by 40%.

Climate change is an international environmental issue. Climate change is an excellent public issue in economic theory, requiring global collaboration to achieve effective results. Since the United Nations Framework Convention on Climate Change (UNFCCC) was first established in 1992, there have been extensive international discussions, known as Conferences of the Parties, or COPs, aimed at reaching a global agreement on emissions reduction (See Table 37.2). The international climate change is discussed as given above in Table 37.2.

37.7 Green Economy

A green economy is defined by its low carbon footprint, reliance on renewable resources, and social equity. Employment and income growth are fueled by investment in infrastructure, assets, and the environment, which enables carbon emissions to be reduced, energy use to be more efficient, and contributes to avoiding the loss of the ecosystem.

A green economy prioritizes environmental protection while also focusing on economic growth. Ecological economics is closely related to economics but places a greater emphasis on society. The 2011 UNEP Green Economy Report asserts that "An economy must achieve both efficiency and equity objectives." UN Environment Programme says that an environmentally friendly economy helps cut environmental risks and ecological scarcities while providing more significant social equity for both people and the environment. Governments, in conjunction with international collaboration, play a critical role in fostering the green economy. According to the reports, an eco-friendly economy grows faster over time than an ecologically brown economy and protects and restores natural capital.

The green economy is a critical component of the broader effort to achieve sustainable development. Resource efficiency is concerned with the most efficient use of resources to create value. It seeks to minimize the number of resources required, waste generated, and emissions generated for each unit of product or service. Investments, employment, and skills are all critical components of a macroeconomic approach to sustainable economic growth. Public investment, policy reforms, and tax and regulatory policy changes are all required to support these green investments. The United Nations Environment Programme recognizes natural capital as critical to a country's economy and as an asset from which citizens have a right to benefit. Despite the widespread belief that there is no need for a distinct focus on sustainable development, a specific category within the green economy aims to improve economic outcomes [12].

Green Economic research focuses on significant areas, viz, establishing regional, sub-regional, and national platforms to promote macroeconomic and microeconomic policies. It prioritizes access to green finance, technology, and investments, exemplifies Green Economy strategies with a decentralized focus. The term "Mainstreaming" refers to providing financial and macroeconomic support to countries transitioning to a Green Economy. UN Environment has expanded its outreach to include the private sector; private companies and non-profit organizations play a critical role in promoting resource efficiency and a green economy. In contrast to the traditional economic order, it is distinguished by an accurate valuation of nature's capital and ecological services. It generates economic value and a comprehensive accounting system that allocates all of the external and societal costs related to ecosystem neglect to the entity responsible for the negligence.

Consumers are becoming more concerned with environmental stewardship. Sustainable development is associated with ecolabels, which have begun to appear on Green Stickers, identifying vehicles that help the environment by reducing pollution in the air. Numerous industries are starting to implement these standards to bolster their efforts to promote environmentally friendly practices in the face of an increasingly globalized economy. Sustainability guidelines ensure that the things you purchase have no adverse effect on the environment or the people who manufacture them. These standards are becoming more prevalent, and they have the potential to usher in a new, more environmentally friendly economy [8].

Economic growth has now become a global priority for the benefit of all. Economic growth is critical for governments, the private sector, and civil society organizations; it is essential for all three. We can progress toward green goals if we can reduce our reliance on fossil fuels while maintaining economic growth. As stated in the UNEP report, we can maintain current production and consumption rates while improving social equity and mitigating environmental impact. At the same time, green economies can be defined as more complex economies that prioritize the environment and social equity. It grows faster than brown economies, misleading, as green economies are a distinct economic category that cannot be moulded into a single form.

There are so many working models for the successful implementation of the green economy. Each model is merely a simplified representation of the underlying system. These are intended to assist us in comprehending the subject. To simplify a model and its output, it is necessary to design and evaluate the model appropriately. However, this model has a flaw: it views the world as a single entity. The model stores all of its data in a single global average. Models will exhibit an overall correlation but no correlation between any two of the model's data points.

While relying on global averages for other purposes, such as global greenhouse gas emissions modelling, international standards can be misleading when it comes to data for local or regional concerns. Because regional emissions increase at varying rates, global emissions are estimated using individual country or region estimates, with significant discrepancies between the two. This analysis concludes that global models, at least when it comes to green versus brown growth, are highly premature and almost certainly inaccurate.

Despite all contradictions, the UNEP investment model favours the green G2 scenario over the brown BAU2 method by allowing for more investment. The assumptions made about the wealth disparity focus on a truly green economy that must balance social equity and environmental concerns.

To summarise, significant reductions in carbon dioxide emissions and adequate control of the wealth disparity (as required by a fully green economy) are highly improbable to occur concurrently.

37.8 The Relationship between the Economy and the Environment

The UN Conference on Sustainable Development was founded in response to grave threats to the global economy. In this context, the conference's selection of "green economy" as one of two primary themes was consistent with the emergence of substantial energy to determine the relationship between the rise of "greening" and the current global financial crisis. The primary objective is to ascertain what distinguishes and differentiates the twenty-four sources on the green economy (which include research papers, government and non-governmental organisation policy documents, and other documents). Three patterns emerge when the relationship between emerging concepts in environmental discourse and Dryzek's classifications are examined. The significance of greening for sustainable development is demonstrated through practices, a shift in language toward economy and polarisation, and socioeconomic paradigms' polarisation.

Environmental policy becomes a subject of price reduction and market incentives by eliminating two sorts of failures. Another recommended alternative is the creation and implementation of new markets, such as pollution licences.

As shown above, neoclassical concepts about environmental policy are of interest as part of a pluralistic strategy. Social economists and feminist economists provide alternate ways in which people and their interactions are thought.

The second ideological trait of the neoclassical economy is its economic assumptions and its emphasis on self-interest. Each actor should act in his interest. This belief legitimises behaviour that shows little or no respect for others. Neoclassical economists often argue that it is unrealistic to expect people to work against themselves. Institutional economics also provides ethical and cooperative issues as a unique economic theory. In response to a single motivation, the actor does not act consistently. Each individual therefore approaches or deviates differently from the concept of sustainable development.

Our third point concerning neoclassical economic theory's ideological aspects refers to the emphasis on monetary indicators. The corporation is the sole entity in neoclassical microeconomics and is supposed to maximise monetary earnings. Corporate accountability is also entirely financial. Investment decisions at the corporate level are determined in economic terms using the cost-benefit analysis (CBA). This might be regarded as an attempt to assess the company's profitability. The GDP is an essential measure in domestic accounting practises. GDP is a component of macroeconomic policy, and GDP is utilised, despite its obvious limits, as a measure of progress.

The examples above show that monetary indicators in neoclassical theory are crucial. Quantification and quantification are preferable in financial terms. From a critical point of view, this is known as "monetary reductionism."

Although neoclassical enterprise theory and economic policy contributions can remain essential and beneficial as part of a pluralist approach to the economy in specific settings, a neoclassical approach alone cannot solve complicated sustainability concerns. For example, a different conceptual framework is required to address corporate social responsibility and environmental systems. Accounting systems must systematically handle non-monetary performance at the organisational and national levels.

Our fourth and final ideological part of neoclassical economics is our belief in reasonable prices for an optimal distribution of national resources. When neoclassic economists advocate the development of infrastructure investments using a cost-to-benefit (CBA) analysis (such as highways, dams and energy systems), they recommend something that equates to value and ideological dictatorships. In a summation process, actual prices on the market and sometimes hypothetical prices for different consequences must be used to reach the best and best alternative.

This concept of fair market prices for the effective allocation of resources is based on a social consensus on the usefulness of the CBA evaluation approach. As Ezra Mishan, a textbook author on CBA argues, assuming such agreement is no longer realistic due to the wide range of views on environmental issues in contemporary society.

37.8.1 Assessing Pollution's Environmental Impact

Industry and the resources found in ecosystems benefit modern societies significantly, as these resources serve as the foundation for the products produced by these societies. Not only does our environment provide us with food, fuel, and fibre, but it also helps to regulate disease and climate spread, provides recreational opportunities, and fosters a sense of well-being. Nonetheless, industrialization and environmental degradation are frequently linked, as industrial expansion may result in pollution and ecosystem degradation, requiring further industrial development.

We must have a rigorous monitoring technique to ensure that pollution remediation is effective over the long term. Numerous advancements have been made in the field of freshwater monitoring; however, some obstacles remain. The first step in determining what "reference condition" means is to define what it is. Numerous statutes, including the WFD, use this phrase to evaluate an ecological community in light of "reference circumstances." On the other hand, it is not easy to find or construct an authentic example because of our long-term and widespread impact. It is widely accepted that the goal of environmental pollution reduction should not be to return to pre-industrial conditions but to ensure the long-term sustainability of water resources. Alternatively, Gell believes it is preferable to establish baselines before industrialization to safeguard against future dangers caused by pollutants.

In terms of ecological consequences, significantly less progress has been made in monitoring terrestrial and marine ecosystems. Rather than using a freshwater filtration system, where the physical environment is very different, it is proposed that sediment toxicity testing be conducted in marine environments to determine the severity of lethal and sub-lethal effects. The concept of sub-lethal effects being critical in other systems is also discussed concerning the development of toxins. There is a need for novel toxicity tests due to the current shortage of subtle endpoints in existing trials. Soil contamination assessment is notoriously tricky due to the extreme diversity of soil and its high variability in bioavailability. Although attempts have been made to develop Soil Guideline Values for various contaminants, the shortcomings of these chemical assessments are severe, and no link to the area's ecology has been established. Recent progress has been made in establishing a more direct connection between land contamination and its communities through a risk-based approach. This advancement will be closely monitored in the years to come.

37.8.2 Ecosystem Recovery and Rehabilitation

Whenever a contaminated site is cleaned and restored, it is critical to specify success criteria for the cleanup work. On the other hand, it is not clear where it should establish its cut-off point. Should the objective be to restore a specific community composition, such as the presence of a rare species, or should it be to restore the environment to its pre-contaminated state? There appear to be two significant impediments to restoring an ecosystem to its pre-industrial state. This first point is that there is a shortage of high-quality data describing the components of an environment, whether biotic or abiotic. While attempts to use palaeoecological data have been made, these efforts have been unsuccessful due to preservation issues and inconsistent techniques and data analysis applications. Until recently, little effort has been made to learn from more recent data. However, it is evident that even with sufficient time, the community will not revert to previous levels of engagement. According to all sources, the physicochemical environment has improved significantly due to either reduced industry or the use of active remediation. Regrettably, this improvement does not result in a complete restoration of the community to its pre-contaminated state. Because colonizing organisms are scarce, the environment lacks physical habitat. Environmental pollutants other than those targeted by remediation play a significant role; these constraints on recovery are almost certainly due to a scarcity of sources of colonizing organisms, a lack of physical habitat, additional pollutants in the environment other than those targeted by remediation and the movement of industrial waste. Additionally, even if the ecosystem's health has returned, there is a shortage of data on the causes of ecosystem dysfunction in recovering ecosystems [15].

When organisms adapted to live on contaminated industrial sites are present, they may present a unique opportunity to use them for site stabilization or active remediation. For several years, the use of metallophyte plants to remediate metalliferous soils has been proposed and implemented; however, due to a lack of understanding regarding the mechanisms of adaptation and metabolic and genetic responses to pollutants, metallophyte plants are generally underutilized. Bacterial remediation of contaminated land has advanced, mainly where organic pollutants or radionuclides are present. The enormous potential for microorganisms to be used in various situations, particularly those involving mixed contaminants, is enabled by a better understanding of the genetics and mechanisms of action of these organisms.

While industrial pollution is decreasing globally, the environmental consequences of industrial activity continue to exist, and changes in industrial processes have introduced new pollutants into the environment. This is critical because a lack of understanding between the contaminants contaminating the ecosystem and the environmental responses at both the individual organism and ecosystem level can create uncertainty about what should be done. Without achieving this, we will not fully appreciate the consequences, devise the most effective monitoring strategies, or design remediation strategies.

The WFD defines ecological status as "undisturbed conditions," which requires the identification of reference states. The WFD reference condition has been defined as a state "in the present or the past, corresponding to extremely low pressure, with only minor changes in physical, chemical, and biological characteristics, and without significant industrialization, urbanization, or agricultural intensification." While such a state may not exist today, the question is whether we have sufficient data to assert that it did exist in the past, and if not, how far back in time should we look, and do we have all the necessary information?

The ideas of biological integrity and ecosystem health are widely debated when it comes to ecosystem quality. Maintaining an ecosystem of balanced, integrated and adaptive organisms requires a variety of species composition, diversity and functional organization similar to the diversity of the habitats of the geographic region. Natural definitions mean that human influence or technological use is not available. The concepts "balanced, integrated and adaptive" do not consistently give suitable criteria for assessing the state of the ecosystem without additional explication. This leaves the idea of naturalness alone. The notion of biological integrity is applied to detect variations of natural reference conditions induced by human activity. Humans impact ecosystems, but this has no detrimental effect on the ecosystems' biological integrity. As a result, these environments serve as a baseline for assessing the magnitude of human-induced changes to ecosystems. Because integrity is defined in terms of a naturally defined natural reference state, it implies that it must either already exist or refer to something historically established.

Ecosystems cannot be defined in isolation from human activity or technology, as they have been influenced by human activity in many parts of the world for centuries. Numerous large temperate lowland river ecosystems, such as the Amazon, no longer exist in their natural state in a large number of parts of the world. Natural reference states can be defined in various ways, including through the use of historical or paleo-ecological data. Excellent historical data are scarce, while excellent paleoecological data are almost nonexistent. Although historical data on many landscapes may or may not be available, humans have altered the landscape significantly over hundreds or thousands of years. Additionally, changes that occur naturally over time and disruptions caused by other factors make it more difficult to distinguish between human-induced and natural changes.

Removing the majority, if not all, of human activity from an area may be impossible, posing a fundamental practical issue: returning to a natural state may require the removal of the majority, if not all, human activity, only in areas where human activity is restricted by nature.

The ability of the worldwide community to progress energy systems to satisfy future growth demands while also decreasing greenhouse gas emissions will eventually depend on the introduction of a significant number of new technologies over the next several decades. According to further microeconomic analysis, key low-carbon technologies could account for up to 11% of projected GHG abatement capacity by 2030 and up to 27% by 2051.

Numerous industry experts believe that one of the most significant contributions technology will make to a low-carbon future in the coming years will be expanding low-carbon options and making them even more affordable. This requires lowering the cost of driving technologies through new research, engineering, and mass-market introduction. We believe that making such complex, long-term investments in cutting-edge research, development, and deployment will require both public and private sector efforts to be stepped up and more coordinated.

37.8.3 Sustainable Development Ideology

Sustainable development can be considered an ideology or a collection of specified ideological norms. Several interpretations of sustainable development can be generated and explored to prevent unsustainable trends and become more sustainable.

In the Brundtland report, sustainable development became a political issue. It enables people to broaden their ideas as players in space and time; it deals with the interests of present and future generations. Although no one can be selfless, everyone can take action by considering the ethical implications of their conduct and judgments. Structures of institutional and incentives can be changed to enable such behavioural shifts.

As the name of the World Commission indicates, the Brundtland report contained a section on the environment and natural resources. In future generations, permanent natural resource deterioration is a problem of inequality that needs to be prevented or minimised. The natural resource basis for future generations should be improved whenever possible. Some points from the report are detailed below.

Substantial institutional reorganisation. The Actor feels that while tiny institutional adjustments are essential in the right direction, major changes must also be addressed in the current political-economic system.

Every society or community has a natural tendency to avoid contentious issues and deal with those in positions of power opportunistically. For example, joining the mainstream makes life easier for economists, and something similar holds for society. Accepting mainstream perspectives and behaviours, on the other hand, may exacerbate rather than resolve sustainability problems.

Public discourse is encouraged rather than shunned in a democratic society. Tensions between supporters of conflicting views (and even narrow-minded conflicts of interest) are respected and are necessary for a functioning democracy.

37.9 Industrial Ecology

Industrial ecologists are concerned with the utilisation of natural resources and waste management in the industry. To understand contemporary society, it is necessary to comprehend the flow of materials and documented industrial processes and their chain of custody. The environmental concerns of industrial ecologists include the use of natural resources and waste management in the industry. The interdisciplinary field of study of industrial ecology is growing rapidly in popularity.

The term "Industrial Ecology" (IE) refers to the study of the interactions between industry and the environment, as well as measures for assessing and mitigating their effects. This fundamental concept demonstrates how complex and involved it is to discover environmental effects, classify them according to their risk level, and then provide the appropriate cure to minimise each impact while taking their relationship into account. Industrial ecology is a concept with applications for both microeconomics and macroeconomics. It can be implemented within a single organisation and expanded regionally/globally.

The content of industrial ecology is derived from an analogy with a biological system: energy and materials are utilised or devoured by the animals in this system, and waste generated by one organism is utilised or transformed by another organism's activities. Industrial systems having linkages to other systems must also be seen as a component of a greater whole.

Industrial ecology is the scientific study of industrial systems, such as factories and their interactions with the surrounding environment. Industrial systems can be compared to natural ecosystems, in which resources and infrastructure coexist with natural capital in an "ecosystem" framework. It aims to inspire sustainable design through the belief that natural systems do not generate waste.

Encouragement of industrial ecology is one of Natural Capitalism's four goals. This strategy discourages amoral purchasing, based on the assumption that what happens in the distant world does not affect the larger world. Furthermore, this strategy places a higher premium on natural capital than on instructional capital.

The linear view of industrial activity is problematic since there are limited natural resources, and the environment cannot manage all waste safely. The logical application of the concept of cyclical evolution inside and between the system components seems to be an analysis of the biology system. That means a system that reuses goods and trash from one element to another so that less material and energy are consumed and less waste is generated. It introduces a new technique to analyse and design industrial processes, with the primary objective of decreasing waste consumption, energy consumption and emissions from all kinds of waste [10].

A critical topic to emphasise is the conflict between industrial ecology and the concept of waste. The industry manufactures commodities by utilising materials and energy and implicitly produces waste and pollution during the manufacturing process. Following the manufacturing process, we must examine the product's end of life or the point when it becomes unusable and must be disposed of. Using an industrial ecology approach, the manufacturing process must avoid converting raw materials into useless goods to the greatest extent possible. Every waste must be viewed as a source of resources and energy.

Businesses are critical in achieving industrial ecology, first and foremost because the company serves as the arena for improvements and innovations, where ideas are executed. Second, as a result of the management process, an increasing number of organisations view control and control processes as inefficient and even detrimental. Industrial ecology is considered a more adaptable and collaborative approach to process management that focuses on balancing economic and environmental goals concurrently.

37.9.1 Ecosystem's Health and Adaptability

A balanced and self-sustaining ecosystem is defined as "stable and long-lasting, with no loss of independence or environmental sensitivity over time, as well as resilience to external stress." The crux of the problem is that ecosystem health is an ever-evolving concept. Satisfying these criteria necessitates only a brief period of change. A critical distinction between integrity and aspiration is that the latter is aspirational rather than existing.

The concept of resilience is critical when considering the health of ecosystems. In resilience assessment, the crucial question is: Which disturbance is the most difficult to overcome? This will indicate whether or not you possess a high level of resilience. Due to the scarcity of known perturbations, it is impossible to use resilience as a proxy for the capacity to absorb or recover from them. Ecosystem resilience cannot be quantified because we cannot predict the dynamics of stressed ecosystems with certainty.

Resilience is irrelevant for ecosystem protection management. While resilience is necessary for success, a high level of resilience may impair the system's ability to transition to another ecosystem state. Numerous critics of the concept of ecosystem health argue that it is impossible to define an ecosystem's ideal state due to the way its various components interact.

When we define an ecosystem's integrity in terms of its resemblance to a state free of human influence, two primary concepts emerge: the quality of an ecosystem can be defined in terms of its ability to maintain its structure over time, or it can be defined in terms of a simplified form of biological integrity. While establishing "natural reference conditions" has proven complex, numerous environmental statutes, including the US Clean Water Act and the EU Water Framework Directive, have adopted the concept of getting as close to the natural state as possible.

A desired or normal state must be defined to apply the ecosystem health concept. When the "natural state" is used as a reference point, the idea of ecosystem health (i.e., ecological integrity) becomes synonymous with the concepts of ecological stability and resilience.

Humans put ecosystems under both direct and indirect pressure. Immediate requirements include food, water, and fuel; indirect requirements include maintaining and regulating ecosystem health. Apart from providing ecosystem services, humans can influence them in various ways, either directly or indirectly. Improved living standards, decreased infant mortality, and increased life expectancy is frequently associated with economic development, also referred to as industrialization. Economic development is a necessary component of financial growth promotion. Fossil fuels and timber harvesting are used to meet energy demands, and waste is disposed of in waterways, landfills, and the environment [9].

Because ecosystem services are interdependent, managing or optimizing one service may have unintended consequences. The importance of comprehending and managing the complex trade-offs between the various demands society places on ecosystems increases as human influence on the world's ecosystems grow. Since previous definitions of ecological quality do not consider these issues, we must develop new ways of defining ecological quality that enable us to address these issues, weigh the benefits and drawbacks of various approaches, and apply it at the most miniature scale possible policy implementation. Essentially, we're going to test these definitions' long-term robustness and generality and then see if there are any operational difficulties. We introduce one possible definition in this article, examine a few properties that a measure of this type may possess, and discuss some of the implementation issues that may arise. We will use the issues surrounding catchment management and river quality to illustrate the concepts at hand.

First, we advise that certain characteristics may be advantageous, or even critical, in developing a high-quality ecological concept that emphasizes ecosystem services. The following is a list.

This allows the notion of quality to depend on the context. In some cases, "good" quality does not always indicate consistency; what is an acceptable and meaningful objective for a rural river might not be appropriate for a rural river in an urban situation. Both systems provide ecosystem services, but the balance of those services differs significantly between the two circumstances we aim to recognize.

When evaluating the quality of something, it should never be relied exclusively on the execution of a particular function or service, or very seldom. If services are required in any system, there will always be explicit demand for many services. If a process or function is altered, unanticipated adverse implications may emerge on other not anticipated or addressed system components.

The evaluation criterion for functions and services is the sustainability of the system and not exploitation potential. Although the fast rates of supply of goods and services in certain types of ecosystems may be relatively high, these rates are unlikely to continue over the long run. Since each service provides evaluation and the following quality measurement are dependent on a sustained level of service provision, each service provides evaluation. The resulting quality measurement must be carried out in as near a sustainable manner as possible.

The ability to incorporate existing ecological assets into the assessment process, but not entirely. Specific existing criteria may be quite dissimilar; for example, in freshwater ecosystems, certain criteria are related to the concentration of certain chemicals or the toxicity of certain species, which are typically quantified in legislation. In the middle of the scale, "naturalness" or pre-human simplicity may still be criteria worth considering, but they must be viewed holistically, not objectively.

The numerous services included here are linked to a facility that incorporates pre-existing criteria and the concept of service demand. Additionally, if demand and supply can be quantified for various services, problematic mismatches in which demand exceeds supply point to potential management locations. Theoretically, correcting such a mismatch would require either an increase in supply or a decrease in demand. The central issue here is that increased demand for something can result in improved quality. Domestic water supply management entails both supply expansion and demand regulation.

Additionally, quality has a third characteristic: it is not absolute. It is critical to note that while there is no single objective that encompasses "good quality," the following goals can be used to help make up the difference: for instance, if a waterway is in high demand as a component of the recreational, conservation, or aesthetic services provided by a system, that feature would be considered when determining the overall system's quality.

It should be possible to create an ESP and use it to manage the environment. Numerous stages are required to deliver such a service profile, and each step presents significant challenges.

Above all, a set of services that meet the system's requirements must be specified. While the list may seem self-evident in some cases, we recognize that creating such a list is frequently a complex problem to solve. It must either be predefined "core" lists applicable to specific system types or be customized for each system. While the involvement of stakeholders is likely to benefit, another critical consideration is that attention should be paid to non-obvious services that could add significant value to the system if added. Additionally, flood regulation serves as a helpful illustration: if the current system of flood regulation works well, the absence of catastrophic flooding may mean that most people are unaware it is in place.

Following this stage, the team would investigate methods for quantifying the services. Concrete measures must be used to quantify various benefits, and they must also take into account data that is already available or can be easily generated. Although direct assessment of flood regulation is limited, it may be simpler to examine factors such as the area's estimated maximum water retention capacity based on land use and floodplain capacity. Similarly, assessing sustainable provision may necessitate a less direct approach. The practical utility of this metric is limited by the expense and time required to collect long-term data. Despite this, these issues apply to a variety of contemporary management scenarios, and the use of indicators is already widespread in various applications, including biodiversity conservation.

37.10 Does Protecting the Environment Harm the Economy?

Green growth is frequently viewed as complementary by government policymakers, with an emphasis on energy efficiency and clean energy technologies, as well as on sustainable development. Green economic growth will not produce sustainable economic development unless the degradation and loss in the environment continue to worsen. The world faces an increasing number of problems related to environmental resources scarcity and the loss of many benefits and services provided by these systems. Overcoming these problems will require addressing additional sustainability and funding challenges. Confronted with massive amounts of the market, policy, and institutional failure that have prevented us from grasping the economic value of this scarcity, the sustainability challenge is to overcome these multiple failures. There is a disconnect between the value we place on ecosystem services and the amount of money we are willing to invest in preserving them. To address the sustainability challenge, it is necessary to improve economic and scientific analysis of ecological scarcity, as well as to more thoroughly value and explain to policymakers the benefits lost. By utilizing a variety of financing mechanisms, ranging from providing ecosystem services via international payments to levying taxes on financial and currency transactions to establishing international financing facilities, the funding issue will be alleviated.

37.10.1 Market and Accounting Mechanism

Climate change can have a mixed effect on businesses while also significantly impacting product prices and market advantages. One of the most significant impediments to the transition to a low-carbon economic growth model is a shortage of comparable, comprehensive, and reliable climate data from corporate emitters. The collaborative efforts of business and environmental non-governmental organizations have resulted in an agreement on preparing entity-level carbon emissions inventories: the Global High Protocol. The Carbon Disclosure Standards Board is working with these communities to develop a general standard for annual reports that includes information on emissions inventories, risks associated with carbon emissions, and company management strategies. The direct emissions component of the framework is based on the GHG Protocol. The most effective method of establishing international climate change policy accounting standards is for governments to direct their securities and to account regulatory bodies to initiate these game-changing processes with the ultimate goal of establishing a set of world-class international accounting standards.

However, it is critical to establish a link between national and regional carbon pricing. These initiatives seek to increase the economic efficiency of our climate change efforts while also encouraging low-carbon investment, particularly in developing countries. We effectively achieve our emissions reduction goals by setting ambitious national targets and linking regional and national emission reduction schemes early and effectively.

International and domestic carbon credits are the most crucial areas of policy convergence, rules to supervise reporting and verification, methods to prevent excessive market fluctuations, and financial intermediaries' role. Other strategies for decreasing emissions not currently targeted effectively by international climate policies could, if correctly conceived, assist in lowering emissions in areas and programmes not presently targeting climate policies.

Carbon markets can also be used to encourage significant investment in sectors and regions outside the emissions trading system (ETS) by enabling compensation to be used as a substitute for compliance. An emission tax, by contrast, mandates that individual polluters pay for every tonne of GHG discharged into the atmosphere. Taxes and carbon markets, which are related to polluting behaviour, impose greenhouse gas emission prices.

In the developing world, those who stand to benefit the most from private financing will benefit the most from a system of state finance tailored toward leverage maximization. The majority of cash could come from institutional investors if a risk-reward balance is struck.

37.10.2 UN Environment Program (UNEP)

The UNEP Finance Initiative and the UNEP Sustainable Energy Finance Initiative are two excellent examples of the UNEP Sustainable Energy Finance Initiative. Recently, there has been a surge in collaboration between UNEP programmes and other private finance networks. At 450 parts per million CO2 equivalents, the United States would need to invest \$550 billion in renewable energy by 2030 to prevent global warming to two degrees Celsius. The UNDP estimates that adaptation will cost \$86 billion. Consequently, private sources of finance or newly developed funding mechanisms will be required in the upcoming years to supplement public funding. Even at current levels of Official Development Assistance (ODA), the costs of long-term development are unlikely to be covered. ODA contributes between \$5 and \$7 billion each year, or less than 1% of the total required. Currently, the global community is building a slew of new financial and alternative financial tools based on the market to incentivize businesses to locate and produce low-carbon and climate-resilient products and services. Firms invested more than \$150 billion in renewable energy technologies as a result of these new regulatory and financial incentives. However, it is unclear if these flows will expand beyond the OECD and a few fast-growing economies in the foreseeable future. Such barriers will need to be overcome for greater access to occur.

Both private and public funding is necessary to sell a variety of new technologies and to keep ahead of global climate change damage. The government must fund the primary research and development (technology research) in the early stages. On the other side, they are producing more market-potential technologies that attract risk capitalists and private equity financiers who desire higher potential risks and better profits from developing technology with greater market potential. When new technology is shown, funding can be secured through public equity markets or procurement. Access to debt and carbon money can help to get further funding for large-scale deployments of well-established technology.

The Kyoto Protocol established a Clean Development Mechanism (CDM) to promote a reduction of emissions of sustainable development and greenhouse gas (GHG) both internationally and in developing countries. A Cap-and-Trade facilitates credit to developing countries for carbon cuts that they then can use to grow their economies or sell at a discount to industrialized countries. There is concern that, despite its capability, the CDM will reach only a few nations and bypass Africa entirely.

The CDM can be quite beneficial in this regard, as it enables developing countries to acquire emission reduction credits and then sell them to developed countries. UNFCCC expects CDM investment in the range of \$10 billion to \$100 billion by 2030, depending on emission cutting targets and CO2 pricing. A recent World Bank analysis has shown that Africa might add up to 170 GW by implementing CDM-eligible projects. It is over four times the current modern energy production in the region.

Many sustainable land-use initiatives and low-carbon technologies are having difficulty recruiting investors due to present market rules. The development of novel market-based and innovative financial instruments must consider the unique market conditions prevalent in developing countries. Numerous CDM enhancements are being considered to accomplish this purpose (programme approaches, etc.). To maximize new financial prospects, developing countries will require assistance in establishing an enabling environment. A new cooperation order is required—one that prioritizes the development needs of poor nations while enabling them to rapidly transition to a low-carbon path that outpaces North American economic trends.

37.11 Creating a Green Economy

Climate models all predict rising temperatures unless we take action to reduce greenhouse gas emissions. Additionally, there is an implicit agreement among economic models, expecting rising costs without effort. Based on an experimental model survey, the Congressional Budget Office estimates that Waxman-Markey will reduce GDP growth by 0.03 to 0.09 percentage points annually between 2010 and 2050.

According to one recent study, the cost of a more aggressive climate policy—roughly equivalent to the policies proposed in current legislative proposals—could range between 1% and 3% of global gross domestic product (GDP). Typically, these figures are derived from a comprehensive model that incorporates engineering and market estimates. Calculating the cost of electricity generation from various sources at various resource prices is just one example of the information that will be included in these reports. After analyzing the data, a breakdown of how the energy industry and customers are expected to respond to carbon pricing policies and how much the system will ultimately cost the country can be deduced.

While it is highly improbable that these models capture all aspects of climate change accurately, we can conclude that they significantly overstate the economic costs of the action. The facts uncovered by the acid rain cap-and-trade programme indicate that prices were considerably lower than anticipated[16].

Unfortunately, those who oppose any form of emissions-reduction policy assert that doing so would be economically disastrous. Cap-and-trade juxtaposes this extreme pessimism with standard conservative rhetoric. Indeed, contemporary conservatives place a high premium on the benefits of market incentives, particularly those favoured by Ronald Reagan in the 1980s, who referred to them as "the marketplace's magic." They believe that the capitalist system can deal with all types of constraints, including resource constraints, and that technological advances, such as those found in renewable energy can help overcome these constraints.

As of now, it is estimated that global temperatures will rise by 9 degrees Fahrenheit by 2100 if business as usual continues. It's a significant difference in terms of the average temperature difference between the two locations. This is such a significant change that it must be disruptive. Additionally, global temperatures would continue to rise, resulting in increased chaos.

37.11.1 Green Project Financing

Green financing's mission is to increase financial flows between the public, private, and nonprofit sectors to facilitate sustainable and environmentally conscious projects. To succeed, it is critical to improve risk management for environmental and social risks, to identify profitable and environmentally friendly business opportunities, and to instil accountability in business practices.

Green financing will be facilitated by regulatory framework adjustments, consolidation of multiple public financial incentives, and increased investment in green technology.

To guarantee that financial systems are linked with the Sustainable Development Goals, the United Nations Environment Programme has been collaborating with nations, financial authorities, and the financial sector. Capital markets underpin today's globalized economy. Through this system's various financial markets, banks and investors allocate capital and funding to multiple sectors of the economy. The capital budget determines how ecosystems, production, and consumption patterns will be shaped over the next decade.

Additional work is required to scale up current green financing efforts consolidating the public sector to enable it to create an enabling environment fostering collaborations between the public and private sectors in areas such as green bond financing. With financial assistance, disadvantaged individuals are trained to become microentrepreneurs.

37.11.2 Natural Capital Sustainably

As long as natural resources are available, economic and social development are contingent. While increased primary, secondary, and tertiary sector dependence on natural resources to power these sectors and meet rising demand, these regions have in the meantime seen a significant increase in demand for natural resources, Asia and the Pacific have seen increasing demand over the last four decades. The region's natural resource use is inefficient and unsustainable, but the process of developing a method for quantifying and valuing that resource use has only just begun.

The 12.2 SDG is to sustainably manage natural resources by 2030. There are assessments and evaluations, raising awareness of data and developing capacity Sensitization. UN Environment Asia and the Pacific launched the Indicators for a Resource Efficient and Green Asia Pacific in 2015 with a regional database and assessment report. This has resulted in materials for communicating about Sustainable Development Goals indicators for natural resources as well as a training programme for those working in the field.

SDG 2.4 assists in achieving sustainable food production by 2030. It is recommended that we enable sustainable food systems while also implementing food production and agriculture practices that contribute to ecosystem preservation, increase resilience to climate change, extreme weather, drought, and flooding, and improve the overall quality of land and soil.

Numerous families in developing countries could benefit from sustainable food value chains, which could pave the way for these families' future development. However, food value chains are intricate and complex systems. While there are numerous reasons for a chain to become unsustainable, there is not always a clear connection between the root causes. The vicious poverty cycle is composed of a series of obstacles that must be overcome in tandem. This requires collaboration among diverse organizations, including farmers, agribusinesses, governments, and civil society organizations, to ensure value chain alignment and cooperation. On every continent, policymakers and practitioners from public, private, and nonprofit organizations collaborate to develop and implement forward-thinking solutions that will sustainably affect the overall system. Significant knowledge is acquired during the process.

37.11.3 Partnerships

In cooperation with a wide range of stakeholders, UN Environment works to help industries adopt new consumption and production patterns. Even though governments and not-for-profit organizations will continue to play a crucial role in expediting the transition to a resource-efficient and green economy, our private sector partnerships are gaining ground in their function as a mechanism for attaining that aim.

All stakeholders—including policymakers, academia, and the private sector—must collaborate to make sustainable lifestyles a reality. To ensure the region's sustainability, future leaders must empower and inspire the next generation. All prominent market actors, including banks, investors, microcredit entities, and the public sector, will promote multi-stakeholder partnerships in the financial market.

Commodity, consumption, and production patterns all change, and multi-stakeholder partnerships are formed to facilitate and accelerate that change. While government agencies and nonprofit organizations continue to play a critical role in promoting resource efficiency and a green economy, the UN Environment recently expanded its role to include the private sector.

37.11.4 Educational Sustainability

Asia and the Pacific are home to approximately 65% of the world's population. This group currently controls 20% of the world's consumer power, but that will change dramatically over the next decade, increasing to 48% by 2030. This is a global scale, not just for Asia. The average person in Asia consumed 4.5 metric tonnes of materials per year in 2000. This figure had risen to more than 9 tonnes per person fifteen years later.

Today's lifestyles are generally better due to significant improvements in food, housing, and transportation accessibility, as well as increased disposable income. However, while pollution and waste emissions have increased in the region, there have also been setbacks due to the area's high levels of pollution and waste accumulation. Citizens throughout Asia and the Pacific and national governments and businesses are interested in a more resource-efficient approach to meeting people's needs, wants, and dreams.

37.11.5 Environment Friendly Technologies

The current sustainable city project's primary focus areas are:

- Access to environmental-friendly technology information.
- Involving the participation of mechanisms for technology transfer and the creation of novel business practises and the strengthening of capacities for innovation.
- Adaptation.
- Management of technologies.

The UN Environment agency is pushing for international agreements and policies that promote environmentally sound technology. UN Environment has established a strong link between science and policy. Reduced waste and increased energy efficiency of materials are critical components of this sub-program on resource efficiency [11].

To effectively use resources, people should use environmental management methods that make the best possible use of resources while also protecting the capacity of natural systems to regenerate and sustain themselves. Adopting and using alternative, environmentally sound development strategies and related technologies is a requirement for this project.

While technology has a wide-ranging impact on the availability of resources (materials and energy), pollution, and waste management across various sectors, it can also bring about resource scarcity, corruption, and waste management issues. Since the 1992 Rio Earth Summit, environmentally sound technology has significantly impacted international environmental cooperation. A cornerstone of the groundbreaking Addis Ababa Action Agenda is having access to cutting-edge technology (2030 Agenda for Sustainable Development). The 193 member states unanimously approved the agreement of the United Nations. A significant role is played by the private sector's involvement in helping to make environmentally sound technology widely accessible for all industries, as it includes both technology providers and homegrown technology developers.

Current development trends are fundamentally unsustainable. What steps can we take to ensure a more sustainable future? It is possible to concentrate on market failure and government failure, but far from sufficient in neo-classical economic theory. Additionally, something else is required. A critical step would be to abandon the current monopoly on neoclassical theory and embrace pluralism. Neoclassical economics and neoclassical environmental economics are not only theoretical constructs but also ideological constructs. While neoclassical theory can continue to contribute as one of several theoretical perspectives, the current monopoly on neoclassical theory runs counter to our democratic values.

Hopefully, there is a global backlash against the concept of economic segregation from politics and democracy. Economic textbooks that are not committed to democracy should be phased out in favour of economic history. Pluralism and democracy as values and principles must be promoted both within and across countries.

References

1. Climate Change 2021: the Physical Science Basis, the Working Group I contribution to the Sixth Assessment Report, 9 August 2021. https://www.unep.org/resources/report/climate-change-2021-physical-science-basis-working-group-i-contribution-sixth
2. Climate Change | United Nationshttps://www.un.org ' climatechange
3. Effect of New Energy Conversion Technologies on Electrical power engineering Education Ebrahim Farjah, 2010.
4. Pogue, David, *How to Prepare for Climate Change: A Practical Guide to Surviving the Chaos* Simon & Schuster, 2021. ISBN13: 9781982134518
5. Gates, Bill, How to Avoid a Climate Disaster Book by Bill Gates, originally published: 16 February 2022, Alfred A. Knopf. Original language: English OCLC: 1122802121 https://web.worldbank.org/archive/website01291/WEB/0__MEN-3.HTM
6. Helen Locher, Andrew Scanlo (2012) "Chapter 1: Sustainable Hydropower – Issues and Challenges".
7. Janusz Nowotnya, John Dodsonb, Sebastian Fiechterc, Turgut M. Gürd, Brendan Kennedye, Wojciech Macykf, Tadeusz Baka, Wolfgang Sigmundg, Michio Yamawakih, Kazi A. Rahmana *Towards global sustainability: Education on environmentally clean energy technologies,* www.elsevier.com/locate/rser
8. Judith Lean, "Cycles and trends in solar irradiance and climate," *Wiley Interdisciplinary Reviews: Climate Change*, vol. 1, January/February 2010, 111-122.
9. Jouni Korhonen (2004), "Industrial ecology in the strategic sustainable development model: strategic applications of industrial ecology", *Journal of Cleaner Production* 12 (2004) pp 809–823.
10. Kathy A. Meney, Ljiljana Pantelic (2020), "Chapter 32 - Decentralized Water and Wastewater Systems for Resilient Societies: A Shift Towards a Green

Infrastructure-Based Alternate Economy", Springer Science and Business Media LLC, 2020.
11. Kanayathu C. Koshy (2018), *Sustainability models for a better world*, Wageningen Academic Publishers, 2018.
12. Peter A. Victor, Tim Jackson (2012) "A Commentary on UNEP's Green Economy Scenarios", *Ecological Economics*, 2012.
13. National Geographic Society https://www.nationalgeographic.org ' encyclopedia ' climet
14. Rosena, Richard A., & Edeltraud Guenther, The economics of mitigating climate change: What can we know? *Technological Forecasting and Social Change*, Volume 91, February 2015, pp. 93-106.
15. Roland Clift and Angela Druckman (2016), *Taking Stock of Industrial Ecology*, Springer Open (eBook), Springer Cham Heidelberg, New York.
16. Olivia Bina and Sofia Guedes Vaz (2011), "Humans, environment and economies: From vicious relationships to virtuous responsibility", *Ecological Economics* 72 (2011) 170–178.

About the Editors

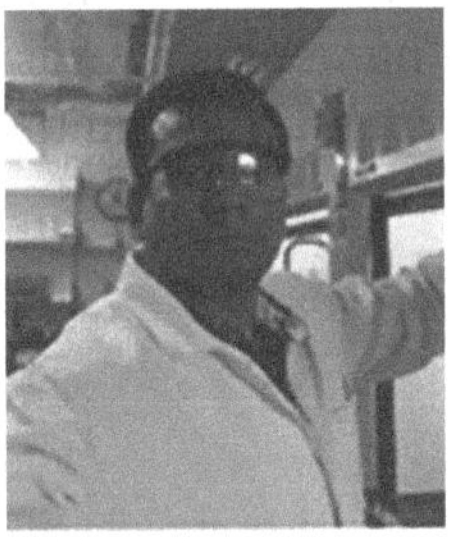

Milind Shrinivas Dangate, PhD, is an assistant professor in the Department of Chemistry at Vellore University of Technology, Chennai, India. He has authored several publications and has a grant and a fellowship to his credit, in addition to several post-doctoral appointments.

W. S. Sampath, PhD, is a professor in the Department of Mechanical Engineering, Colorado State University, Director for Next Generation Photovoltaics (NGPV) Laboratory at Colorado State University, and Site Director at NSF I/UCRC for Next Generation Photovoltaics. With over 30 years of industry experience, he has contributed significantly to the science of renewable energy.

O. V. Gnana Swathika, PhD, is an associate professor in the School of Electrical Engineering at VIT Chennai, India. She earned her PhD in electrical engineering at VIT University and completed her postdoc at the University of Moratuwa, Sri Lanka.

Sanjeevikumar Padmanaban, PhD, is a faculty member with the Department of Electrical Engineering, IT and Cybernetics, University of South-Eastern Norway, Porsgrunn, Norway. He received his PhD in electrical engineering from the University of Bologna, Italy. He has almost ten years of teaching, research and industrial experience and is an associate editor on a number of international scientific refereed journals. He has published more than 300 research papers and has won numerous awards for his research and teaching. He is currently involved in publishing multiple books with Wiley-Scrivener.

Index

Also of Interest

Check out these other related titles from Scrivener Publishing

INTEGRATED GREEN ENERGY SOLUTIONS VOLUME 1, Edited by Milind Shrinivas Dangate, W. S. Sampath, O. V. Gnana Swathika, and Sanjeevikumar Padmanaban, ISBN: 9781394193660. This first volume in a two-volume set presents the state of the art for the concepts, practical applications, and future of renewable energy and how to move closer to true sustainability.

RENEWABLE ENERGY SYSTEMS: Modeling, Optimization, and Applications, edited by Sanjay Sharma, Nikita Gupta, Sandeep Kumar, and Subho Upadhyay, ISBN: 9781119803515. Providing updated and state-of-the-art coverage of a rapidly changing science, this groundbreaking new volume presents the latest technologies, processes, and equipment in renewable energy systems for practical applications.

Encyclopedia of Renewable Energy, by James G. Speight, ISBN: 9781119363675. Written by a highly respected engineer and prolific author in the energy sector, this is the single most comprehensive, thorough, and up to date reference work on renewable energy.

INTELLIGENT RENEWABLE ENERGY SYSTEMS: Integrating Artificial Intelligence Techniques and Optimization Algorithms, Edited by Neeraj Priyadarshi, Akash Kumar Bhoi, Sanjeevikumar Padmanaban, S. Balamurugan, and Jens Bo Holm-Nielsen, ISBN: 9781119786276. This collection of papers on artificial intelligence and other methods for improving renewable energy systems, written by industry experts, is a reflection of the state of the art, a must-have for engineers, maintenance personnel, students, and anyone else wanting to stay abreast with current energy systems concepts and technology.

POWER ELECTRONICS FOR GREEN ENERGY CONVERSION, Edited by Mahajan Sagar Bhaskar, Nikita Gupta, Sanjeevikumar Padmanaban, Jens Bo Holm-Nielsen, and Umashankar Subramaniam, ISBN: 9781119786481. Written and edited by a team of renowned experts, this exciting new volume explores the concepts and practical applications of power electronics for green energy conversion, going into great detail with ample examples, for the engineer, scientist, or student.

INTEGRATION OF RENEWABLE ENERGY SOURCES WITH SMART GRIDS, Edited by A. Mahaboob Subahani, M. Kathiresh and G. R. Kanagachidambaresan, ISBN: 9781119750420. Provides comprehensive coverage of renewable energy and its integration with smart grid technologies.

Green Energy: Solar Energy, Photovoltaics, and Smart Cities, edited by Suman Lata Tripathi and Sanjeevikumar Padmanaban, ISBN: 9781119760764. Covering the concepts and fundamentals of green energy, this volume, written and edited by a global team of experts, also goes into the practical applications that can be utilized across multiple industries, for both the engineer and the student.

Microgrid Technologies, edited by C. Sharmeela, P. Sivaraman, P. Sanjeevikumar, and Jens Bo Holm-Nielsen, ISBN: 9781119710790. Covering the concepts and fundamentals of microgrid technologies, this volume, written and edited by a global team of experts, also goes into the practical applications that can be utilized across multiple industries, for both the engineer and the student.

Energy Storage, edited by Umakanta Sahoo, ISBN: 9781119555513. Written and edited by a team of well-known and respected experts in the field, this new volume on energy storage presents the state-of-the-art developments and challenges in the field of renewable energy systems for sustainability and scalability for engineers, researchers, academicians, industry professionals, consultants, and designers.

Biofuel Cells, Edited by Inamuddin, Mohd Imran Ahamed, Rajender Boddula, and Mashallah Rezakazemi, ISBN: 9781119724698. This book covers the most recent developments and offers a detailed overview of fundamentals, principles, mechanisms, properties, optimizing parameters, analytical characterization tools, various types of biofuel cells, edited by one of the most well-respected and prolific engineers in the world and his team.

Biodiesel Technology and Applications, Edited by Inamuddin, Mohd Imran Ahamed, Rajender Boddula, and Mashallah Rezakazemi, ISBN: 9781119724643. This outstanding new volume provides a comprehensive overview on biodiesel technologies, covering a broad range of topics and practical applications, edited by one of the most well-respected and prolific engineers in the world and his team.

Energy Storage 2nd Edition, by Ralph Zito and Haleh Ardibili, ISBN: 9781119083597. A revision of the groundbreaking study of methods for storing energy on a massive scale to be used in wind, solar, and other renewable energy systems.

Hybrid Renewable Energy Systems, edited by Umakanta Sahoo, ISBN: 9781119555575. Edited and written by some of the world's top experts in renewable energy, this is the most comprehensive and in-depth volume on hybrid renewable energy systems available, a must-have for any engineer, scientist, or student.

Progress in Solar Energy Technology and Applications, edited by Umakanta Sahoo, ISBN: 9781119555605. This first volume in the new groundbreaking series, Advances in Renewable Energy, covers the latest concepts, trends, techniques, processes, and materials in solar energy, focusing on the state-of-the-art for the field and written by a group of world-renowned experts.

A Polygeneration Process Concept for Hybrid Solar and Biomass Power Plants: Simulation, Modeling, and Optimization, by Umakanta Sahoo, ISBN: 9781119536093. This is the most comprehensive and in-depth study of the theory and practical applications of a new and groundbreaking method for the energy industry to "go green" with renewable and alternative energy sources.

Nuclear Power: Policies, Practices, and the Future, by Darryl Siemer, ISBN: 9781119657781. Written from an engineer's perspective, this is a treatise on the state of nuclear power today, its benefits, and its future, focusing on both policy and technological issues.

Zero-*Waste Engineering 2nd Edition: A New Era of Sustainable Technology Development*, by M. M. Kahn and M. R. Islam, ISBN: 9781119184898. This book outlines how to develop zero-waste engineering following natural pathways that are truly sustainable using methods that have been developed for sustainability, such as solar air conditioning, natural desalination, green building, chemical-free biofuel, fuel cells, scientifically renewable energy, and new mathematical and economic models.

Sustainable Energy Pricing, by Gary Zatzman, ISBN: 9780470901632. In this controversial new volume, the author explores a new science of energy pricing and how it can be done in a way that is sustainable for the world's economy and environment.

Sustainable Resource Development, by Gary Zatzman, ISBN: 9781118290392. Taking a new, fresh look at how the energy industry and we, as a planet, are developing our energy resources, this book looks at what is right and wrong about energy resource development. This book aids engineers and scientists in achieving a true sustainability in this field, both from an economic and environmental perspective.

The *Greening of Petroleum Operations*, by M. R. Islam *et al.*, ISBN: 9780470625903. The state of the art in petroleum operations, from a "green" perspective.

Printed in the USA/Agawam, MA
September 16, 2026

910661.106